D1175644

Principles
of Companion
Animal Nutrition

Principles of Companion Animal Nutrition

John P. McNamara

Scientist and Professor
Animal Sciences and Nutrition
Washington State University

PEARSON

Prentice Hall

Upper Saddle River, New Jersey 07458

Library of Congress Cataloging-in-Publication Data
McNamara, J. P. (John P.)
Principles of companion animal nutrition /John P. McNamara.—1st ed.
p. cm.
Includes bibliographical references and index.
ISBN 0-13-151258-7
1. Pets—Nutrition. 2. Pets—Feeding and feeds. I. Title.

SF414.M39 2006
636.088'7—dc22

2004024505

Executive Editor: Debbie Yarnell
Assistant Editor: Maria Rego
Production Editor: Ginny Schumacher, Carlisle Publishers Services
Production Liaison: Janice Stangel
Director of Production & Manufacturing: Bruce Johnson
Managing Editor: Mary Carnis
Manufacturing Manager: Ilene Sanford
Manufacturing Buyer: Cathleen Petersen
Creative Director: Cheryl Asherman
Cover Design Coordinator: Christopher Weigand
Marketing Manager: Jimmy Stephens
Composition: Carlisle Publishers Services
Printing and Binding: R.R. Donnelley

This book was set in Palatino by Carlisle Communications, LTD. and was printed and bound by
R.R. Donnelley.

Pearson Education Ltd.
Pearson Education Singapore, Pte. Ltd.
Pearson Education Canada, Ltd.
Pearson Education—Japan

Pearson Education Australia PTY, Limited
Pearson Education North Asia, Ltd.
Pearson Educacíon de Mexico, S.A. de C.V.
Pearson Education Malaysia, Pte. Ltd.

10 9 8 7 6 5 4 3 2 1
ISBN: 0-13-151258-7

Table of Contents

Preface

This book is primarily meant to serve the newcomer to nutrition of companion animals at the college or university level. Some high school courses may find it appropriate as well. A thorough background in chemistry and biology should not be necessary. My own experience, advice of many colleagues and reviewers of the book, and the changes occurring in animal sciences programs led to the consensus that it is better to provide a text at a level closer to the newer student, yet having enough information that teachers of more advanced students can build on the framework by adding more detail.

These decisions were not made without much review. Departments of animal sciences have taught courses such as "Animal Nutrition," "Feeds and Nutrition," or "Feeds and Feeding" for eighty years or more. The tradition was to cover agricultural animals: cattle, sheep, pigs, chickens, and sometimes horses. Dogs, cats, and "exotics" (birds, fish, reptiles, rabbits) were not considered appropriate. Yet, even as much as thirty years ago, many departments began to admit and even recruit students from the wider population, students who had little or no experience or interest in agricultural animals. The number of families living on agricultural enterprises has been declining for seventy years to the point where it is now less than 1 percent of the population (U.S. Census, 2000). There remain few land-grant university departments of animal science that teach agricultural students as a majority of their enrollment.

It was, in fact, thirty years ago, that my friend and I were two of the first members of the Companion Animal Club at the University of Illinois. Under the guidance of Dr. Jim Corbin, one of the first animal scientists who concentrated on companion animals, we started talking about careers as "small animal" nutritionists, veterinarians, and other specialists. In growing numbers, faculty realized the importance in social, economic, and even environmental terms, and simply desired to educate the next generations as to the biology of these important species.

This certainly was the case for me. As a boy from the little farm town of Chicago, Illinois, at that time, several faculty wondered if I had wandered into the wrong department (probably several still do!). I have worked, taught, and done research in the dairy and pork industries. Yet when I was offered the chance to teach "Nonruminant Nutrition" seventeen years ago, I jumped at it. One reason was that there were many colleagues who could teach ruminant (cattle) nutrition, but not so many on the "chickens, pigs, and horses" side. The importance of agricultural animals will vary from region to region.

As time passed and students changed, I had the choice of cancelling the course or tailoring it to the students who were coming into our department. I chose the latter. I have taught this senior-level class, primarily dedicated to companion animals, with necessary examples from poultry and pigs, for twelve years. Many successful books preceding this one, along with my own experience, demonstrate that the need for a companion animal biology text is no longer seriously questioned.

Yet the target audience for this book remained undetermined. Traditionally, animal science departments teach their first nutrition class in the junior year, after students have had some biology and chemistry. Yet many more people than animal sciences majors are interested in companion animals, and the level of biology and chemistry knowledge can vary tremendously (even among majors). In addition, many community colleges teach students interested in companion animal nutrition, many in veterinary technician curricula. Our department has taught a course in companion animal nutrition, open to all majors with a minimal scientific background. The class has been successful, now one of the larger ones in the college. Over many years I had garnered facts from advanced texts, scientific literature, and experience to help me teach.

It was never a problem to find good, detailed, advanced sources for dogs, cats, horses, rabbits, and some other species, as you can find in the "Further Reading" sections in each chapter. These would be appropriate primarily for senior students in curricula including biology, animal sciences, or students in colleges of veterinary medicine. I was, however, left with a void for a simple, introductory text covering the major companion species that could be used in a number of different courses for those interested in the basis of nutrition of companion animals. Existing resources were too narrow in scope; too broad (also included other nonnutritional subjects such as animal management); too advanced (requiring college-level chemistry or beyond to fully understand); not as sound scientifically as they might be; or written by authors from a commercial firm. This latter issue was not a problem of inaccuracy or even bias; in fact, most "corporate" publications contain quite adequate scientific accuracy. The problem was more of completeness and purpose. There are a lot of pamphlets and minibooks from companies on practical feeding management containing good scientific explanations, yet not what most would call a complete introduction.

I decided to write a new book, then, and my goals included having sufficient scientific explanation of underlying chemistry and biology so that the beginner could understand the basis of nutritional-management practices. The framework is similar to most nutrition texts: the early chapters cover the basic nutrients and their uses (Chapters 3–7), then we cover feed chemistry and ration formulation (Chapters 8 and 9), and third, chapters on specific species information for dogs, cats, horses, llamas and alpacas, rabbits, birds, fish, rodents, and reptiles (Chapters 10–18). There are two chapters in the text (Chapters 1 and 2) that are not often included in traditional texts; one covers a little bit of the history of nutrition, from very early days to now, and the second covers the overall scheme of life-cycle management. I thought it critical to help teachers and students to put the present research and applications in companion animal nutrition into an important context. Companion animal nutrition is a living science, with quite a bit of professional and personal research into best practices. Although we know many specific chemical and biological facts, there is a lot of gray area and differences of opinion on companion animal nutrition. My goal is to show that this is, in fact, part of the scientific process and a good thing!

> *I have tried to provide the key frameworks necessary to understand principles and applications, allowing teachers and students from more advanced classes to fill in specific detail from other sources as needed.*

This is not a senior-level or veterinary student text. From another perspective, in the species chapters, there is much more information that is not included in any other text. I must agree with my friend and colleague, Phil Senger, who has written a very good text on reproduction and whose guidance I have sought throughout this process. In his preface, he states:

> This book is written for students, not my colleagues. In general, college texts are too lofty. They often target the instructor and assume students have a higher level of knowledge than they really do. I have tried to strike a balance by presenting . . . concepts without unessential detail. At the same time, I have tried to provide a format from which instructors can incorporate additional information according to their own specific priorities and expertise. I am a firm believer that learning should be stimulating, interesting, and fun, not a task. In this light, I have broken textbook tradition and included some humor . . . (Senger, *Pathways to Pregnancy and Parturition*, 2003, Pullman, WA: Current Conceptions, Inc.)

Another concern of mine was the mass of text now available on thousands of Internet sites, most of which is never checked for accuracy, and the vast majority of which are trying to sell someone something. Students and the lay population often now use these as their first, if not only, source of information. I thought there should be an accessible reference for those who wanted to check the information on the World Wide Web. There will be few direct references to Internet sites in this book for several reasons, some of which include: Internet sites have never been meant to be authoritarian references for scientific information; there are far too many to even highlight a few; and they change daily. However, general references will be made throughout the book to some Web sites I have found helpful and some general guidelines for using and interpreting Web sites concerning nutrition of companion animals. I am not referring to the sites from reputable nutrition companies and universities, as they are almost without exception scientifically accurate and understandable.

I hope that this book can serve a wide range of interested parties, university students interested in companion animals as part of their lives but not necessarily their careers, as well as students interested in the animal sciences and veterinary medicine fields. At the beginning of each chapter on nutrients, there will be a "Take Home and Summary" section that highlights the key points of each chapter. A little history will help you understand how basic concepts and practices have developed. Then we will go through description of the topic, including the main concepts and new vocabulary, and provide sufficient biological information to make the topics clear. Finally, at the end the first several chapters there will be a small section, "Practical Applications," to help the reader extrapolate from the scientific topic. In the chapters that cover each animal, we will then focus on practical applications. This format will help you understand why we do things certain ways to optimize the nutritional health of an animal, and why there may be a variety of good and acceptable ways to feed a given animal.

There are study questions at the end of each chapter. They include vocabulary that students are expected to be able to use (one level would be to use the word in a sentence; the advanced student would be able not only to use the word in a sentence, but also to explain the sentence). In addition, there are questions of interpretation and application. They are not simple regurgitations of fact, but meant to help the student integrate the basic biology into the animal as a whole, to provide a practical application of principles, and to be able to explain the principle behind the application.

I hope that this book provides the reader with an easy to understand, maybe even fun and a little humorous, scientific background of the art and practice of feeding our companion animals. Nutrition as a field of human endeavor includes the scientific, provable, quantitative aspects as well as practices based on philosophies of science, sociology, personal preference, and even religion. That is as it should be in the beauty of the world, yet we must never fail to appreciate the "truth" of scientific knowledge, though we may apply it in various ways. We should not allow personal or other preferences to deny the scientific truth, even though we might freely choose to practice in various ways. I stand by the scientific facts and proven, documented nutritional practices herein that have been shown to ensure a good life for animals and simultaneously I recognize the wide variety of applications, interpretations, and nutritional philosophies. My hope is to give students the tools to make informed decisions on companion animal nutrition and to manage the feeding of their animals to provide a healthy and long life.

Reviewers for *Principles of Companion Animal Nutrition*

Dana Call
Oklahoma State University—
    Oklahoma City

Richard Heitmann
University of Tennessee

Gita Cherian
Oregon State University

Keith Cummins
Auburn University

Harold W. Harpster
Penn State University

Kelly Swanson
University of Illinois

Anne Duffy
Kirkwood Community College

Chapter 1

Introduction: Nutritional Terms and Definitions

Take Home and Summary

The primary objective is that students will gain an appreciation of the nutritional principles governing growth, health, and performance of animals kept as companions. The secondary objective is to supply practical guidelines for nutritional management. If students understand some basic chemistry and biology, they will be able to nurture their companions with more competence and confidence. This text emphasizes the biological basis of nutrient requirements and therefore helps the reader to interpret and evaluate pet food labels. It provides an introduction to nutrients, their primary plant and animal sources, and their uses, including carbohydrates, fats, proteins and amino acids, and vitamins and minerals. Students will be able to recognize basic feed ingredients in order to evaluate the worth of rations for specific animals. Students will be able to recognize proper and improper feeding of companion species, including general deficiencies in macronutrients, vitamins, and minerals. In doing this, students will be introduced to the metabolic basis and practical preventative management of nutritionally related problems.

> *Primary objectives:*
>
> Appreciate principles of nutrition.
> Supply basic practical guidelines for nutritional management.
> Have fun with your companion.

Achieving these objectives will allow you to have more fun with your companion animals. This book is meant to be fun to read, so for the sophisticated practitioners attempts at humor and perhaps even some simplification may at first seem out of place. However, I hope that the style will help the majority of readers on their first foray into the "whys and hows" of good nutritional management to keep the perspective that one major purpose of having an animal companion is to *enjoy life*! When readers are done with this book, they may or may not proceed to higher study in companion animal nutrition and health, but they will be able to manage and care for the nutritional needs of their companion animals with confidence.

The Study of Nutrition

Nutrition: sustenance of the basic processes of life.

The *study of nutrition* is scientific investigation that provides us with knowledge to apply sound principles of biology and chemistry to help ensure health, welfare, and longevity. Nutrition has been a subject of formal and informal study almost since "civilization" began. Aristotle in his "Ethics" discusses *fasting* (periods of reduced food intake, but not starvation, which is a lack of food) and *gluttony* (overeating for long periods of time) and how they affect life and health. People learned early on that if they did not drink water or eat sufficient food, they sickened or even died, and that if they ate too much, other problems arose over time. Further, if they ate sufficient quantities of food, but without a wide mix of foods, or certain types of food, health and longevity would similarly deteriorate. Early in the written history of civilization, including the Bible, there are accounts of the consequences of gluttony: obesity, various maladies (sicknesses), and often, early death. That genetic variance in many body organs and systems also plays a role in obesity, and this is still a topic of much debate among scientists, behaviorists, and general folk. This is just one example of the scientific, personal, sociological, and even religious interpretations of nutrition.

People have actively investigated the various nutritional relationships to health since the ancient Greeks and Egyptians, and since the advent of "modern science" in the 1700s, nutrition has developed into a major field of its own. The feeding and care of companion animals was not given the same attention as that to human health until the latter part of the nineteenth century. Research activity in the nutrition of humans and then to food-producing species expanded into that for dogs and cats in the early 1900s, and began in earnest after World War II. Since then, many nutritionists have devoted their careers to studying nutrition of only one companion animal species, and our knowledge today is truly vast. Figure 1.1 shows a general schematic of the progression of feeding of companion animals. It was through trial and error, observing and changing practices, that we eventually came to have improved concepts and practices in nutrition. With a more sophisticated scientific process of investigation, the concepts become more detailed and the good concepts became more widely practiced. Eventually, a level of knowledge was disseminated to the general population, leading to better health and longevity of companion animals. This process is never

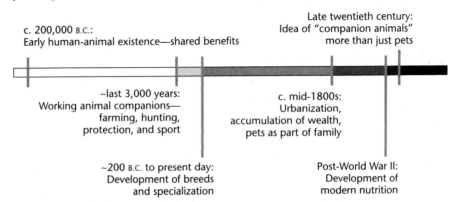

Figure 1.1   Evolution and domestication of companion animals.

"done," and is proceeding right now. The time span from when these words were written to the time you read them has included new observations, investigations, and discoveries, some of which may eventually improve the way we feed our animal friends.

Myths, legends, philosophies, and ideas of many kinds still cloud the scientific picture. The goal of this book is only to provide sound scientific facts and principles, and parts of chapters will dispel myths with proven facts. (A more advanced reader may want to refer to this treatment in Case, Carey, Hirakawa, and Daristotle, 2000.) It is also true that the understandable goal of an easy life with little effort has certainly been applied to companion animal feeding. The abundance of food in the Western Hemisphere has led to an increase in obesity and nutritionally related problems for pets unheard of in history. We have gone from basically throwing scraps out to the dog and cat in the farmyard or backyard to providing them with all they want of scientifically balanced foods, and all manner of various snacks. The problems of too much have replaced the problems of too little.

However, the reader is asked to consider a broader context here, in that what is important in Western Europe, parts of the Americas, Australia, and Asia is still basically unknown in most of Africa, Asia, and Central and South America. That is, nutrition of companion animals is of much less importance to them than it is to us here, or in comparison to the nutrition of humans there. Sustaining human life remains of utmost importance, and the thought of even keeping companion animals that cannot fend for themselves is a luxurious dream. We might at least keep in the backs of our minds, lest we take some matters too seriously, what a wonderful privilege it is to enjoy the companionship of a healthy human-animal bond.

History of Research and Development in Pet Foods

Domestication

Dogs and cats have been recorded as pets since the Egyptians. For about the first 3,000 years or so of *domestication*, "nutrition" has been primarily "assisted foraging." In more affluent cultures, various types of dogs were kept. The ownership of these animals increased for purposes ranging from utility, such as sheep herding and guarding, to more leisurely pursuits such as prizing the "beauty" of different animals. Basic genetic selection was practiced for centuries, eventually resulting in dozens of breeds of dogs, cats, birds, rabbits, fish, and rodents of all types of sizes and shapes.

Historical records would suggest that dogs have been domesticated for at least several thousand years (Cheeke, 2003). Yet they are all still the same Genus: Canis. There are records of domestication in Asia, Europe, Africa, and North and South America. The foundation populations were all wolves.

> North American wolf: gave rise to Eskimo dogs
>
> Asia: Chinese wolf became the Pekingese
>
> India: the dingo, greyhound, and mastiff
>
> Europe: shepherds, terriers, and spaniels
>
> Indian and European wolf hybrids (mastiffs): bulldogs, bloodhounds, Great Danes, Newfoundlands, and St. Bernards (Cheeke, 2003)

Early on, it was not expected that you had to feed dogs or cats directly much of anything, as they would forage on their own, similar to their wild "cousins." An occasional scrap, bone, or dish of milk was sufficient. This worked pretty well for most animals up until the early twentieth century. As animals were bred, changed, and kept for various reasons in various environments, sooner or later, observations on practices of nutrition began to be passed down orally or in written form. Often, practices used by humans in their own nutrition were tested for animals, with varying degrees of success. As urbanization grew and scientific research in nutrition developed, people started to develop and sell specific pet foods. Some of the earliest recorded were in the mid-1800s.

In the 1900s, with greater growth in nutritional research for both humans and farm animals, companions also benefited. Many foods were on the market by the 1920s, such as Gaines Meal and Alpo (Hand, Thatcher, Remillard, & Roudebush, 2000). The majority of these foods were canned, wet foods. The development of canned pet food followed closely the development of canned foods for distribution to humans. The industrial revolution was in full swing in the mid-1800s and mechanization of many processes was taking place. In addition, the techniques for safely processing and storing dry foods had not

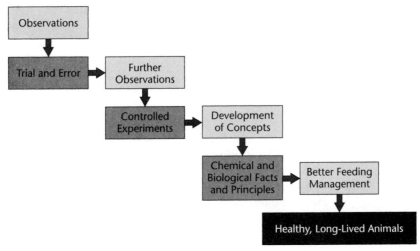

Figure 1.2 Science, knowledge, and practice.

yet been perfected. The modern concept of companion animal nutrition really began in the post-World War II era, as "dry food" was developed. As with many major developments in history, the development of nutritionally complete, dry pet foods was not necessarily born of a great scientific discovery or a visionary following a dream, but primarily due to necessity: a shortage of tin available for cans during the war.

Figure 1.2 summarizes the general flow of knowledge leading to the present scientific practice in nutrition of companion animals; this is also applicable to many areas of biology. Through trial and error, successful strategies become common practices. After some time, these practices become recorded, discussed, and formulated into general concepts that are used and taught to others. These concepts are usually tested in scientific experiments to develop basic principles that are then developed further into feeding-management practices, with the end result of healthy, long-lived animals. Note that each event was dependent on the observations of people who came before, and that these observations had to be recorded and passed down, or else they were lost and had to be rediscovered.

Scientific Research in Pet Foods and Nutrition

After World War II, direct research into the nutrient requirements for dogs and cats began in earnest at private companies and at universities. This was a time of great discovery and improvement in our knowledge of nutrient use by humans and animals. The *National Academy of Sciences (NAS)* is a body of scientists from all fields who are elected to the Academy based on the quality and importance of their work. The *National Research Council (NRC)* is made up of members of the Academy and does research on various topics at the request of (but not necessarily the expense of) the federal government. Those scientists whose specialty is nutrition contribute to the Committee on Nutrition of the NAS. These scientists oversee the compilation of knowledge and production of guidelines for nutrition of all animals, both agricultural and companion. The first NRC publications on nutrient requirements for dogs and the nutrient requirements for cats were published in the 1950s and 1960s, respectively. They were revised in 1984 and 1985.

Since then, thousands more research trials have increased our understanding of nutrient use in companion animals to an impressive level, although research continues into the areas we do not understand. In 2004, the first revision in eighteen years was published as a combined book on the nutrient requirements for dogs and cats (NRC, 2004, see *www.NAP.edu*). It simply cannot be overstated how much time and effort and money have been used, and are necessary even today, to discover and teach good scientific knowledge on nutrition of companion animals. This work will continue only as much as you and I, the owners of animals and buyers of animal foods, support both public and private research and teaching in animal biology.

The Scientific Process

Before we can discuss in more detail the present state of nutritional knowledge for companion animals, it is important to have a short discussion of the scientific process. The *scientific process* is the generally agreed-upon set of steps that must take place to truly test an idea to determine its correctness in describing the world around us. These steps have been described in many different ways over the centuries by many people. The four basic components of the scientific process are generally referred to as observation, hypothesis, experimentation, and analysis and interpretation. In the last sixty years, development and research in pet nutrition has followed, for the most part, the scientific process quite rigorously. From the first early attempts at simply mixing cooked meat and grains and putting it in a bag or canning it, through informal "testing," up to modern scientific experimentation and analysis, the pet food industry has developed into one based on sound scientific principles. Major-brand foods on the market today are based on excellent application of the scientific process. In addition, conscientious pet owners, knowingly or not, follow most of the scientific process in selecting foods for their animals. Let us review the basic principles.

The scientific process:

Observation—really notice something

Hypothesis—think about your observation and form a question

Experimentation—test your question in reality

Analysis and interpretation—check your answer; what does it mean?

Observation

An *observation* is simply something one sees *and* records. We may see a dog roaming down a street and pay no attention—that is not truly an observation. But if we notice that there is a collar on the animal or certain coloring or behavioral characteristics, we make an observation. We may observe that a particular animal appears to like one food more than another. Anyone can and does make observations all the time—you do not need to be a scientist (although we hope that scientists are better trained in making observations). We might observe how an animal grows, or that it seems more healthy over another. We may notice that food intake increases during cold weather or after an exercise program has started. We also can make valid observations by reading a column in the newspaper, a magazine, or by reading a scientific article or textbook. We do have to be careful about believing everything we read—on one level we trust what we read, but at a higher level, we should do research to follow up on the story in the actual scientific literature. If we see and take note of something (and by implication, begin to think about it), that is an observation.

Hypothesis

When one makes an assumption or has an idea or a thought, that usually can be stated as a hypothesis. A *hypothesis* is an assumption, idea, or theory, usually made up to test its "logical or empirical consequences" (Merriam-Webster Dictionary, New York: Simon & Schuster, 1974). So, a hypothesis really is a subset of assumptions made in the scientific process to organize one's thoughts and test the validity of one's theories. After making one or more observations, a person may be curious enough to ask a question about that thing. A question is potentially a hypothesis. For example, a pet owner notes that her dog prefers one dog food over another. Further, they then ask a friend—*does your dog prefer it too?* That question really is the beginning of a hypothesis. If the friend's dog displays the same behavior, it might lead to a statement such as: "Dogs prefer this food over that one." That implies a general assumption about dogs and food preferences—it is now a hypothesis (not a fact!). It can be tested to determine "logical or empirical consequences." Taste-preference tests are often done to test a specific hypothesis of whether many dogs prefer one food over another. Critical elements here include the initial observations and thoughts followed by enough thinking to formulate an idea or hypotheses. Again, anyone can and should do this—not just scientists.

Experimentation

The next step in the scientific process is to test the hypothesis. Put differently, upon making observations and hypotheses, we "test the empirical consequences"—this is often called *experimentation.* Such tests are done under strict conditions with careful recording of observations; this is usually referred to (somewhat redundantly) as a scientific experiment. In the previous example, we might test our hypothesis by asking six friends if their dogs prefer that same food over another. This is a test, but it is not an experiment. We might find that all eight dogs (ours, our first friend's dog, and the other six) all prefer one food over another. That is consistent with our hypothesis. We might find it to be four and four—some dogs prefer the one and others prefer the other. We might find that only the first two dogs prefer the first food. We have made an observation, formulated a hypothesis, and done a test. Yet we still have not done an experiment.

An experiment requires controlled conditions, so that in fact we are sure we are actually testing our specific hypothesis. The key element of an experiment is that it is controlled—if some change is made to determine some effect, there must be a clear frame of reference (control) to compare it to, so that it is not confused with some other effect. If we set out to test that food A is preferred over food B, and we always offer food A in a well-lit corner and food B in a dark corner, we do not know if we are actually testing the taste preference, or the preference for being able to see what is being eaten. Other key requirements include **replication**: the experiment is repeated a sufficient number of times such that we have some confidence that our results are representative of the system we are testing. The experiment must also be clearly recorded and it must be *repeatable*; that is, someone else must be able to do it the same way to make an independent test. We might find with 100 dogs that 95 preferred one food over another in many different situations. Then we might do that same experiment on different types of dogs, dogs in different environments, and so on. There are many tested and accepted statistical methods with which we can help determine our level of confidence in our experiment that are beyond the scope of this book. Once an experiment is repeated sufficient times, we can make an interpretation that the results are applicable to the larger population (of dogs, for example). That does not and cannot mean that it always applies to every single dog. Scientifically we can say "most dogs prefer this food," and if the experiment has been done sufficient times with sufficient animals, we can even say "95 percent of dogs prefer this food," but we can usually never say that *all* dogs will.

Although anyone can and should make observations and hypotheses and conduct tests and even experiments, a truly scientifically valid experiment requires at least some serious thought and planning. In-depth experimentation and understanding in biology and nutrition today requires sophisticated, expensive equipment and procedures. This is a critical reason as to why there is not more direct testing of pet foods, and more details and examples are given as we cover specific subjects.

Analysis and Interpretation

The final critical element of the scientific process is **analysis** and **interpretation.** It is really not enough that we have made observations and hypotheses and have done experiments. We must think about the results in relation to the conduction of the experiment as well as previous knowledge; this process is called analysis. We may need to make statistical comparisons, or compare the results to other similar experiments. We need, basically, to make sure that our results are what we think they are. Were the controls correct—did we make a valid comparison? Was there some factor during the experiment that affected the results without us realizing it? (One real-life example would be that dogs offered food A were not allowed the same access to water as those offered food B—thus some of the difference in the performance of the two foods really was inadvertently due to a difference in water consumption.) In short, we must *think* about our results. How do they compare to similar experiments? Are the results consistent with what we know? This is the hard part—we must first realize that the results of one experiment that contradict everything ever thought before, by themselves, without repeating the experiment, cannot disprove all prior knowledge. But we must also realize that they have the potential to, and that only further repetitions of the experiment, or similar experiments, can help us decide. Sooner or later, then, we accept or reject (or more likely, refine) our initial hypothesis. That is science.

The scientific process thus is defined broadly so that we can think freely and observe the world around us without unnecessary restrictions. But at the same time, the key elements of a soundly based hypothesis, a well-controlled experiment, and a well-thought-out interpretation are essential to the scientific process. Science has often been called a search for the truth. In biology, and in the small subset of biology that is companion animal nutrition, we do search for truth. But one truth is that there is still a wide variation of characteristics, even among dogs within a breed, that can affect the results of scientific experiments. At the same time, there is enough commonality among dogs, even across all breeds, so that we can eventually come to some "scientific truths." The reader should keep these general truths in mind as we explore the scientific evidence for sound application of nutritional principles to our animals.

Present Focus in Companion Animal Nutrition

So if we already know all these facts and principles about nutrition, why do we still study it? Even though we do have command of many facts and principles, there is still a lot to learn. Recent clinical data would suggest that dental calculus, dermatitis, allergy, diarrhea, obesity, and pruritus are presented as significant problems at clinics in the United States. No longer are gross deficiencies the major problem, but more subtle interactions of nutrition and health are now our focus. In addition, scientists and teachers must continue to inform the next generation so that the knowledge is not lost. As we learn the basic principles and practices of nutrition, we will focus on identifying and learning how to prevent several nutritional problems. Nutrient excesses or deficiencies can cause many diseases, either directly or indirectly. Deficiencies, excesses, or imbalances of nutrients will also interact with the genotype of the animal and the environment to cause diseases.

As society has changed, nutrition has progressed from a science of defining requirements and formulating sufficient rations to presently include study of the interaction of nutrition and disease, as well as learning how to optimize total animal health and longevity through better nutrition. Examples include refining vitamin and mineral requirements to provide optimal health and defining the roles of specific nutrients in the function of the immune system and resistance to disease. These and other areas are presently being researched at universities, nutrition companies, and private clinics. Many exciting research areas in nutrition are highlighted throughout this text. Some practical problems of our companions in present times in the Western world are a result of nutrient excesses. Much present research is thus centered around several situations or problems that have nutritional causes or can be prevented or improved by good nutritional management.

Present-day focus in nutrition:
Obesity
Feline urinary tract problems
Renal dysfunctions
Diabetes
Pregnancy and fetal development
Skeletal dysfunctions
Heart disease
Food sensitivities
Immune function

There continues to be a growing understanding of the role of the genome in nutrient requirements. We know that several genes are controlled by specific nutrients to direct nutrient metabolism, reproduction, and immune function. Food is not medicine, but specific nutrients do interact with the genome to produce different effects. It is easy to note large differences in the nutrient requirements among different classes such as mammals, birds, fish, and reptiles; it is also easy to note them among different mammals (ruminants, nonruminants, carnivores, herbivores). But it is also the case that there are genetic differences even among closely related species, and sometimes even among breeds that have been selected for certain traits that alter use of nutrients in subtle but real ways. This does not mean that we need to have many breed-specific products, but we do understand that there are some feeding differences among breeds due to their genetics. Swanson, Schook, and Fahey's (2003) article is a good starting point into this literature.

Nutritional Principles

This short section is intended to introduce you to some of the core concepts and terminology that we will rely on and use throughout the book. Occasionally, you might come back and re-read this section to help keep some of the detail in a simple perspective. It is the author's strong belief and experience that nutrition can and should be simple; bringing large amounts of detailed information back to the core concepts can help accomplish this simplicity.

> The principles of nutrition describe the chemical basis for nutrient requirements and optimal health.

Certain principles apply to all animals; some apply to only a specific species, or even a specific species at one point in the life cycle or in one environment. There tends to be as much or more similarity in nutrition and nutritional problems among species than there are differences. Yet there are differences among species and among individual animals in a species, the study of which makes up the core of population biology. Attention to detail is important. However, if you remember the general principles, it will make many complicated situations appear simpler, and make many problems easier to prevent or solve.

In this information age, remember that "A little knowledge is a dangerous thing." In fact, information is quite dangerous without *knowledge*—defined as the basis and ability to obtain, judge, evaluate, and act on information. This text will supply you with both information *and* knowledge. If you keep your outlook simple, and remember some general principles, you will not be swamped by information, but will be able to obtain, decipher, interpret, evaluate, and act on the information. You will be knowledgeable.

Conservation of Matter and the First Law of Thermodynamics

The *First Law of Thermodynamics* simply states that matter (anything that exists) and energy (the ability to do work) are always conserved. They cannot be created or lost. For an animal, this can be stated as: Energy out always equals energy in. Stated differently, everything the animal consumes is accounted for—it is either never digested and thus lost in the feces, or it is absorbed and used by the body, with by-products excreted in the urine, feces, and respiration (breathing) or lost as heat. The animal neither creates nor destroys matter, although it does change the *form* of the matter. Thus we can define how animals use various foods for their different requirements such as growth, pregnancy, or exercise.

Metabolism, Growth, and Reproduction

Metabolism is the chemical interconversion of nutrients to supply energy for life or to build cellular and organ structures. It encompasses all the processes necessary to sustain life or to grow body components. The body *grows* by combining the catabolism (breakdown and restructuring) of ingested nutrients to supply energy and substrates for the anabolism (synthesis of new compounds) of body structures and components. *Catabolism* is the breakdown of nutrients to supply energy or other compounds, and ***anabolism*** is the building or synthesis of body components such as muscle and fat. In general, catabolism supplies the energy and the substrates for anabolism, the building of new cells and organs. Nutrients are supplied by a simple or complex mixture of foods, either in a natural state or after some type of processing. Animals must take in nutrients in sufficient amounts and in the proper balance to support growth. At maturity, the body may no longer be growing, so metabolism then supports life processes to maintain body size and composition. In addition, an animal might be reproducing—females may conceive and thus the fetus and placental tissues are growing.

Metabolism: conversion of one chemical to another chemical in cells.

Catabolism: breaking down chemical nutrients to intermediates, generates chemical energy and heat.

Anabolism: making new chemicals for structure and function from intermediates, uses chemical energy from catabolism.

Required Nutrients

What the body cannot synthesize in sufficient amounts for normal or optimum performance are ***required nutrients***. These are categorized into the various functions they perform in the body.

Energy

Energy is defined as the ability to do work. Energy is a concept that humans have developed to measure and describe different processes including physical motion, creation of loss of heat, and chemical and biological reactions. The concepts and uses of energy in nutrition are described in more detail in Chapter 7. Objects and chemical compounds may possess different amounts and forms of energy, which simply means that they can do work, which might be movement of an object or transformation from one chemical form to another, or raising the temperature of some chemical or system. This can be confusing sometimes, but let us stick with the simple for now: It is the energy contained in carbohydrates, fats, and amino acids—their ability to do work—that drives the processes of respiration, movement, nervous activity, and chemical syntheses (growth, lactation, pregnancy) in animals. Energy is required for all life functions. But the student must appreciate that it is the actual chemical nutrients (water, oxygen, carbohydrates, fats, and amino acids) that the body requires and uses. The concept of energy is useful for keeping track of feed use and animal performance. The concept of energy inspired many scientists to do experiments to disprove, prove, or define the various elements of energy. A major portion of what we now understand has come from these studies throughout the last 300 years. The concept of energy provides a common, measurable basis for comparing the value of foods or the performance and health of animals, from the nucleus and mitochondria to the cell, the organ, the animal, and the population.

Carbohydrates

Carbohydrate is the name given to a variety of chemicals of a certain common structure containing carbon, oxygen, and hydrogen only. With few exceptions, most carbohydrates contain these three critical elements in the ratio of 1 carbon, 1 oxygen, and 2 hydrogen. The most common carbohydrate is known as *glucose*, commonly referred to even today as "blood sugar" (table sugar, called sucrose, is another carbohydrate, made up of two molecules: one glucose and one fructose). Glucose can come from sucrose, but they are not the same compound. There are many polymers (long chains made from different sugars) that are also carbohydrates; two common ones are starch and cellulose. These are found only in plants (and a few bacteria) and provide energy (starch) or structure (cellulose) to plants. When consumed by animals they can be broken down to sugars that are then used for energy. This energy drives many important biological reactions and is essential for life.

Fats

Fats are chemicals made of carbon, hydrogen, and oxygen in a structure different from that of carbohydrate, usually of the ratio 1 carbon to 2 hydrogen. Oxygen makes up a small proportion of the molecule, called the carboxylic acid part. This acid part is important in helping to make fats more soluble in water solutions, but the important part is the highly energy-rich carbon and hydrogen. Animals use the majority of fats for energy, oxidizing them in the mitochondria to supply ATP. However, several fats have specific functions in the body. There are only a few specific fatty acids that are required to be in the diet. Most animals can make all the fat they need from carbohydrates and proteins. Those fats that are required to be in the diet of most species are called *essential fatty acids.* As is the case with many nutritional terms, this is a nutritional definition, not always a strict biological constant. Some animals do not need all of the three types of essential fatty acids; some need a greater amount than other animals. The three major essential fatty acids are *linoleic acid, linolenic acid*, and *arachidonic acid*. We will explain the functions of these in detail in Chapter 4 and in the specific chapters for each species.

The requirement for specific fats, the role of fatty acids in body functions, and the actual requirements in the diet are confusing and still somewhat controversial issues in nutrition. Carnivorous animals by their nature have evolved while consuming significant amounts of fat and protein and little carbohydrate. There is not a physical, chemical, or biological rule or law that they must do so. That is, animals can survive without fats in the diet (with the exception of the essential fatty acids), even though this is not usually what happens in nature. Most fats that the animals require are made in various body cells. Finally, the student should remember that fats are not only used for energy, but are important for membrane function, nervous tissue function, for production of important hormones, and in production molecules that control immune functions and inflammation. Nutrition is truly still a "living science."

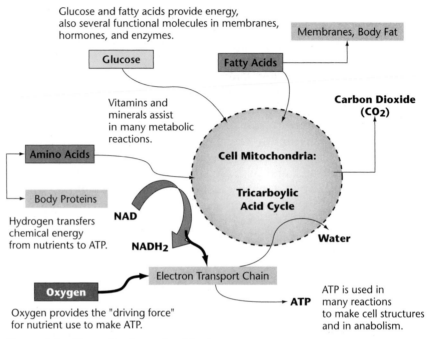

Figure 1.3 Generalized use of nutrients.

Figure 1.3 shows a summary of nutrient use to generate energy in animals. The nutrients glucose, amino acids, and fatty acids are broken down to carbon dioxide and form a compound called nicotinamide adenine dinucleotide ($NADH_2$), which is then broken down further in the electron transport chain, which requires oxygen and makes water. In this process, energy is trapped in the chemical known as ATP, adenine triphosphate. This compound is used in many reactions to keep the body alive, to grow, and to reproduce. Note that it is the chemical changes in the nutrients that transfer energy into the reactions that make ATP and $NADH_2$ to supply energy to make the proteins and other structures that make up animal life.

Amino Acids

Amino acids are the individual components (often referred to as "building blocks") of proteins. Amino acids contain carbon, hydrogen, and oxygen in ratios sometimes similar to carbohydrates and sometimes similar to fats. The major difference is that amino acids also contain nitrogen, which provides them with some unique chemical properties. Proteins are polymers of a few up to many hundreds of amino acids. Proteins are truly an essential element of life in all but a very few simple life forms. It is the proteins that provide structure (cell walls, muscles, bones, connective tissues) as well as function (enzymes, membranes, hormones) for life. Genes code for proteins through a system of connecting amino acids in a determined sequence. Even the simplest bacterium contains over 1,000 genes for different proteins, while dogs and cats possess over 30,000. Amino acids are provided nutritionally either as plant protein or animal protein, which is then broken down in the gut by other proteins (enzymes) to the amino acids. These are then absorbed by the body and used to make the body's specific necessary proteins. There are about twenty amino acids central to making proteins. Most animals can make many of these amino acids and do not really require them in their diet (although they usually do consume them). But several amino acids (eight to twelve depending on the species and time of the life cycle) cannot be made by the animal, and these are known as *nutritionally essential amino acids.*

Vitamins

Vitamins are compounds that are not usually part of the structure of the body or used directly as energy or for proteins. Rather, they are chemical cofactors in metabolic reactions that provide for the life functions—synthesis of tissues, breakdown of nutrients, and conversion of nutrients to energy to breathe, walk, and reproduce. They were originally named "vital amines"—recognizing that they were vital for life, and some were amines (nitrogen-containing compounds). We now know of over fifteen vitamins that are required for many different metabolic functions, and we know that they are not all the same type of chemical. From the making of bone and muscle, to reproduction, and even for the basics of life, vitamins are absolutely essential. The body, with few exceptions, cannot store them so they must be consumed fairly regularly (daily or every few days). Scientists working throughout the last century have so thoroughly defined vitamin requirements, sources, and uses that vitamin deficiencies are almost never really seen in practice. Vitamins are added to almost all prepared foods, and this has prevented the scourge of vitamin deficiencies that claimed many human and animal lives not so very long ago (and, unfortunately, still do in some areas of the world).

Minerals

There are twelve to fifteen known required minerals in the nutrition of animals. These range from calcium and phosphorous, used as structures in the body (bone, teeth) to chromium and cobalt, which are only needed in tiny amounts but are necessary for specific metabolic functions. *Minerals* are similar to vitamins in that they are cofactors in metabolic reactions. Mineral deficiencies are actually metabolic abnormalities. Similar to vitamins, few are stored in any great amounts in the body. There are exceptions—calcium, phosphorous, iron, and copper can be stored and a deficiency may not develop for many weeks. Some minerals such as sodium, chloride, and potassium are so important in cell life that a deficiency may develop in only a few days. Most others are needed in such a tiny amount that, in fact, only when animals were raised in a completely mineral-free environment (no mineral in the food, no metals in the room) were deficiencies actually seen. Although minerals are required, for many of them a deficiency would almost never occur in natural life.

Practical Application

This section comes back to the general topic we started with, the interaction of normal observations and practices with scientific research. Most traditions and general practices have some foundation in the truth of long and consistent observation. Most folks do not feed their animals on a daily basis consciously thinking about scientific methodology. They often follow general practices, common knowledge, or traditions based on things that have worked in the past. This is known as the useful, though sometimes inefficient, practice of "trial and error." The practices we use often come from the observation principle noted earlier—people have observed how well individual animals or groups of animals have performed on various feeding systems. An example is that if an animal eats too much and does not exercise much, it gains weight. Another example is that garlic may prevent flea infestations. The first example has been "proved" over and over in practice and research—and thus "common knowledge" is consistent with "scientific truth." In the second example, many people may have observed, or thought that they observed, a decrease in fleas when feeding garlic to dogs. Yet in controlled studies, this has never been confirmed. Sometimes, what people observe may be two unrelated events: feeding garlic could have coincided with a change in the weather or season that in turn resulted in fewer fleas. Thus, many advances in scientific nutrition have risen from simple observations by people over time. Thus, we must never have a closed mind regarding new ideas. However, ideas must be tested scientifically before we can understand them completely or pass judgment one way or another and make recommendations.

> *Common sense or knowledge:* consistent with scientific evidence.
>
> *Myth or perception:* commonly held belief not consistent with scientific evidence.

After reading and discussing the material in this book, you should at least be able to tell the difference between nutritional advice and advertisement, and be willing to recognize when (1) advertisement is either good or bad nutritional advice and (2) nutritional advice is advertising. I have no axes to grind or products to sell. I personally have experience and believe that the modern pet nutrition industry is excellent, with a track record of providing a variety of good products for many purposes. There are also several other good alternatives, if good scientific principles are followed. I hope that the facts and concepts in this book help to make you more knowledgeable, increase your confidence in making good purchasing and feeding decisions, and allow you to have more fun with your companions.

Words to Know

During the reading, I hope you have noticed words that are in bold and italicized type. Other important words may be italic, in which case they are introduced now and then defined in more detail in later chapters. The bold words are those for which you should learn the definition and be able to describe. In each chapter, I will list vocabulary words defined in the text that you should understand and be able to use.

amino acids	*First Law of*	*National Research*
anabolism	*Thermodynamics*	*Council (NRC)*
analysis	*gluttony*	*nutritionally essential*
carbohydrate	*hypothesis*	*amino acids*
catabolism	*interpretation*	*observation*
domestication	*knowledge*	*replication*
energy	*metabolism*	*required nutrients*
essential fatty acids	*minerals*	*scientific process*
experimentation	*National Academy of*	*study of nutrition*
fasting	*Sciences (NAS)*	*vitamins*
fats		

Study Questions

Further chapters will list helpful study questions for you to answer.

Further Reading

There are literally thousands of books on biology and nutrition, and hundreds of thousands of scientific papers and books on nutrition of companion animals. For example, Case et al. (2000) list 245 references just for the sections on nutrient use. This present text is not appropriate for long lists of such papers. Many recent ones may be easily found with online searches, and most university libraries (at least those with a veterinary school) will have almost all of the references relating to companion animal nutrition. However, at the end of each chapter there will be a brief "Further Reading" section that can take the student or professional either deeper into the science through next-level texts (texts in more detail) or out into the general application through books written primarily as "how-to" books. Some more specific points will have specific book or journal article citations. I will rarely cite an Internet site, due not only to the rapidly changing nature of the Internet but also to the wide range of quality of many sites. With full knowledge that I may come off to some as a "pawn of corporate pet nutrition," I can nevertheless state that the reader's best start at helpful and scientifically correct Web sites are those of the well-known pet food manufacturers. From there you can often find further bibliographies that can help you understand a topic at a deeper level. The larger company Web sites, while admittedly attempting to help sell products, do provide accurate information at a practical level. The books that I cite are written by experts in the field and they contain many hundreds of very detailed research papers. Finally, as mentioned earlier, The National Research Council (2004) has recently released the newest version of *Nutrient Requirements of Dogs and Cats*, which is probably a first best resource for more detail on the topics we cover.

Burger, I., ed. 1993. *The Waltham book of companion animal nutrition*. Oxford, UK: Pergamon Press.

Case, L. P., D. P. Carey, D. A. Hirakawa, and L. Daristotle. 2000. *Canine and feline nutrition*. St. Louis, MO: Mosby, Inc.

Cheeke, Peter R. 2003. *Contemporary issues in animal agriculture*, third edition. Upper Saddle River, NJ: Pearson Prentice Hall.

Hand, M., C. D. Thatcher, R. L. Remillard, and P. Roudebush, eds. 2000. *Small animal clinical nutrition*, fourth edition. Topeka, KS: Mark Morris Institute.

Kelly, N., and J. Wills. 1996. *Manual of companion animal nutrition and feeding*. Gloucester, U.K.: British Small Animal Veterinary Association.

Lund, E. M., P. J. Armstrong, C. A. Kirk, L. M. Kolar, and J. S. Klausner. 1999. Health status and population characteristics of dogs and cats examined at private veterinary practices in the United States. *Journal of the American Veterinary Medical Association*, 214: 1336–1340.

Merriam-Webster Dictionary. 1974. New York: Simon & Schuster.

National Research Council. 2004. *Nutrient requirements of dogs and cats*. Washington, D.C.: National Academy Press.

Swanson, K. S., L. B. Schook, and G. C. Fahey, Jr. 2003. Nutritional genomics: implications for companion animals. *Journal of Nutrition*, 133: 2033–3040.

Chapter 2

The Life Cycle and Nutrient Requirements

Take Home and Summary

The term ***life cycle*** describes the continuous development and growth of an animal from birth to the birth of the next generation. This term captures the essence of development and life over time, and reminds us that we must provide the optimum nutrient supply for each phase of life and into the next generation. Nutritional companies have embraced this philosophy, as evidenced by several examples of "life-cycle feeding" products and specific age- and development-related food products. In nutrition, we generally break the life cycle down into some discrete parts: the ***neonate,*** the newborn animal living solely on milk from the mother; the growing animal, the physiologically maturing animal (young adult); the adult animal maintaining its body weight; the pregnant and lactating female; and the ***senescent*** (aged) animal. We are starting to understand that the nutritional needs of the aged animal are different than for the mature adult. In this chapter we introduce the life-cycle stages and describe their unique developmental and nutritional needs. In the later chapters we will fill in the details.

The Neonate

The neonate, or newborn, requires energy-yielding nutrients (fat and glucose) immediately after birth, first for survival outside the mother and then for development and growth. The body heat naturally supplied by the mother, by littermates, or by the neonate itself through shivering or nonshivering thermogenesis (body-heat production) is critical to surviving the first several hours and days. If needed, we "environmental managers" can supply external heat so that the young do not need to use so much energy to keep warm and can apply it to growth and development. The energy-yielding nutrients are supplied from the milk as *lactose* (milk sugar) and fat.

Figure 2.1 shows the animal life cycle. Starting with birth, the animal develops through life in different stages, each with its own nutritional priorities. The suckling phase demands an easily digestible diet concentrated in high-quality protein, fat, vitamins, and minerals supplied by the mother's milk.

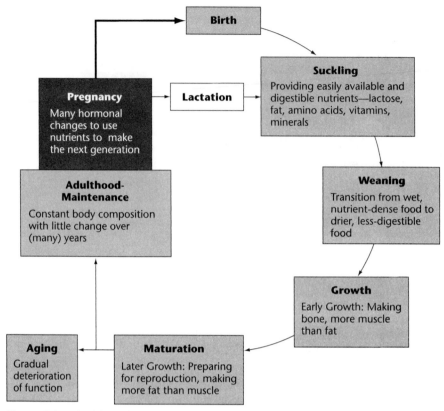

Figure 2.1 The life cycle.

Growth still requires a concentrated form of protein, energy, minerals, and vitamins. Requirements decrease per unit body mass and remain constant through most of adulthood. Pregnancy and lactation increase demand for all nutrients. The aged animal, like the neonate, requires highly digestible, high-quality protein, energy, vitamins, and minerals, but in lesser amounts. Note that each phase proceeds over time and requires constant observation and adjustment of diets and feeding management.

> Glucose is always required by the brain and nervous system, and it is only supplied in milk by lactose, but can be made from amino acids if needed.

After the first several hours or days of adapting to the new environment, the suckling animal begins to grow. Lactose is made up of two sugars, glucose and galactose (see Chapter 4), and galactose is converted back to glucose for the neonate to use. Amino acids are required to make bone matrix, connective tissue, muscles, and cell proteins of all types. Energy is required to run these

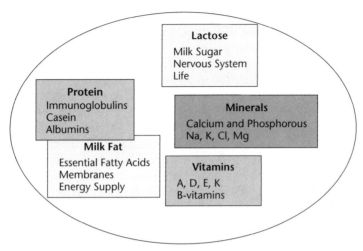

Figure 2.2   Milk—nature's "most perfect food."

processes, and milk fat and lactose are excellent sources of energy. (This is a major reason why there is no better substitute for milk—the young body cannot make sufficient glucose for survival without it.) Vitamins and minerals are needed to supply all the cofactors and precursors for the growth reactions. Minerals such as calcium and phosphorous are needed directly for bone growth.

Figure 2.2 shows the complex components of milk. Milk is often called nature's "most perfect food." Because the baby requires a highly digestible, high-quality ration, it is not surprising that milk meets almost all the animal's needs. The proper form and balance of amino acids, sugars, fats, vitamins, and minerals are supplied almost without exception.

In general, the younger or smaller animal requires more energy, amino acids, vitamins, and minerals per unit body mass and per unit diet (*nutrient density*) than at any other stage except lactation. Younger animals need more amino acids, calcium, and phosphorous for structural growth, and more energy to supply that growth. Also, younger animals are generally not as developed in regulating body temperature, and thus require more nutrients to generate body heat.

Weaning

The critical element to understand about *weaning* is *adaptation,* a coordinated response or reaction to a change in conditions. The young animal must adapt from a wet, highly digestible diet, to a drier, more varied, and less-digestible one.

Weaning is the act of the mother ceasing to make milk and to allow the young access to suckling.

Weaning also means the period of making the transition from a highly nutritious, wet, and digestible diet to one of widely varying nutrient content of poorer nutritional quality and less moisture. The milk-based diet of fat, lactose, and the milk protein *casein* is 85 to 90 percent water. Eventually, the diet becomes much drier (60 to 70 percent water if diet is actually composed of prey-type animals [in the wild] to only 10 to 12 percent water in dry diets). The chemical makeup changes to a wide variety of proteins of varying digestibility and quality, starch, fibers, and different types of fats.

The young animal adapts to this change by making and secreting different enzymes from the mouth, stomach, intestine, and pancreas. An *enzyme* is a protein that allows a chemical reaction to take place that otherwise would not, or would only take place at a rate much too slow to be useful. We will discuss enzymes in more detail later, but for now simply note that these proteins are absolutely essential to life and are really a critical function of the body. Changes in the regulation of the genes to make different enzymes takes time (a few to several days) to avoid a period of digestive upset or damage and reduced growth and increased risk of infection and disease. In nature and in most managed situations, suckling animals begin to eat other food before they stop drinking milk to help this adaptation. In the final stages, weaning is always a rapid process, initiated by the mother. Because of the supply of other food, the young have now adapted and can survive without any milk. In managed situations, the weaning process can be extended; however, it is critical to understand that it is natural that the mother will one day say "enough." It is our responsibility to make sure the young are ready.

During weaning, as during any shift in diet, the chances of digestive upset, disease, and infection increase. This period is one of physiological and emotional stress on the animal. The stress of separation may alter the functions of the body for some time, including eating behavior (up or down) and digestive function, heart rate, and metabolism in the liver. Therefore, it is our goal to provide a system to minimize the stress. We can do this with a planned and steady change from the wet, milk diet to the drier food diet. The body will normally adapt by changing the type and amount of digestive enzymes. For example, milk sugar, or lactose, is broken down by one enzyme, *lactase.* Plant starch and amylopectin (a more complicated polymer) must be broken down by two enzymes (amylase and amylopectinase) that are not needed for milk lactose, so those enzymes must now be made by the animal. Milk protein, primarily casein, is highly soluble and made up of an almost perfect balance of amino acids for the young; it only requires two or three enzymes to break it down for use. Nevertheless, the wide variety of plant and animal proteins must first be physically broken into small pieces (by teeth or stomach action), then acted upon by several different *proteases* (protein-breaking enzymes) from the stomach and pancreas. Plant and animal fat, unlike the already-soluble milk fat, must be broken down into small globules and then acted upon by the enzyme *lipase* (lipid-breaking enzymes). It is primarily the physical form and the wide variety of chemical forms in plant and animal foods that require the greatest adaptation during weaning.

If the weaning transition does not go well, and animal or plant food is introduced too rapidly, it can lead to upset or actual damage of the digestive system. It is a normal process for the bacterial populations in the intestines to change when the animal consumes a different diet (we have all experienced this firsthand). The rapid introduction or removal of a major food source will almost always lead to some kind of digestive malfunction—diarrhea, gas, constipation, or worse forms of *gastroenteritis* (inflammation and damage) or infection. The sharp edges on broken plant pieces, or bones, can physically irritate the tract lining, leading to inflammation and further damage. In mild forms, these are no problem, but unchecked can lead to serious problems.

> Feeding transition in weaning exemplifies a nutritional management rule: The slower the better. We make changes slowly so that the animal adapts safely and comfortably.

Growth

During the growth stage, the need for protein and amino acids remains high to build muscle, and energy is required for tissue growth, primarily protein synthesis, as well as activity of the animal. Early growth after weaning is primarily in the order of mass: muscle, internal organs, bone, and then fat, or *adipose*, tissue. Growth of an animal includes development of organs such as the *liver*, which carries out hundreds of important metabolic reactions to make what the animal needs; the *kidneys*, which help remove toxins and wastes from the body; and the *spleen*, which makes blood cells. The primary increase in mass is in muscle and organs, requiring a constant source of amino acids. Bone growth requires amino acids to build the matrix and calcium, phosphorous, and magnesium for the mineralization. Energy is required to supply all of these metabolic reactions. Growth of fat (adipose tissue) is a minor aspect in early growth.

> The primary purpose of feeding management in early growth is to establish good eating habits for the lifetime of the animal.

Figure 2.3 examines the stages of cell growth. Tissues grow by cells dividing and increasing in number (*hyperplasia*), and in the case of muscle and fat, by *hypertrophy*, or growing in size. Nutrients must be supplied to provide for both processes. There exists a theory, supported by some good research data, that animals born with more fat cells, or that have fast rates of hyperplasia early in life, are at greater risk of obesity. This is because it is thought that the hyperplastic state affects the hormonal systems leading to more energy being supplied to the fat tissue. This can lead to hyperplastic obesity, which tends to be more important in younger animals. Unfortunately, once the condition exists, it is difficult for the animal to remove the fat and lose the fat cells without severe dieting in many cases. Although still a theory, there is significant scientific evidence for it; this line of work has focused attention on the problem of obesity and ways to prevent it. The student should note that there are both genetic (cells, hormones, and enzymes) and environmental (diet, temperature, and exercise) factors that control adipose tissue growth and obesity.

Figure 2.4 shows an example of growth curves. All bodies grow following a general allometric equation such as the two shown in Figure 2.4. The curves show growth of different animals. Some animals take longer to reach mature weight than the rapid growth of a smaller animal that reaches a smaller mature weight in a faster time. The larger animal requires more food because of its larger size and, in addition, more nutrients per unit of food to

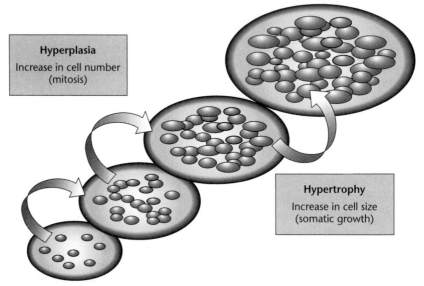

Figure 2.3   Cell growth: hyperplasia and hypertrophy.

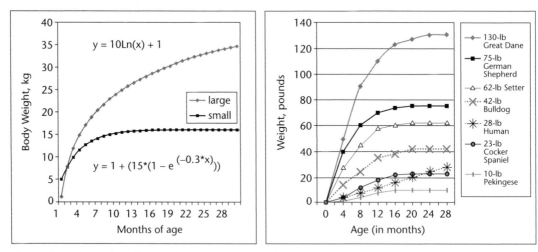

Figure 2.4   Examples of general growth curves and growth of dogs.

supply the growth. By about one year in this example, smaller dogs or cats are already adults and can be fed a maintenance diet with a lesser concentration of nutrients per amount of diet.

As the animal reaches maturity, the rate of muscle growth slows and most growth is as body fat. In addition, this is a time of development of the reproductive systems to start the next generation. The nutrient requirements for this development are not great, and the hormonal changes taking place change the conformation, metabolic rate, and feeding behavior of animals. During growth we manage feeding to avoid a lack of energy, which can decrease growth and negatively affect maturation, or to avoid an excess of energy that leads to obesity.

Adulthood and Maintenance

Most animals spend the vast majority of their lives in adulthood. Adulthood means in one sense that the body has ceased to grow. Nutritionally, this can be rather dull and boring—years and years of consuming about the same thing simply to maintain body composition. Unfortunately, that long, dull, and boring period is exactly when several problems may arise. In recent history this has become a matter of too much or the wrong kind of food. Normal adulthood will be characterized by maintenance of body weight or a slow gain in weight, followed by a slow loss in weight, followed by some length of degeneration (aging). During adulthood (compared to growing animals), feed intake per unit body weight starts to decrease because the metabolic rate of the body slowly decreases. This can be countered by regular exercise, which will help maintain metabolic rate and, thus, body composition. If properly managed, animals will maintain a constant weight and body composition for years and, in some cases, decades. In the United States and Western Europe, there tends to be a problem of too much energy, too much protein and, in some cases, too much or too little of specific minerals.

In adulthood, energy and protein needs are primarily a function of body weight—the bigger the animal, the more they will need. The requirements will also be affected by activity and exercise level, environmental temperature, and disease. Although body weight and composition are roughly constant and thus seemingly boring, the student must remember that muscle, bone, and fat, as well as all organs, are constantly recycling—a synthesis of new cells and tissues and a breakdown of old ones is always happening. We often call this recycling by another term: *turnover*, either muscle protein turnover or fat turnover in the adipose tissue. This turnover, in addition to basic functions such as breathing, blood pumping, and cell life, is really what requires a significant amount of nutrients in adult animals.

In the adult animal at maintenance, this turnover is hard to notice, as the body from the outside changes little. Yet it is this turnover and its affects on nutrient requirements that accounts for most of the energy, amino acid, glucose, and vitamin and mineral requirements. Protein, fat, and bone are constantly being made, requiring amino acids, energy, and vitamins and minerals. The brain and nervous tissue continue to require glucose.

The astute student might now be thinking, "Well, if the body just breaks down the tissues and makes them right back, why do they need any new nutrients?" This is a great question with several good answers. One is that the chemical reactions taking place require inputs of energy as they never proceed at 100 percent efficiency. The amino acids released from the breakdown of muscle are not of the same exact balance as those that went in—some are oxidized (broken down to carbon, nitrogen, and oxygen) at different rates than others. A few are metabolized into slightly different compounds that cannot be used again. Therefore, there is always a need for amino acids from protein in the diet, although it is much lower than during growth or lactation. In fact, in the maintaining adult, the need for protein in the diet is often lower than the content of the animal's diet (this is certainly the case in carnivores), so in nature animals often actually consume more protein than they need. This may lead to problems in kidney function later in life, as we will learn. The student should never forget that even though the animal is "maintaining," that does not mean that there isn't a whole lot of metabolism, both catabolism and anabolism, occurring.

Figure 2.5 shows an example of the breakdown and remaking (synthesis) of a tissue, in this case fat. Fat is constantly being made from glucose and fatty acids from the diet. Fat is stored as a chemical called triacylglycerol. This molecule is constantly being broken down. If synthesis is greater than breakdown, the total amount of fat increases; if the opposite is true, the amount of fat decreases. You should note that our feeding-management goal in the adulthood stage is to provide just the amount of energy to keep these two processes proceeding at the same rate. This might seem like a waste of energy, but this constant turnover gives the animal a great advantage in survival during periods of rapid stress (fright, flight) or lack of food (fat can be easily released from the adipose to supply other organs).

Variations in genotype, age, physiological states such as pregnancy and lactation, disease, inactivity, or other causes can lead to changes in nutrient requirements that we don't often notice until some serious symptoms occur. For an example, providing even a small amount of food over the requirements leads over time to mild obesity; this in turn changes turnover rate and metabolic rate, and this changes nutrient requirements more, leading to advanced obesity. Now the animal is in serious trouble and often "we didn't even notice." Small changes over time can be occurring long before the advanced symptoms occur. This is the single biggest cause of many problems with the health of companion animals in the United States today, and is the reason we need to be preventative managers in our animal-feeding programs.

Adult Females

During adulthood, the major physiological states that affect nutrient requirements occur in the reproducing female.

The nutrient needs of growing fetuses in late pregnancy and during lactation are the greatest demand on the animal in the life cycle.

Maintenance—Body Fat

• Body triacylglycerol storage is also "turning over."

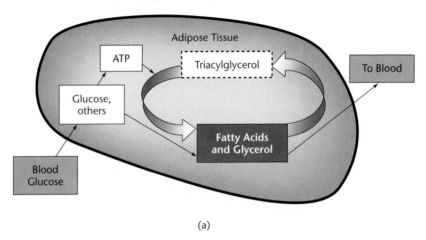

(a)

Maintenance—Muscle (Organ) Protein Turnover

• Muscle (and all protein) is always being made, broken down, and remade, or "turnover."

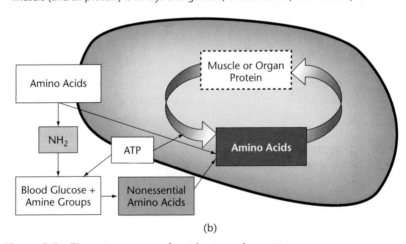

(b)

Figure 2.5 Tissue turnover and nutrient requirements.

The Breeding Season (Prior to Actual Pregnancy)

The term *breeding season* is often used to describe the total period of breeding, pregnancy, and lactation, which is fine. But breeding itself (the act of ovulation of an egg and fertilization by the male) does not really affect nutrient requirements. Many myths and practices abound that people swear by to "get animals ready to breed" but, in fact, if the females have been properly cared for, they are ready. No special tricks are needed to control or affect ovulation other than supplying a properly balanced ration in the correct amounts.

Pregnancy

Pregnancy begins after the egg is fertilized and the embryo implants into the uterus. This period is also called *gestation.* From the frame of reference of the fetus, the proper balance and amounts of nutrients is absolutely critical. However, we really do not need to make any changes (from a proper diet) during this time. The metabolic system of the mother can easily adapt to supply the small amounts of amino acids, fats, glucose, vitamins, and minerals that the developing fetus needs. In fact the metabolic and hormonal regulation of maternal adaptation to study the fetus(es) is a beautiful and amazing example of biology at its best. Normal feeding management during early pregnancy is to provide a properly balanced adult ration. About the last trimester (one-third) of pregnancy (or about week three in cats) is usually when the actual weight gain of the fetus or fetuses increases to the point where more nutrients must be supplied. At that time, an increase in protein, glucose, fats, vitamins, and minerals is required to allow for the growth of the fetuses in the mother.

Lactation

Lactation is the synthesis and secretion of milk. During this period, the mother needs to maintain her own body, and to provide for the maintenance and growth of one to several others. It is common that the total need for each nutrient increases from two to as much as five times more than for maintenance. The maternal system adapts in two ways, by increasing feed intake and by using nutrients from the mother's body. There is a wide range of adaptations from species to species, but all mammals adapt by these two means, some more by feed intake, some more by using body reserves. A small litter-bearing animal such as a rat usually increases feed intake four to five times more than normal and, in addition, loses most of its body fat and a large amount of body muscle to supply energy and amino acids for the milk. By the time she weans the litter, the litter's total mass and nutrient requirements far exceed that of the mother! For a horse nursing the usual single young, with a much slower growth rate, a smaller (20 to 30 percent) increase in food intake plus a moderate contribution from her body fat and muscle can easily do the job.

Recovery and Rebreeding

The demands of pregnancy and lactation require a period of recovery. In some species in the natural state, it is normal to actually initiate one pregnancy while still nursing the last young. For most species we deal with, we delay rebreeding to allow the dam (mother) to rebuild the lost body fat and protein from the last reproductive cycle. This is another example in which our management of these animals actually optimizes their health and extends their life. The usual goal is to feed a normal adult ration to allow the dam to return to her normal body condition (composition). Depending on the species, number of young, and body condition of the dam, this might be from a restricted amount to full feeding.

Adult Males

For most species, adult males are primarily kept at maintenance, although an allotment should be provided for exercise. No extra food or nutrients are needed for breeding, unless services are unusually many or close together. Unless one is dealing with a large breeding colony or herd in which each male is used to "maximal effect," this is not the normal case. Myths and practices notwithstanding, the normal breeding male does not need anything other than access to a properly balanced adult ration.

Aging

We now understand much more about the changes occurring in the body during the process of aging then we did just a few short decades ago. I use the term *decade* specifically to impress upon those who may only be in their second or third decade of life that scientific knowledge and understanding often takes a long time to develop. Life-cycle feeding of dogs and cats has really only been practiced widely for about a decade or two. The thirty years of intensive research activity on many fronts finally allowed visionary nutritionists to apply the knowledge in commercial products. This may seem obvious to young folks growing up now, but I guarantee you, it took a lot of people doing a lot of work to reveal this "obvious" information. It actually speaks well of the scientific research and application of scientific practices in companion animal feeding over the last forty years. How else could one explain the tremendous increase in longevity of our companion animals over that time? Not all would agree with this inference; some are worried that commercial diets, especially those with significant amounts of plant materials, are not appropriate and, in fact, even life-shortening for dogs and cats. I will not cite specific Web sites here, but a good activity either now, or later in the class, would be to search for "raw" and "foods" and "dogs" on a search engine and see what you find. The evidence available to me and those that I trust (Hand, Thatcher, Remillard, & Roudebush, 2000) does not support the concept that commercial foods have been bad for dogs and cats. Yet the scientific process is always one of reexamination, and perhaps with more research activity in this area we may learn more and better ways to feed.

> Available scientific evidence does not support the claim that commercial foods containing carbohydrates have been bad for dogs and cats when fed at correct amounts.

Aging animals generally have reduced total requirements for nutrients at the same body weight, as the metabolic rate slowly decreases over time. Because the energy needs slowly decline, this means the animal must either consume less or consume a diet with less energy-yielding components in it. Muscle protein turnover slows, and the ability of the kidney to synthesize and excrete the by-products of amino acid metabolism normally declines with age. Therefore comes the oft-repeated guideline of "higher-quality protein" for aging animals. Few people who have watched television, read a magazine, or bought dog or cat food in the last ten years have not heard this term, but we often don't really understand it. As described earlier, a *high-quality protein* is a protein that is highly digestible and that matches well the amount and balance of amino acids that the individual animal needs. It is critical to provide a good balance of amino acids in aging animals, primarily because they are no longer as efficient in their metabolism. It may also lessen the amount of work for the aging kidneys, but scientific evidence does not necessarily show that protein intake moderately above requirements reduces kidney function.

The concept is the same for several minerals: too much sodium or potassiam stresses the kidneys. Also the type of protein, mineral and pH balance, digestibility, palatability, and timing of feeding affect organ function. This is when normal aging can lead to liver, kidney, or other organ malfunction if nutrition is not properly managed. Generally, the ability of the animals to digest foods or to adapt to changes in diets lessens with age. Immune function changes and resistance to disease may also decrease. We truly do not fully understand the amino acid needs in elderly animals or the full extent of the interaction between immune function and nutrition. However, research activity continues strongly in these areas. Recent research has demonstrated that by adjusting the nutrient content we can enhance immune function and improve quality of life in our companions by this means.

Practical Application

The practical application of the basic information in this chapter is to match the nutrients you are supplying to the true needs of that animal in its current stage of life. Neonates' best food is mother's milk; in the case where that cannot happen, there are milk replacers available for many species. The best advice in feeding young growing animals is to establish good eating habits. They should be fed a properly balanced ration without allowing excess consumption, and a regular exercise program should be implemented in a timely fashion. Growing animals need food with a greater concentration of protein, energy, vitamins, and minerals to supply growth of bone and other organs. Adult animals need a proper mix of nutrients, but usually in lesser concentrations, as the goal is to only facilitate normal tissue repair, not growth. Adults should be fed responsibly and exercised regularly to maintain weight and health. Reproducing females in the last trimester of gestation should be fed a diet more concentrated in protein, energy, vitamins, and minerals, but again should not be allowed to gain or lose significant amounts of weight (other than fetal tissues). Aging animals are not just old adults, and care should be taken to ensure they have a highly digestible diet sufficient in a higher-quality protein (proper amino acid balance, see Chapters 4 and 7) and vitamins and minerals to ensure sufficient intake of these even if total food intake diminishes.

Words to Know

adaptation	*hypertrophy*	*neonate*
adipose	*kidney*	*nutrient density*
casein	*lactase*	*protease*
enzyme	*lactation*	*senescent*
gastroenteritis	*lactose*	*turnover*
gestation	*life cycle*	*weaning*
high-quality protein	*lipase*	
hyperplasia	*liver*	

Study Questions

1. Why do we separate phases of life for nutritional application?
2. What is the most important thing we should ensure for the newborn, and why?
3. What is the phase of weaning, and what simple feeding-management practices should we follow?
4. What are the general changes in nutrient requirements during growth, gestation, and lactation?
5. How are aging animals different in their general nutrient use?

Further Reading

Burger, I., ed. 1993. *The Waltham book of companion animal nutrition.* Oxford, UK: Pergamon Press.

Case, L. P., D. P. Carey, D. A. Hirakawa, and L. Daristotle. 2000. *Canine and feline nutrition.* St. Louis, MO: Mosby, Inc.

Hand, M., C. D. Thatcher, R. L. Remillard, and P. Roudebush, eds. 2000. *Small animal clinical nutrition,* fourth edition (Chapter 1). Topeka, KS: Mark Morris Institute.

National Research Council. *Nutrient requirements of dogs and cats.* 2004. Washington, D.C.: National Academy Press.

Taylor, Robert E., and T. G. Field. 2004. *Scientific farm animal production: An introduction to animal science.* Pearson Education, Inc.

Warren, D. M. 2002. *Small animal care and management,* second edition (chapter 18). Albany, NY: Delmar, Thomson Learning.

Chapter 3

Glucose and Fatty Acids: Providers of Body Structure and Function

Take Home and Summary

Carbohydrates and fats supply energy to cells. They also provide several molecules that are important in basic cell functions, such as DNA and RNA, several hormones and molecules involved in immune function. The major carbohydrate required by cells is glucose, the sugar required by red blood cells, nervous tissue, the liver, and the kidneys. Glucose is the most important nutrient for animals after water and oxygen and is supplied in the diet through starch from plant seeds and certain leaves. Carbohydrate (cellulose and hemicellulose) from plant stems and leaves can also supply, after digestion, some glucose to omnivorous and herbivorous animals (dogs and cats use very little, if any, fiber). Glucose is used to supply precursors for genetic material and some amino acids. Fatty acids are energy-dense carbon chains that provide energy. Some also play a role in membrane function, immune response, and inflammation. They are the major type of energy stored in animals, and have 2.25 times as much energy as glucose. Fatty acids and other lipids (sterols, cholesterol) are also used for cell membranes, nervous tissue, and several hormones. Only a few specific fats are actually required by the body; these are called essential fatty acids (linoleic acid, linolenic acid, and arachidonic acid), and they are not all required by all species. Animals can make all the other fats, if they need to, from glucose or amino acids. Once fats are formed in animals, they cannot be converted back into carbohydrates or amino acids, but must either be stored or oxidized for energy.

Types and Functions of Dietary Carbohydrates

Carbohydrates are the major form of chemical energy in the plant world, and the most important one in the animal world. Even though fats contain more energy, carbohydrates are absolutely essential for cellular life in animals. Therefore, animal systems have developed extremely complex and redundant control of glucose use through the nervous system and endocrine organs. First let us go through the types of carbohydrates in plants (these are the ones used for animal food) and then their function in animals.

Structural Carbohydrates

Carbohydrates make up the hard structure of stems, wood, and leaves. The generic nutritional term for *structural carbohydrates*, that is, those found in **plant cell walls,** is *fiber.* Because there are so many kinds of fiber, we'll be a little more sophisticated than that. There are many different terms in use to refer to these compounds, depending on whether you are a chemist, botanist, or nutritionist. We will only use the basic nutritional terms in this chapter.

> Glucose is the most important nutrient after water; it is required by the brain, nervous system, kidneys, and liver. If glucose does not come from the diet, animals must make it from certain amino acids and excrete the excess nitrogen.

Plants have cells. Cells have **cell wall constituents (CWC)** and **cell constituents.** It is the CWC of the cells that provide structure to stems, leaves, and the wood of trees. The major carbohydrates that make up CWC are **cellulose, hemicellulose,** and **lignin.** Lignin is actually a different kind of molecule that has little to no nutritional value and it is not really a carbohydrate. We will forget about it for now except to remember that when we use the term *dietary fiber,* we do include lignin. Because lignin has no nutritional value, this part of the fiber is actually unavailable to the animal and can be used to reduce the energy content of the food while keeping the digestive tract fuller. This strategy is sometimes used in weight-reduction diets.

> Plant cell walls are the source of *dietary fiber,* which is primarily cellulose and hemicellulose. Other fibers include pectins, arabinosides, and xylans, all common in fruits and vegetables.

The basic units of carbohydrates are **hexoses,** sugars that have six carbons, and the major one of these is glucose. **Glucose** is a 6-carbon chain (hexose) that usually forms a ring in nature (Figure 3.1). The six carbons each carry their own water molecule, broken up into what we call a hydroxyl group made of one oxygen and one hydrogen and then an additional hydrogen. Thus, the name *carbohydrate*: or "hydrated" (watered) carbon (chemistry really is simple stuff if you give it chance). We have seen in the last chapter that we need carbon, oxygen, and hydrogen to drive metabolism, and here is glucose, bringing in carbon for the necessary structures and its own two hydrogen to supply energy.

Glucose is the primary carbohydrate **monomer** (single unit) in nature. When we speak of **blood sugar** we are talking about glucose. *Sugar* is really not a strict chemical term, but it sounds like the German "sacher," which means sweet, because the German chemists were working hard in the early 1800s to determine which chemical was the "sweet" in sacher. The name stuck, and carbohydrates are still called sugars. Glucose is required by brain

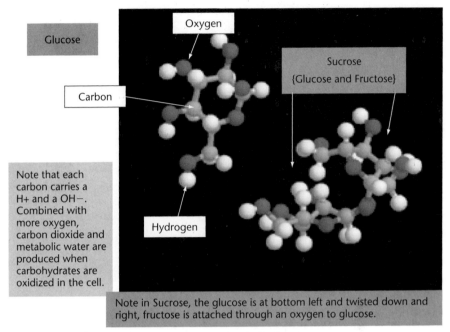

Glucose

Oxygen

Sucrose
{Glucose and Fructose}

Carbon

Note that each
carbon carries a
H+ and a OH−.
Combined with
more oxygen,
carbon dioxide and
metabolic water are
produced when
carbohydrates are
oxidized in the cell.

Hydrogen

Note in Sucrose, the glucose is at bottom left and twisted down and
right, fructose is attached through an oxygen to glucose.

Figure 3.1 Sugars, the building blocks of carbohydrates.
Source: Molecular models used with the kind permission of Dr. William McClure.

and nervous tissue cells as well as by the liver and kidneys. This is an absolute requirement (except in adaptation to severe starvation) for these organs, making glucose the most important nutrient after water and oxygen. Remember, we do not have to feed glucose all the time, because animals can make it from some of the amino acids if they must. But that does *not* mean that glucose is not required by the cells.

Figure 3.1 shows the chemical structure model of glucose (left) and sucrose, which is made up of fructose and glucose. Each carbohydrate carries oxygen and water. As the molecule is oxidized, oxygen is added and the carbon and oxygen are released to carbon dioxide (CO_2), and the hydrogen is transferred to the electron transport chain (Chapter 7) to make ATP for energy to make other cellular components. This process makes metabolic water in the proportion: 1 glucose plus 6 oxygen = 6 carbon dioxide and 6 water. Figure 3.2 shows the structure of fructose, a component of sucrose.

The supply of glucose is so important that animals have evolved complex hormonal and nervous regulation of glucose production and use by the body. The term for low blood sugar is *hypoglycemia*; high blood sugar is *hyperglycemia.* Both conditions are dangerous to cells and the animal. If the animal is not consuming enough carbohydrate in the diet, that can lead to hypoglycemia. At that point, the animal then must make glucose, and it can only make it from amino acids. Carnivorous animals, such as cats, make most of their glucose from the extra amino acids they consume, yet they can also consume carbohydrates to obtain glucose.

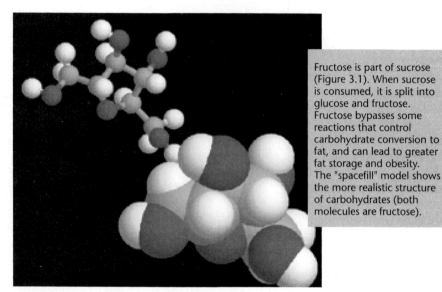

Fructose is part of sucrose (Figure 3.1). When sucrose is consumed, it is split into glucose and fructose. Fructose bypasses some reactions that control carbohydrate conversion to fat, and can lead to greater fat storage and obesity. The "spacefill" model shows the more realistic structure of carbohydrates (both molecules are fructose).

Figure 3.2 Fructose, the "deadly sugar."
Source: Molecular model used with the kind permission of Dr. William McClure.

There are some problems with glucose metabolism that have genetic as well as dietary causes. *Diabetes* is the name for the situation in which there is a malfunction in the production or use of the hormone *insulin.* Insulin is made by the pancreas and is required for glucose entry into cells of the muscle and fat tissue. Diabetes can be caused by a genetic problem with synthesis of insulin in the pancreas, or by a malfunction of insulin brought on by age and obesity. We will cover diabetes in more depth when we discuss dogs and obesity in Chapter 10.

> *Diabetes:* from Greek, to "stand with legs apart" (as in urinating).
>
> *Mellitus:* "sweet."
>
> Sweet urine, or diabetes mellitus has been recognized for 3,000 years and has genetic and dietary causes.

Glucose can be attached together (several hundreds of glucoses) in long chains generally called *polymers.* The type of bond connecting the two glucose units dictates whether it is a simple, carbohydrate-energy-storage form (starch) or a complex, tough, linear, cell-wall-constituent structure form, cellulose.

Starches are usually referred to nutritionally as *nonstructural carbohydrates,* which are primarily energy-storage forms in the seeds of plants. The primary ones are starch (*amylose*) and *amylopectin* (a more complex glucose polymer)(Figure 3.3). In starch, the glucose units are connected by an alpha-1, 4 linkage (different than the beta-1, 4 linkage). This linkage, instead of providing tight, linear chains for structure, gives a more granular structure. This

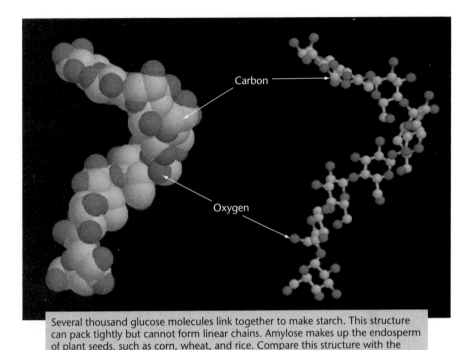

Several thousand glucose molecules link together to make starch. This structure can pack tightly but cannot form linear chains. Amylose makes up the endosperm of plant seeds, such as corn, wheat, and rice. Compare this structure with the more linear and tightly wound cellulose in Figure 3.4.

Figure 3.3   Amylose is starch, the major plant energy storage polymer.
Source: Molecular models used with the kind permission of Dr. William McClure.

allows close packing for dense energy storage. Plants store a lot of starch in seeds, and the seed uses this energy while it is germinating and beginning to make leaves and roots, before the leaves and roots make or gather the nutrients needed. Amylopectin is a looser structure than starch that allows water molecules in between the polymer chains. It is found commonly in fruits and vegetables such as apples and potatoes, usually between the cell walls. It is still energy storage, but not tightly packed like in corn and wheat.

Animals use amylose and amylopectin for the same thing—energy. Animals make the enzymes *amylase* (breaker of "amyl" chains) in the mouth and intestine and amylopectinase in the intestine to break down the chains and absorb the glucose for energy. Therefore, nowadays, plant starch is a major energy source for all animal life, even for our more carnivorous friends, the dog, and our very carnivorous friend, the cat. It is easy to come by, inexpensive, easily digestible, and a great source of energy.

Cellulose is a linear structure of glucose units connected by a beta-1, 4 linkage between the first carbon of one glucose and the fourth carbon of the next. Without understanding the chemistry, all you need to understand now is that this provides for a tight interlocking association between the strands of cellulose. This makes up the cell walls and provides rigid structure. Cotton and wood are made up of cellulose as are all stems and leaf structures. Most plant cellulose also contains other sugars such as arabinose and xylose. In fact, cellulose is so strong

Fiber is a generic term for plant cell walls. These are made primarily from cellulose and hemicellulose, which can be digested by some bacteria, usually in the large intestine, rumen of ruminant animals (cattle, camelids), and specialized cecum of herbivores such as horses and rabbits.

Note from the models that several glucose molecules are bound in a very tight formation. The stick model shows how the glucose units relate to each other, while the spacefill model represents the actual space used. Several of these strands can wrap around each other to make tough cell walls.

Figure 3.4 Cellulose: plant cell walls, food for herbivores.
Source: Molecular models used with the kind permission of Dr. William McClure.

that many polymers in use in clothing and several packaging films are actually made up in part or based on the structure of cellulose. Figure 3.4 shows carbohydrate molecules linked together in large fiber polymers, cellulose and hemicellulose. Cellulose and hemicellulose, along with lignin, make up the structural parts (stems, leaves) of all plants, including wood. The bacteria in the large intestine of mammals (or the rumen of ruminants such as cattle, deer, elk, llamas, alpacas, and camels) can break down most plants.

The properties that give cellulose such strength are exactly those that make it such a poor nutrient source for many animals: In the first place, it is poorly soluble, so it does not dissolve well in the intestines, so the enzyme that breaks it down (cellulase) cannot get at it easily. But even more important, the beta linkage that makes it so strong also means that it cannot be broken down by enzymes made by animals. Only some fungi, yeasts, and *anaerobic bacteria* (bacteria that do not require oxygen) can break down cellulose.

The anaerobic bacteria that make the enzymes to break down cellulose must have an oxygen-free atmosphere to work. The major oxygen-free environments are in the bottoms of swamps, compost piles, and in the rumen of ruminants and large intestines of animals. So you see, although cellulose is the most abundant biological substance on earth, it is not very available to use as a food source by nonruminants such as dogs, cats, and rodents. Humans, dogs, and cats can eat all the fiber desired, and although a little energy is gained from some types of fiber, it is not much. That is why humans sometimes eat a high-fiber diet to reduce energy consumption while still keeping the digestive tract full. The same is also true for dogs and cats. Their lower intestines do not support enough of the type of bacteria that can break down fiber well. So although we sometimes increase the dietary fiber in rations for dogs and cats to reduce energy intake, it is not a good idea to increase it too much as intestinal problems can result.

Hemicellulose is another carbohydrate polymer, but is a little bit different from cellulose. Instead of being only glucose units, it is made up of several different kinds of simple sugars, including glucose, galactose, arabinose, xylose, and ribose. Basically, hemicellulose has the same functions as cellulose with one major difference: It is much more digestible by animals and can be used for an energy source by several species. However, it is almost always found in a tight complex with cellulose and lignin, and so if you want hemicellulose, you get the other two, too.

Lignin is actually a complex network of rings of carbon with small carbohydrate side chains that makes a tough, strong, almost indigestible structure. Great for holding up plants, but impossible to live off of, unless you are an aerobic fungus. Anaerobic fungi live off dead plants because they can break down lignin. Intestinal bacteria cannot use lignin, so adding this fiber will provide "fill" to reduce energy, but will not provide any nutrients to the animal.

> Glycogen is the animal form of starch: long chains of glucose stored primarily in muscle cells and the liver to use as source of glucose for short-term needs.

Well, we have been discussing how plants store glucose for energy, and if animals eat it they can get the glucose for energy. But what if the animal eats more than it needs—can it store glucose for energy later? The answer is *yes.* Some animal cells can store large amounts of *glycogen*—glycogen is the animal equivalent of starch used to store glucose. It is a chain of glucose with alpha-1, 4 and alpha-1, 6 linkages and it is a little like amylopectin but with a lot more space for water. Thus glycogen does not pack well and uses a lot of space, so animals cannot store very much. Primarily it is stored in the liver and muscle. It exists to supply glucose when the body is running low. Because it takes up so much space, there isn't much, and it doesn't last for long. So the animal really requires a fairly regular supply of glucose for key life functions, and actually converts most of the excess consumed glucose into fats, to supply energy.

Figure 3.5 shows a schematic summary of glucose use by animals. Glucose is broken down to carbon dioxide to derive energy (ATP and $NADH_2$) to keep the animal alive, to grow, and to reproduce; a small amount is also used to make genetic material. However, note that if glucose is consumed in amounts more than the cells can use, some is stored as glycogen, and more excess glucose is converted to fat, a dense form of energy storage that cannot be converted back to glucose. Glucose is routinely used in combination with nitrogen to make some amino acids to make protein (see Chapter 4). Note, following the arrows in the figure, amino acids can be converted to glucose and can be converted to fat if there are too many amino acids consumed. Nitrogen is removed from amino acids that are used for making glucose. In mammals,

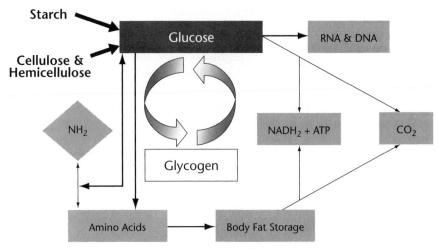

Figure 3.5 Summary of glucose use by animals.

this forms urea; in birds, uric acid; and in fish, ammonia which take energy to remove from the body, as well as affecting the environment. For this reason, feeding of protein in quantities significantly greater than requirements (even though it might be "natural") is not always the best strategy.

Types and Functions of Dietary Fats

Fats are the major energy storage form in the animal body. They are also used in the plant world, but not to the same extent. Fats consist of energy-dense hydrogenated carbon chains. When oxidized in the cells, they can provide tremendous amounts of chemical energy to allow for biosynthesis, work, or heat generation. To a certain extent, fats have gotten a bad name in nutrition. There has been some wonderful research on the role of fats in the diet and the effects of different types of fat on metabolism, immunology, and lifetime health. Unfortunately, there has been so much new knowledge that a lot of the meaning has been confused and garbled, and the general public is thus also often confused. Fats do contain a tremendous amount of energy, and this makes it easy for humans and animals to eat too much energy and become obese. But that fact does not make fats *bad*; many fats are essential to life and perform amazing and important functions.

Fats are essential to the survival of animals; certain types of fat are necessary for normal function. Most companion animal species do not metabolize fats the same way as humans and they are much less susceptible to the negative effects of certain types of fats and amount of fat intake. The primary problem remains the potential of too much total energy intake, leading to obesity and related problems.

The chemical name for the whole group of chemicals known as fats is **lipids.** A major process in lipid metabolism is *lipogenesis*—the synthesis of new fat (lipo) and creation (genesis). When animals consume foods such as carbohydrates and proteins, they can convert the excess glucose and amino acids into fat by lipogenesis in the *adipose tissue* (see Figure 1.3), which leads to the axiom "an instant on the lips, a lifetime on the hips." Excess glucose and amino acids will be converted to fats unless the energy expenditure of the animal increases to oxidize them to carbon dioxide. Dietary fats in excess of requirements do not need to be made into fat—they already are, so they simply get stored in the adipose tissue directly. It is easier, of course, to eat too much energy on a higher-fat diet as fat is so energy-dense.

The other major process of fat metabolism that we need to remember is *lipolysis,* the breakdown (lysis) of fat (lipo). Once fat is stored, it must be released, or *mobilized*, from the adipose tissue to be of use to the other organs such as muscle and heart. By the process of lipolysis, the fatty acids are cleaved off of the ***triacylglycerols,*** leave the fat cell (*adipocyte*), and can be used by the tissues as needed. Tissues can also use fatty acids to make specific lipids that they need for their function, such as the transfer of fat from adipose tissue of the mother to the milk to provide energy to the young. Now let us define some basic fat molecules and what they do in the animal body.

Simple Fats

Fatty Acids: Introduction

Simple fats are those existing as single molecules. Of these, fatty acids are the simplest. Fatty acids are to fats as sugars are to carbohydrates—one of the simplest forms of the chemical class. Fatty acids are chains of carbon, with a minimum length of two carbons. Some fatty acids in nature can reach up to sixty carbons. At one end is a *carboxylic acid* group made up of a carbon and two oxygen, thus the term ***fatty acid.*** This means that fatty acids can be fat soluble because of the long fat chain and to a certain extent water soluble because of the ionic acid at the end. We know some of these fatty acid molecules, when combined with some ions, as soaps. That is how people first cleaned things: making soap out of animal fat and *lye*, which is a strong chemical base that makes the fats soluble in water (you may ask your grandma about lye soap). The fat part sticks into the grease and dirt and then the carboxylic acid part stays soluble in the water and is diluted into the wash water, and *viola,* the clothes are clean.

The simplest fatty acid is *acetic acid,* and you may know this molecule as vinegar—vinegar is a fat. Because it is so small, it is *volatile:* it actually has a boiling point not much greater than room temperature. When you open a bottle of vinegar, you smell acetic acid. We already introduced acetate as a key nutritional intermediate. Much of carbohydrate, fat, and amino acid metabolism ends up producing acetate that can then be used to produce chemical energy in the tricarboxylic acid (TCA) cycle and electron transport chain. Or, acetate can be used to make many different compounds the cell needs, such as other fatty acids, cholesterol, sex hormones, and amino acids. Other fatty

> ### Chemistry of Fats and Hydrocarbons: Our Energetic Friends
>
> By the way, just a small chemical change in acetate (removing one oxygen) makes another chemical, which I am sure none of my readers knows anything about, called ethyl alcohol (better known as just "alcohol," although that term really applies to many compounds). So the smallest fat molecule is vinegar and its close cousin is alcohol. Both compounds carry a lot of energy and can do a lot of good or bad in the body based on their concentration. The 3-carbon fatty acid, *propionic acid*, is formed by microbial action in the hindgut, as well as by bacteria in the rumen (foregut) of cattle, sheep, and other ruminants. Propionate is converted back to glucose in the rumen or intestinal cells and in the liver. If we take the acid and turn it into an alcohol, it is *propanol*. A close cousin is *isopropanol*, also known as *isopropyl alcohol*, which your mother used to rub on your back or tummy to cool you off when you had a fever (and which you could use to treat acne). And if you take the alcohol off completely, you have three carbons and eight hydrogen, called *propane*, that you may use to burn in a camp stove or heat your house. Want more chemistry fun? The next biggest fatty acid has four carbons and is called *butyric acid*. It is also produced by bacteria, and can be converted to longer fats (it is a normal breakdown product of fatty acids used for energy in cells). Butyric acid produced by bacteria in cattle and horses is one major reason why cattle and horses smell like they do. Well, if we take the acid off and make *butane*, four carbons, ten hydrogen with no oxygen, you can use this in a butane stove or butane lighter. And one more: Octanoic acid is a fatty acid of eight carbons. It is made in the body in small amounts. Take the acid off, you have *octane*. That's right, octane, a major component of the gasoline you can't afford to put in your car. So, these fatty acids, alcohols, and alkanes have in common that they have a lot of energy.

acids can then be made by attaching molecules of acetate to each other to make longer chains—this is the process we earlier defined as lipogenesis. Acetate made from carbohydrate, amino acids, and other fats can be used to make a variety of fatty acids as the cell requires. We need to talk a little chemistry now, so hang in there. The use of fatty acids in the body and as foods is very much a function of their chemical properties, so we need to learn them. Two characteristics that are critical to fatty acid properties are *chain length* and *saturation*.

> *Chain length* is the number of total carbons in the fatty acid molecule.
>
> *Unsaturation* is the presence of one or more double bonds between carbons that give different shapes to fatty acids.
>
> Longer chains and more saturation means more solid fats. The amount of unsaturation is the important factor in fluidity.

Fatty Acid Chain Length

Remember glucose had six carbons, but fats can have from two up to forty, fifty, or more. In general the longer the chain, the more water-insoluble the fat becomes. It also becomes more solid at room temperature. The three simplest fatty acids, *acetic acid, propionic acid,* and *butyric acid* are two, three, and four carbons long, respectively, and we know these as the *volatile fatty acids.* If you have stood too close to the south end of a horse (or a cow, or sheep, or deer) facing north, you have almost certainly smelled these three compounds. They are made by anaerobic bacteria in the rumen of ruminants or the large intestines of many species. They each have melting points close to room temperature, so they become volatile and can be turned into gasses in a low percentage at body temperature. We will talk more about their unique use as nutrients for horses and other nonruminant herbivores in Chapter 11.

As chain length increases, fatty acids become less volatile and less water-soluble. From about five to twelve carbons or so they are liquid at body temperature. Therefore these are not found in large amounts in nature, and when they are, they are usually combined with longer fats to stay solid. These are usually intermediates on the way to longer fats. At about chain length fourteen to sixteen, fatty acids become solid at room temperature. In general, the larger the molecule, the more solid it is at room temperature, which is just another way of saying it has a higher melting point. The most common and important fatty acids are shown in Tables 3.1 and 3.2 with their primary functions.

The fatty acid that is sixteen carbons long is *palmitic acid* and this is widely distributed in nature. It is usually the fatty acid that plants and animals make in

Table 3.1	Characteristics of Fatty Acids through Sixteen Carbons			
Chemical Name	Common Name	Chain Length	Primary Purpose	Major Sources
Ethanoic acid	Acetic acid	2	Energy, cholesterol lipogenesis	Fermentation
Propanoic	Propionic	3	Energy, glucose	Fermentation
Butanoic	Butyric	4	Energy	Fermentation
Pentanoic	Valeric	5	Energy	Fermentation
Hexanoic	Caproic	6	Energy	Fermentation
Octanoic	Caprylic	8	Energy	Synthesis
Decanoic	Capric	10	Energy	Synthesis
Dodecanoic	Lauric	12	Energy	Synthesis
	Myristic	14	Energy	Synthesis
Hexadecanoic	Palmitic	16	Energy	Synthesis

Most fatty acids less than 8 carbons are only found in significant amounts after bacterial fermentation. Fatty acids from 10 to 14 carbons are not present in large amounts in plant or animal tissues, but do tend to be higher in some nuts and palm. Synthesis signifies that this compound is made by the animal.

Table 3.2	Characteristics of Important Long-Chain Fatty Acids			
Chemical Name	**Common Name**	**Chain Length**	**Primary Purposes**	**Major Sources**
Octadecenoic	Oleic	C18:1, Ω 9	Energy, membranes	Vegetable oils
Octadidecenoic	Linoleic	C18:2 Ω 6	Energy, membranes	Vegetable oils
Conjugated linoleic acids	CLA	C18:2, Ω 7	cis-9, trans-11 cell division, inhibits cancer growth	Rumen fermentation
		C18:2, Ω 6 t-10, c12	Reduces fat synthesis	Rumen fermentation
Alpha-linolenic	ALA	C18:3, Ω 3	Energy, membranes "pro-inflammatory"	Soybean, flaxseed
Gamma-linolenic	GLA	C18:3 Ω 6		Evening primrose Black currant
Arachidonic	AA	C20:4,Ω 6	Membranes, immune function, inflammation	Plants, animals
Eicosopentenoic	EPA	C20:5 Ω 3	Inflammation, prostaglandin	Synthesis
Docosapentenoic	DPA	C22:5 Ω 3	Inflammation, prostaglandin	Synthesis
Docosahexenoic	DHA	C22:6 Ω 3	Inflammation, prostaglandin	Synthesis

the greatest amount during lipogenesis. Then the cell can elongate that chain by adding more acetate units to make fatty acids of eighteen, twenty, twenty-two, or more carbons. The fatty acid of eighteen carbons long is *stearic acid*, and is found in greater amounts in animal fat (which, along with palmitic acid, is one reason why animal fat is more solid than plant fat). Stearic acid is found in even greater percentages in the fats of ruminants (such as cattle and sheep, this fat is called *tallow*), which is one reason why those fats are much harder at room temperature than pork fat (*lard*), or chicken fat.

Fatty Acid Saturation

We now know that the longer the fat the more solid it is and the less water-soluble it is. What about plant oils that are liquids at room temperature? Is it because they are a lot shorter than animal fat? Well, no. *Saturation* refers to how the carbons in the fatty acids are saturated with hydrogen. If every carbon is saturated with (bonded to) two hydrogen, then this is a fully **saturated fatty acid.** However, if the average number of hydrogen in a fat is less than two per carbon, it is an **unsaturated fatty acid.** This is done by the carbons forming what is called a double bond (Figure 3.6). If one hydrogen is removed from each of two touching carbons, then the carbons can form two bonds between them, or a *double bond*, carbon bonded to another carbon with two bonds. This major characteristic of saturation level dictates much of the function of fats. The saturation of fats with hydrogen makes major differences in chemical properties and will dictate whether or not a fat is a *fat* (solid) or an *oil* (liquid) at room temperature. In practical nutrition and food preparation, we call lipids that are solids at room temperature *fats* and those that are liquid at room temperature *oils*.

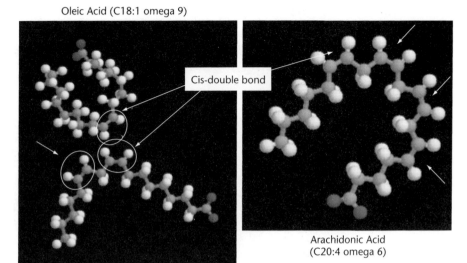

Oleic Acid (C18:1 omega 9)

Cis-double bond

Arachidonic Acid
(C20:4 omega 6)

Linoleic Acid (C18:2 omega 6)

Figure 3.6   Fatty acids.
Source: Molecular models used with the kind permission of Dr. William McClure.

If a fatty acid has two or more double bonds we say it is *polyunsaturated,* and you have probably heard or read about polyunsaturated fatty acids (PUFA) if you have watched much television or read a magazine or two in the last twenty years. **Linoleic acid** is eighteen carbons long and has exactly two double bonds, so the chemical formula is $C_{18}H_{32}O_2$. Linoleic acid is also found widely in many plants such as corn, soybeans, nuts, and sunflowers.

As the number of double bonds in fatty acids *increases,* the melting point *decreases.* It might not seem like this little difference in chemical formula or structure could make such a big difference in biology, but it does. Stearic acid (C18:0 as in beef fat) is a solid at room temperature, and oleic acid (C18:1, Ω 9 corn oil), just by removing two hydrogen and forming a double bond, is a liquid. Linoleic acid (C18:2, Ω 6), with one more double bond, is a liquid with an even lower melting point. **Alpha-linolenic acid** (C18:3 Ω 3) has one more double bond. All of these fatty acids have eighteen carbons, but the melting point *decreases* from stearic to oleic to linoleic to linolenic because the double bonds form a bend in the chain of the molecule, giving it more options to bend which makes it bigger in space, so that it cannot pack as tight (see Figures 3.6 and 3.7). Alpha-linolenic acid (C18:3, Ω 3) is also eighteen carbons, but has three double bonds, this decreases the melting point and makes it more fluid. This fatty acid is not common, being found primarily in soybeans and flaxseed, but it has important functions in immunity, which we will discuss in Chapter 11.

Figure 3.6 depicts fatty acids and double bonds. Oleic acid is the primary plant fat, and makes up most of common oils such as corn oil, soybean oil, and the like. Oleic acid has one double bond, so is called monounsaturated and is a liquid at room temperature. Linoleic acid is the same length

The length, number, and type of double bonds dictate the structure and function of fatty acids. Shorter length, more double bonds, and cis-double bonds mean more bends and the ability to fill more space, making membranes more fluid.

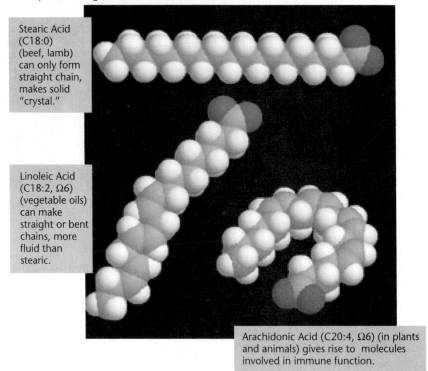

Stearic Acid (C18:0) (beef, lamb) can only form straight chain, makes solid "crystal."

Linoleic Acid (C18:2, Ω6) (vegetable oils) can make straight or bent chains, more fluid than stearic.

Arachidonic Acid (C20:4, Ω6) (in plants and animals) gives rise to molecules involved in immune function.

Figure 3.7   Fatty acids—"solid" and "liquid."
Source: Molecular models used with the kind permission of Dr. William McClure.

of eighteen carbons but has two double bonds, thus it melts at an even lower temperature. Arachidonic has twenty carbons and four double bonds. It is often classified as a nutritionally essential fatty acid for most animals; however, in most cases it is only strictly required for cats and some birds and fish. A cis-double bond usually puts a "bend" in the molecule, which makes it harder to "pack down" and thus more fluid. A trans-double bond usually forms a straight chain, allowing closer packing of molecules and less fluid. Note that these chemical characteristics affect membrane function, including heart and blood cells, and play a role in immune function and inflammation.

Figure 3.7 shows how fatty acids can differ in shape. The space-filling model of Figure 3.7 demonstrates that fatty acids can take different shapes. The different shape can alter how the fatty acids interact with genes to alter gene expression (whether or not a protein is made). In addition, they can affect enzyme function to increase or decrease some reactions in energy metabolism or immune function. Fatty acids with "more space" provide a more fluid membrane, which may have a role in arterial and cardiac health (see Figure 3.8).

The varied fatty chain length and number and type of double bonds allows a number of membrane formations from "more stiff" to "more fluid." More fluidity generally relates to better overall health and longevity. Some of these fatty acids the animals consume, and some they make on their own.

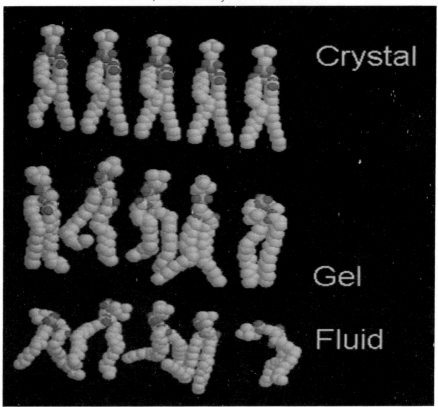

Figure 3.8   Phospholipids in bilayer.
Source: Phosphatidyl cholines taken from model lipid bilayers simulated by H. Heller, M. Schaefer, and K. Schulten, "Molecular dynamics simulation of a bilayer of 200 lipids in the gel and in the liquid-crystal phases," *Journal of Physical Chemistry* 97:8343-60, 1993. See also *http://www.umass.edu/microbio/rasmol/bilayers.htm* by Eric Martz for other treatments. Molecular models used with the kind permission of Dr. William McClure.

Figure 3.8 is an illustration of the effects of the placement and type (cis or trans) of double bond on fatty acid conformation in membranes. The fewer double bonds (greater saturation) or the more trans-configuration double bonds, the straighter the fat, the more it packs, and the more crystalline, or solid, it will be. More double bonds, double bonds closer to the middle of the chain, and cis-configuration double bonds will cause fatty acids to pack more loosely and be more fluid. This affects membrane function, ease of transport in blood, and immune function and inflammation response.

Fatty Acid Double Bond Location

The point at which the double bond is in the carbon chain is referred to in relation to the last, or *omega*, carbon in the chain. The first carbon is the carboxylic acid, and the next one is the *alpha*, or first after the acid, the second is *beta,* then *delta, gamma, epsilon,* and so on. The last carbon is *omega,* for the last letter in the Greek alphabet. There are three general classes of unsaturated fatty acids, the *omega-3, omega-6,* and *omega-9,* named for where in the chain the double bond is made. Arachidonic acid (C20:4 Ω 6) is longer, but has four double bonds, with the last one three carbons from the end (omega-3). Figures 3.6 and 3.7 show different fatty acids with double bonds at different points in the chain, giving different bends and shapes to the molecule. This all affects how they pack in crystals, membranes, and changes their metabolic function.

Gamma-Linolenic acid (C18:3 Ω 6) is also eighteen carbons with three double bonds, just like alpha-linolenic acid, but the double bonds are in a different position. These small differences in structure result in different fluidity and, more importantly, may lead to different molecules that control inflammatory response and immunity. We will discuss the potential role of these fatty acids in inflammation and immune function in Chapter 11. When two molecules have the same chemical formula, or molecular weight, but have different structures, they are called *isomers*. We understand that these molecules, although they are only present in small amounts, may have significant effects on health.

There is one more last bit of chemistry we need to know. The double bond can form two different configurations that dictate how the molecule is shaped in nature, either a relatively straight chain, or a bent chain with a bend at different angles. Figures 3.6 and 3.7 show a trans double bond (such as the one circled in arachidonic acid) that makes a structure that is relatively straight, just like normal saturated fatty acids. A cis double bond, as in oleic and linoleic, forms a bend in the molecule, so that it fills more space because the carbon chains cannot pack together tightly. So in general, unsaturated fatty acids with trans double bonds pack tighter and are more solid than unsaturated fatty acids with cis double bonds. This is one reason why you may have heard or read that *trans fats* are bad: they are simply much more like saturated fats than unsaturated oils. They also may directly affect heart function, inflammation, and immunity; this is still under intense study and we are not fully sure of all their effects. Many trans fats that we or our animals eat are in fact manufactured, not naturally occurring. They are made by chemical saturation (adding hydrogen) to plant oils to make them solid, for example, to make margarine and baking shortening.

One relatively new finding exemplifies the importance of these small chemical differences. There are two molecules that we call **conjugated linoleic acid.** They both are eighteen carbons long and have two double bonds. However, one has a cis-double bond at carbon 9 from the acid end (also 9 from the omega carbon) and a trans double bond at the omega 7 position, so called C18:3 cis 9, trans 11 (numbered from carboxyl end). This isomer has been found to reduce tumor growth, and is the only naturally occurring nutrient to be declared anti-carcinogenic by the American Cancer Society. The other isomer has a trans double bond at the tenth carbon from the front (omega 8) and a cis double bond at the omega 6 position (C18:2 trans 10, cis 12). This isomer has been shown to reduce triacylglycerol synthesis, which can reduce body fat. There are supplements on the market already containing these fatty acids, even though research is not complete. Although there is still not a tremendous amount of scientific evidence for these effects, the evidence is growing. These molecules are produced only in the rumen of ruminant animals such as cattle and sheep, thus these molecules are present in larger concentrations in fats from cattle, sheep (lamb), and wild ruminants. Research is ongoing to determine if there is any benefit to greater consumption of these fatty acids by humans and animals. This is another example of the living science of nutrition.

> The debate rages: Which is worse, natural animal fat, like butter, with saturated fats and cholesterol, or that "healthy" plant oil that has been chemically saturated, with its trans fat? The major issue is in the amount used; these are not truly *bad*, unless consumed in excess of what the body needs. There is some evidence that these saturated fats may have harmful effects, but this is not clear. Moderation in all things (including pet nutrition) is a philosophy that has been somewhat pushed aside by a lot of "scientific knowledge" in the press, but is still a good philosophy.

Triacylglycerols

A triacylglycerol is a molecule that is a combination of three (tri) fatty acids connected to a *glycerol* by *ester* linkages. Yes, we have to think chemistry again for a minute. Glycerol is a 3-carbon compound that is half of glucose with a few other changes made. The carboxylic acid end of the fatty acid can make what is called an ester bond with the carbon on the glycerol. Since there are three carbons on the glycerol, up to three fatty acids can bind to it, thus: *triacylglycerol*.

> Triacylglycerols are to fats what starch is to carbohydrates: the primary storage form. However, they are highly insoluble in water (hydrophobic).

Some are liquid at body or room temperature; some are semisolid at body or room temperature, depending on the chain length or unsaturation of the composite fatty acids. Because there is no free acid group on these compounds, they are basically insoluble in water. This gives them their natural advantage—they can be packed very tightly in dense globules. This is the real reason why oil and water don't mix. The triacylglycerol has no ionic group to interact with the water. It is this chemical fact that makes triacylglycerol so useful as energy storage, but that also takes a tremendous amount of metabolic regulation to make it work, both in plants or animals. Much of what is "bad" about overconsumption of fats has to do with the simple fact that they are water insoluble and thus can "plug up," in a very oversimplified sense, the "water flow" of animals.

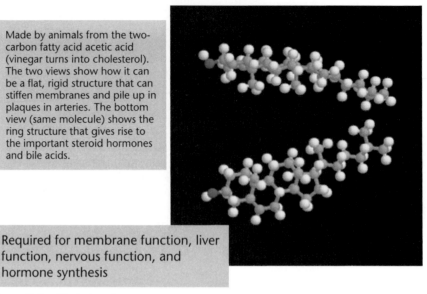

Made by animals from the two-carbon fatty acid acetic acid (vinegar turns into cholesterol). The two views show how it can be a flat, rigid structure that can stiffen membranes and pile up in plaques in arteries. The bottom view (same molecule) shows the ring structure that gives rise to the important steroid hormones and bile acids.

Required for membrane function, liver function, nervous function, and hormone synthesis

Figure 3.9 Cholesterol.
Source: Molecular models used with the kind permission of Dr. William McClure.

Sterols

Now we move on to really a completely different kind of fat. It is not based on fatty acids, but has a structure of rings of carbon we call *sterols.* They are a complex structure of three rings of six carbons each and one of five carbons arranged in a specific pattern (Figure 3.9). These cyclic compounds are only soluble in other fats or oils. They make up a variety of structural and functional lipids in nature; in fact, life as we know it could not exist without sterols. Common sterols include cholesterol, the sex hormones *estrogen*, *progesterone*, and *testosterone*, *corticosteroid* hormones (which control water use and energy use), and compounds such as vitamin D. Their commonality is the compound ring structure—four 6- or 5-carbon rings attached together to make a rigid planar structure. Various side groups containing carbon, hydrogen, or oxygen give them different properties.

Cholesterol

Cholesterol is only made by animals. Other sterols are made in plants, but only animals make cholesterol. Due to its high degree of insolubility, too much can tend to gather in wounded spots on blood vessels and cause atherosclerotic plaques. Too much of this can block arteries and lead to coronary artery disease. Does this mean our pets and us do not need cholesterol and should avoid it at all costs? No. It means it is needed in the right amounts at the right times. Cholesterol is a critical component of membranes in cells; without cholesterol, membranes would not function and cells would not exist as we know them. Animals are not required to eat cholesterol because they can make it. Neither does cholesterol need to be avoided, as the body has a

Cholesterol is the parent compound of most sterols in animals, and is essential to animal life. Too much, however, may lead to problems. Dogs, cats, and horses tend to be resistant to excess cholesterol buildup, but that is not true of some rodents, birds, and reptiles.

regulatory system to reduce synthesis of cholesterol as consumption increases. It is only when repeated, long-term overconsumption of cholesterol and other fats happens that problems such as obesity or arteriosclerosis (in humans and a few companion species) can occur. Cholesterol is a major functional fat in the animal world. From this comes important hormones that regulate water use, kidney function, lipid digestion, and reproductive traits. Cholesterol is a precursor for many other important compounds. The most closely related are *bile acids*. These are the major transport forms to help digest fats and remove some excess cholesterol from the body. They also make complexes with some other toxic compounds to remove them from the body. Their major function is in digestion of fats: They aid in the intestine to keep the fats in solution in water so that they can be digested by the enzyme *lipase* (lipid-breaking) and absorbed into the body. Major bile acids are taurocholic and glycoholic acid. Bile acids are made in the liver stored as bile in the gall bladder and transported to intestines though the bile duct.

Steroids

Steroids encompass a large class of sterols built upon the cholesterol ring structure. They make up many important regulatory hormones in the body, most of which affect nutrient use. The first group is sometimes called *corticosteroids*, steroid hormones made by the cortex of the adrenal gland, next to the kidney. Sometimes they are referred to as *adrenal steroids*. The first are the **glucocorticoids**; these are essential in the regulation of body glucose and fatty acid use. **Cortisol** and **corticosterone** help regulate how the body uses glucose to store as energy. They are required for normal adipose tissue function and synthesis of body fat. A deficiency in these hormones can be caused by some genetic mutations, and in those animals a marked abnormality in body fat deposition is noted. Some corticosteroids can reduce inflammation, and are often used in prescription or over-the-counter products to do so. *Mineralocorticoids* are also made by the adrenal gland and are essential in the regulation of body water and mineral balance.

The last major category of steroids includes the sex hormones: estrogen, progesterone, and testosterone (Figure 3.10). These are made from cholesterol and are chemically very similar. It is their minor differences that dictate how the whole animal kingdom "goes around"—without them, we would not have sex differences and reproduction. The circles in Figure 3.10 show the small chemical difference between estrogen and testosterone. **Estrogen** is the major female hormone needed for ovulation and milk production. Estrogen and related compounds help control the estrus cycle of making new eggs for fertilization and also help the mammary gland grow to get ready to feed the young.

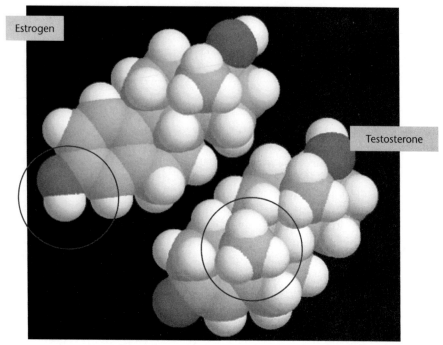

Figure 3.10 The major sex steroids, estrogen (female), and testosterone (male). *Source:* Molecular models used with the kind permission of Dr. William McClure.

Progesterone is the other major female hormone needed for pregnancy and milk production. Once the egg is fertilized, progesterone is made from cholesterol in the ovary and this helps the new embryo and placenta to be maintained by the mother and tells the mother not to start any new reproductive cycles. Progesterone is also required for mammary gland growth to get ready for lactation.

Testosterone is the major male hormone needed for spermatogenesis. In addition to this, testosterone in the male and estrogen and progesterone in the female direct what we call secondary sex characteristics. These are the different body shapes, hair and feather coating, coloring, and musculature that differentiate the male animals from the female animals. Early in development, while the animal is still a fetus, it is the absence of testosterone that allows estrogen to develop the ovaries and uterus in the female. If the testosterone gene is not "turned off," the fetus develops as a male.

Complex Fats

Well we are now done with the list of what are referred to as simple fats. You've probably guessed by now that we aren't even close to done with fats. Complex fats are really where a lot of the action is in metabolism and nutrient use. Complex fats are molecules and compounds that are a mix of fats and other chemicals, usually proteins or mineral groups. They have important functions in many body systems.

Phospholipids

Phospholipids, as you might guess, are fats complexed with the phosphoric acid group. We have actually learned a little about this important chemical already—we have seen phosphoric acid, or phosphate, before in ATP and $NADPH_2$. It is the strong chemical force of the bonds that holds the phosphate groups together and that provides the energy to drive other chemical reactions. Most phospholipids take the form of a triglyceride, but with one fatty acid of the triglyceride replaced by a phosphorous-containing compound. Thus, they have a very strong water-insoluble fat group as part of the structure and one very strong water-soluble group at the other end. This gives phospholipids tremendous flexibility in the world—they live in "both worlds at once." They are absolutely necessary for life and are important in membrane function, nervous transmission, and activity of many hormones.

Sphingolipids

The next class, which is also extremely critical to animal life, is the *sphingolipids,* complex combinations of long-chain fatty acids with other compounds. Some of these are known as *sphingomyelin,* which makes up the *myelin sheath* of nerves. This water-insoluble fat does not conduct electricity, but provides insulation around the nerve cells that do. Electrical impulses move very quickly inside the brain, up and down the spinal cords, and along all the nerves to direct all the functions of the body without interference from or interfering with these functions. All parts of the nervous system, including brain, membranes, and neurons, contain this essential compound.

Lipoproteins

Animals use fats for many purposes such as energy transformation, nervous system function, membranes, and hormones. But you may be wondering how they get from place to place in the body through the blood; they are insoluble in water. In come the *lipoproteins,* complexes of fats with proteins. These large structures allow the hydrophobic fats to be carried through the blood. These proteins have regions that are both hydrophobic (which bind to the fats) and hydrophilic (which are soluble in water). Because fats are insoluble in water, they must be complexed with other compounds in order to move about in the body. There are many different lipoproteins, but they all function to get fats from the digestive tract to the body tissues or from the body tissues to other body tissues or back to the liver and bile for excretion.

Good or Bad? The adjectives *good* and *bad* for cholesterol should never be used.

High-density lipoproteins carry cholesterol to the liver for removal from the body, thus the label "good."

Low-density lipoproteins carry cholesterol into other tissues, thus the label bad. They are neither good nor bad; they each have a specific function.

The high-density lipoproteins (HDL) in general carry cholesterol *away from* tissues to the liver for removal from the body and the low-density lipoproteins (LDL) in general are designed to carry fats and cholesterol *to* the tissues for normal function. Thus, in humans, elevated concentrations of LDL in the blood have been associated with atherosclerosis and arterial disease and are also correlated with overeating of sugars and saturated fats. The HDL tend to be inversely correlated with heart and arterial disease and tend to be increased by regular exercise. It is not the cholesterol per se, but in fact the transport for lipoproteins that can tend to cause excess storage of cholesterol in arterial walls, which can lead to problems in humans. Luckily for dogs and cats and most other animals, their physiology is a little different. Although they definitely do have cholesterol and lipoproteins, for some reason they are very resistant to arteriosclerosis. Thus they can get fat, but arterial and heart disease are not the major problems associated with obesity in these animals.

So, fats are important in energy generation and storage as fatty acids and triglycerides; fat digestion and removal of excess cholesterol as bile acids; insulation as triglycerides and myelin; membrane structure and function as triglycerides, phospholipids, and cholesterol; lipid transport as lipoproteins; hormone action as cholesterol, steroids, phospholipids, and vitamin D; brain and nervous function as phospholipids, sphingolipids, and essential fatty acids; skin function as essential fatty acids; and many other metabolic regulatory actions as steroids, phospholipids, essential fatty acids, and their derivatives the prostaglandins.

Practical Application

Carbohydrates and fats are essential elements of life. Carbohydrates and fats should be supplied in the diet of most animals as a mix of plant seeds, sometimes stems and leaves, and animal fats. When balanced correctly and fed in amounts appropriate to the animal being fed, this mix will provide sufficient glucose for energy and other functions, and the right amount and balance of the various fats for energy, membrane structure, hormones, and fat-soluble vitamins. Commercial foods are well-designed to provide the right amount of starch, energy, and essential fats for the animals that they are made for. Choosing the right food for the life stage of your animal and feeding it according to label directions and personal observation will meet the needs of your animal the best.

Words to Know

acetic acid	*blood sugar*	*cellulose*
alpha-linolenic acid	*butyric acid*	*cholesterol*
amylopectin	*cell constituents*	*conjugated linoleic acid*
amylose	*cell wall constituents*	*corticosterone*
anaerobic bacteria	*(CWC)*	*cortisol*

diabetes	*isomer*	*polymer*
estrogen	*lignin*	*progesterone*
fatty acid	*linoleic acid*	*propionic acid*
fiber	*lipids*	*saturated fatty acid*
glucocorticoids	*lipolysis*	*sphingolipids*
glucose	*lipoproteins*	*sterols*
glycogen	*monomer*	*testosterone*
hemicellulose	*nonstructural*	*triacylglycerol*
hexose	*carbohydrate*	*unsaturated fatty acid*
hyperglycemia	*phospholipids*	*volatile fatty acid*
insulin	*plant cell walls*	

Study Questions

1. Describe the characteristics and makeup of starch versus cellulose.
2. What are the major roles of glucose in the body, and what dietary components provide glucose?
3. What are the differences between sucrose and glucose?
4. What are the major uses of glucose in the body?
5. What are simple fats?
6. How does the double bond and chain length alter the properties of fats?
7. What are trans-fats and cis-fats, and why is this difference important in nutrition?
8. What are several complex lipids and their functions?

Further Reading

Case, L. P., D. P. Carey, D. A. Hirakawa, and L. Daristotle. 2000. *Canine and feline nutrition.* St. Louis, MO: Mosby, Inc.

Hand, M., C. D. Thatcher, R. L. Remillard, and P. Roudebush, eds. 2000. *Small animal clinical nutrition,* fourth edition. Topeka, KS: Mark Morris Institute.

Heller, H., M. Schaefer, and K. Schulten. 1993. Molecular dynamics simulation of a bilayer of 200 lipids in the gel and in the liquid-crystal phases. *Journal of Physical Chemistry,* 97:8343-60.

Martz, Eric. 1993. Molecular models. http://www.umass.edu/microbio/rasmol/bilayers.htm.

McClure, William. 2003. Biochemistry 1. Fall Term 2003 Website. *http://www.bio.cmu.edu/courses/BiochemMols/BCMolecules.html.* Images of sugar, fatty acid, amino acid and protein molecules courtesy of Dr. William McClure, Carnegie Mellon University, with permission.

McMurray, John, and Robert C. Fay. 2004. *Chemistry,* fourth edition. Englewood Cliffs, NJ: Pearson Prentice Hall.

National Research Council. 2004. *Nutrient requirements of dogs and cats.* Washington, D.C.: National Academy Press.

Petrucci, Ralph H., William S. Harwood, and F. Geoffrey Herring. 2002. *General chemistry: Principles and modern applications,* eighth edition. Englewood Cliffs, NJ: Pearson Prentice Hall.

Chapter 4

Amino Acids and Proteins: Providers of Body Structure and Function

Take Home and Summary

Proteins provide the structure and function of life in the animal world. Amino acids are made of carbon, hydrogen, oxygen, and nitrogen. Proteins are integral parts of cell membranes; they are the enzyme catalysts that allow metabolism to proceed; they are the framework of bone, joint, ligaments, and tendons; and they make up the structure of muscle and several important hormones. Proteins digested by acids and enzymes in the stomach and intestines are converted to amino acids. The amino acids are absorbed by the body to make its own proteins. The type and pattern of amino acids in each protein dictates the structure and function of the protein. Many proteins are complexed with carbohydrates, fats, minerals, and vitamins to carry out the various cell and body functions. Plant and animal proteins have varied amino acid compositions, so we feed a mix of plant and animal proteins so that the animal can match them up to make their own proteins. No protein or protein mixture can match exactly the amino acid needs of an animal. The amino acids that animals cannot make are dietary-essential amino acids and we must provide them in food. Any excess amino acids must be oxidized to carbon dioxide or stored as fat, making ammonia as a waste product. Ammonia is converted to urea in mammals, uric acid in birds, or excreted as ammonia in fish.

Amino Acids—Building Blocks of Life

We have learned about the monomers and polymers of carbohydrates and fats. Now let us turn to *amino acids* and proteins. Basically, proteins are the reason we need carbohydrates and fats: Proteins actually make up "life as we know it." Carbohydrates and fats in large part provide the energy needed to make new protein. Many proteins have carbohydrate and fat molecules bonded to them to provide more variety of structure and function. Proteins form the beginnings of bone and help to create movement in muscle. Proteins are a key part of cell membranes for movement of ions and nutrients into and

- Methionine (Met)
- Valine (Val)
- Arginine (Arg)
- Phenylalanine (Phe)
- Threonine (Thr)
- Tryptophan (Trp)
- Histidine (His)
- Isoleucine (Ile)
- Leucine (Leu)
- Lysine (Lys)

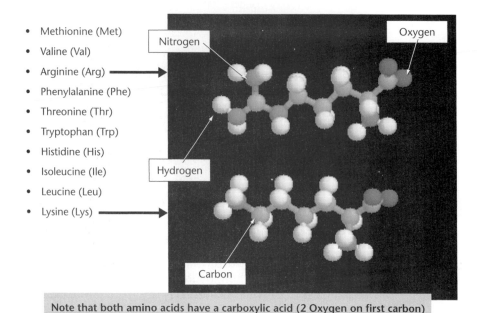

Note that both amino acids have a carboxylic acid (2 Oxygen on first carbon) and an amine group (NH₂) on the next (alpha) carbon. These two both have 6 carbons, but Arginine, on top, has 3 more amine groups on the other end.

Figure 4.1 Essential amino acids. These are the amino acids that most animals cannot synthesize; thus, they are called essential for the diet.
Source: Molecular models used with the kind permission of Dr. William McClure.

out of cells. Proteins make up many critical hormones that regulate processes of respiration, ion transport, nutrient movement in blood, blood clotting, growth, and reproduction.

> The most important function of proteins is **enzyme** action. An enzyme is a chemical **catalyst** that allows a reaction to take place without a change in the catalyst itself.

Most metabolism proceeds because there is a protein *enzyme* that catalyzes a reaction. Without protein enzymes, life as we know it would not exist, and you would be studying something else in biology.

Figure 4.1 shows the structure and listing of the essential amino acids. Dietary essential amino acids are those that animals cannot make in sufficient amounts to maintain normal life, so they must be consumed in the diet. Every amino acid has a carboxylic acid made up of carbon and two oxygen. On the first (alpha) carbon from the carboxylic acid is an amine group made of nitrogen and two hydrogen. The essential amino acid shown here, arginine, also contains three more amino groups at the far end of the structure. These are broken off in the urea cycle to make urea, the waste product that removes excess amino acids from the body in the urine.

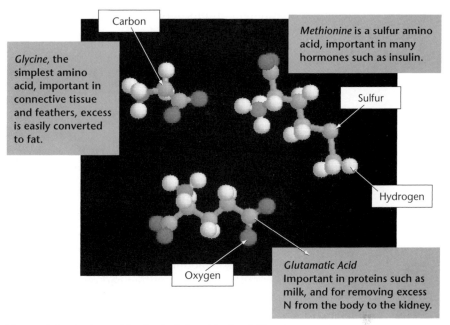

Figure 4.2 Amino acids, the building blocks of life.
Source: Molecular models used with the kind permission of Dr. William McClure.

Amino acids, similar to carbohydrates and fatty acids, have a skeleton of carbon molecules linked together. Like fatty acids, they start with a carboxylic acid group. However, the major difference is that they have an *amine* group (made of nitrogen and two hydrogen) attached to the alpha carbon, right after the carboxylic acid group. Thus, most amino acids are referred to as *alpha amino acids*. Amino acids can range from two to six or seven carbons in length (Figure 4.2), so they are much smaller than most fats but about the same size as glucose. However, the carbon chains can vary tremendously in their chemical makeup. The simplest amino acid, glycine, has just two carbons: one is the carboxylic acid and the other is the alpha carbon with the amine group attached. The amino acid serine is glycine with a hydroxyl group (oxygen and hydrogen attached together). To form other amino acids, we add to this core structure a **side chain:** Some amino acids have sulfur molecules attached, some have a straight or branched chain of carbons, others have a carbon ring structure, and others have another carboxylic acid or other amine groups attached further down the carbon chain (Figure 4.2). It is the chemical property of the side chains that gives great variety in protein structure and function.

A simple *bio-math* exercise can help you to understand just how powerful this variety in chemical side chains can be. If there are twenty amino acids, how many ways can just two of them be combined? Close the book for a minute, go to a computer, calculator, or a piece of pencil and paper and calculate how many combinations can be made from twenty amino acids arranged in different groups of two (called a dipeptide). Then do it for three, four, five, and ten amino acids.

> The tremendous variety of proteins is a simple mathematical function. Twenty different amino acids can form 400 (20 × 20) different *dipeptides*, 800 (400 × 20) different *tripeptides*. The hormone *oxytocin*, which functions in milk removal from the mammary gland, has only eight amino acids, one of the smallest proteins known. But oxytocin is then just one out of 20 × 20 × 20 × 20 × 20 × 20 × 20 × 20 = 25,600,000,000 (that is 25 billion, 600 million) different possible combinations for a peptide chain of eight amino acids in length. For a protein of fifty amino acids in a chain, there are possible combinations of 1.126×10^{65} different proteins! Is it any wonder, then, that animal life is so amazingly varied?

Figure 4.2 shows an example of the structure of three other amino acids, glycine (the simplest), methionine (a sulfur-containing amino acid) and glutamic acid, which contains another carboxylic acid and is used in many important metabolic reactions. In all cases, they have the beginning carboxylic acid (carbon, two oxygen, one hydrogen), with an amine group (nitrogen and two hydrogen) attached to the alpha carbon. The different chemical side chains give the different amino acids their chemical structure and function.

A critically important concept in nutrition is that of dietary *essential amino acids* (Figure 4.1). An essential amino acid is one that the animal cannot make or cannot make enough of to live well or even survive for long periods. Therefore, they must be consumed in the diet. If an animal can make enough of the amino acid in its own body, it is a *nonessential amino acid* in the diet, meaning the animal can make it from other amino acids and glucose. Nonessential amino acids for most species (see Table 4.1) include glycine, alanine, cystine, cysteine, serine, glutamate, aspartate, tyrosine, glutamine, asparagine, proline, ornithine, and citrulline. There are species differences; for example, birds need more glycine (Chapter 15).

> An essential amino acid is required *in the diet* for a specific animal and life stage. However, the cells of the body require all the amino acids; they are all *essential to life*. The dietary essential amino acids are *less* important to the body because the animal evolved over time to survive, only survive, without them. But this is only bare survival, not true life. Practically, these amino acids must be in the diet.

The distinction between nonessential and essential relates to the diet, not the animal. In practical application, most diets provide both essential and nonessential amino acids. You might think of the ability to make the nonessential amino acids as a survival skill: It is good to have, but you only use it when you need to.

Table 4.1	Amino Acid Chemistry and Uses	
Name	**Chemical Type**	**Special Purpose**
Arginine, Lysine	Basic amino acids	Lipoproteins, urea cycle
Methionine, Cysteine	Sulfur amino acids	Cell structure, hormones
Valine, Leucine, Isoleucine*	Hydrophobic, branch-chain proteins	Lipid transport, muscle and milk
Alanine, Glycine	Simple amino acids	Glucose synthesis, energy
Glutamate, Glutamine	Dicarboxylic amino acids	Energy metabolism, urea cycle, milk
Aspartate, Asparagine	Dicarboxylic, diaminic amino acid synthesis	Nitrogen exchange, transport
Threonine, Serine	Hydroxyl amino acids	Mucus, intestinal lining
Histidine, Tryptophan, Proline	Cyclic amino acids	Lipid transport, membranes, connective tissue, milk
Phenylalanine, Tyrosine	Phenolic amino acids	Thyroid function neurotransmitters
Ornithine, Citrulline		Special amino acids used in the urea cycle

*Essential amino acids are in bold type; they must be contained in the diet in sufficient amounts to supply the optimal needs of the animal. Animals may survive for long periods with a small amount of essential amino acids, but it will not be an *optimal* life for longevity and health.

Proteins—Structure and Function of Life

So, how do we know which amino acids are needed and which proteins to feed? Well, it just so happens that a lot of experimental work has been done, and in other books there are tables of hundreds of protein and amino acid compositions of different feedstuffs. For less commonly used feeds we might need to do new analyses. We do experiments to determine amino acid requirements. Thousands of experiments to determine amino acid requirements have been done and we have good estimates of amino acid requirements for many species. There is still a lot we do not know, however. In this text, we are going to learn some more about protein structure and function and the different types of proteins in nature. In Chapters 8 and 9, we will discuss how these are used in food formulations and application to protein nutrition with concepts such as first limiting amino acid and protein quality. For now, let us finish up on protein chemistry.

Proteins are linear chains of alpha amino acids linked by peptide linkages of the amine group of an amino acid to the carboxylic acid of the next. As cellulose is a long chain of glucoses that can form tough, stemy structures in plants, proteins are long chains of amino acids that can form many billions of different proteins for structure or function of cells. The combination of the acid and amine allows a powerful bond to be formed between the amine group of one amino

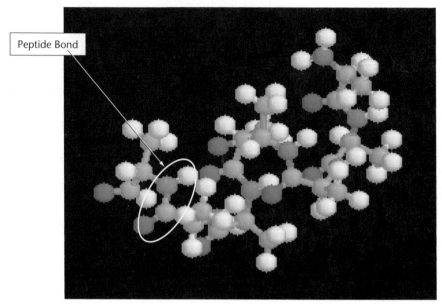

Peptide Bond

Figure 4.3 Chains of amino acids bound by peptide bonds make peptides and proteins, often with a structure called an alpha-helix.
Source: Molecular model used with the kind permission of Dr. William McClure.

acid and the carboxylic acid group of another, which we call a ***peptide bond*** (Figure 4.3). As Figure 4.3 shows, the peptide bond between the amine group of one amino acid and the carboxylic acid group of another, repeated many times, makes a chain of amino acids, which is called the primary structure (Figure 4.3). The secondary structure is caused by folding of the chain on itself, usually caused by chemical interactions (ionic forces) between side chains. A major secondary structure is an alpha-helix.

There are four general levels of protein structure, each building on the one that comes before. The actual sequence of amino acids in the chain is the ***primary structure.*** Then, the amino acids in the chain can interact chemically with other amino acids (without forming tight bonds), causing the chain to fold or bend on itself, and this is the ***secondary structure. Tertiary structure*** is the final shaping of the protein, usually caused by ionic interactions among side chains as well as direct sulfur linkages between cysteine molecules in what is called a disulfide bridge, which locks the protein in a specific shape (think of two poles or pipes bracketed together). Insulin is an example of a simple protein that derives its function from having two disulfide bridges to lock it into shape. Finally, complete proteins can interact between each other to form large "macromolecules," and this is ***quaternary structure.*** Figure 4.4 shows the primary, secondary, and tertiary structure of a protein. At the base of all this structure is the actual amino acid sequence. From this it is decided what secondary, tertiary, and quaternary structure can exist. This provides the basis for the variety of structure and function from enzymes to hormones to huge proteins found in connective tissues and muscles.

The coil is an alpha-helix. Note in the stick model it is an open structure.

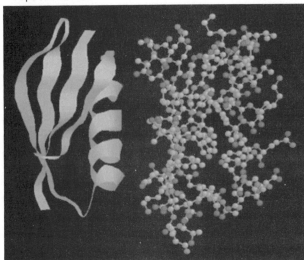

Secondary structure is the alpha-helix or beta-pleated sheet. These are the basic chains of the protein.

Tertiary structure is the folding and arrangement of the chains with each other.

Quaternary structure is the arrangement of different proteins with others to make up large enzymes, receptors, or structural proteins.

The ribbon shows a beta-pleated sheet. Note in the stick diagram this forms a tight structure.

Figure 4.4   Full protein structure, showing secondary and tertiary structure. *Source:* Molecular model used with the kind permission of Dr. William McClure.

We have discussed the structure and function of proteins in the bodies of the animals we feed. But what about nutrition? Protein nutrition is simple: Proteins ingested are broken down to their constituent amino acids in the digestive tract. The amino acids are absorbed into the bloodstream. The amino acids are used to make the proteins required by that animal in that environment at that stage of the life cycle. Thus, simply put, protein nutrition consists of the consumption of plant or animal proteins, their breakdown into constituent amino acids, and the synthesis of new animal proteins. Thus, this is critical to your understanding: The animal's requirement is for amino acids, not proteins.

Proteins are fully digested to amino acids and a few small *peptides* in the small intestines and the amino acids are absorbed to be used for protein synthesis. Even in the extreme case in which an animal would consume exactly the same amount and balance of proteins from a diet as existed in itself, they still would be broken down into the amino acids before use.

Understanding amino acid nutrition is also helpful to evaluate nutritional claims. Proteins in and of themselves have no value—as they are not absorbed whole, and are not available to the body. You may have seen some ads about how we should eat or feed enzymes to help with this or that problem. The only problem with this idea is that any enzyme is going to be broken down to amino acids, so whatever activity it might have had as an enzyme is gone. Thus, feeds that contain enzymes, growth factors, or miracle proteins' should be critically examined before use.

Two possible exceptions to this rule are in neonates (twenty-four to forty-eight hours postpartum), and animals experiencing digestive diseases. In the first case, the tight junctions between intestinal cells are not yet closed, so that milk-borne (colustrum) immunoglobulins can be absorbed intact to provide "passive immunity." In the second case, bacterial, viral, or physical damage can open gaps in the membranes that allow passage of whole proteins or peptides, which then may or may not cause an immune response or infection.

Types and Functions of Proteins

Proteins come in several major types. Some are only found in the plant kingdom, some only in the animal kingdom, and some in both. All are made up of amino acids, or combinations of amino acids, carbohydrates, or fats. Just as we had simple carbohydrates and simple fats, we also have simple proteins, chains of only amino acids without other chemicals. We will briefly go through the types of proteins and touch upon how their chemistry affects their biological function and nutritional value.

Globular Proteins

These proteins make up most of the proteins in nature. They are *globular*, which in rough translation means they form globs when purified in solution— it actually means they form "globe-like" structures. As silly as that sounds, it means that they are *globulins*, fairly water insoluble, and they need to be associated with various ions like sodium, potassium, phosphates, or others to dissolve in water. Globulins include muscle proteins and some plant proteins, among others. They can be structural or enzymatic in nature.

> **Simple Proteins**
>
> Globulins
> Albumins—animal proteins, blood, muscle, many enzymes
> Glutelins—plant proteins, in all seeds, each has a characteristic texture in a food
> Fibrins
> Collagen—basic connective tissue protein; elastin—important in skin, ligaments, tendons

Albumins are usually water soluble, usually of animal origin; their function is as a temporary "storage" of amino acids, as hormones or binding proteins, or transport proteins. Blood protein and egg protein are primarily albumins. Blood proteins perform many different functions (transport oxygen and lipids, carry minerals and vitamins). Egg protein is of course used to supply the developing chick in the egg. Most enzymes are globular proteins.

Glutelins are primarily plant proteins. They vary widely in solubility and are usually part of the seed of plants as storage and function during germination. Because each plant is a little different, they will be quite diverse in chemical properties and amino acid content. Those of you who have ever baked cookies, pies, breads, or pizza dough have seen firsthand the properties of plant glutelins. Along with the starch, the variety of plant proteins gives the characteristic textures to the different foods made with these cereal grains. The formulation of pet foods uses these characteristics to make a food that is not only nutritionally complete, but of a texture that is pleasing and acceptable to the animal, and will handle and store well.

Fibrins

Fibrins are those proteins found only in the animal kingdom that make up the various connective tissues of the body. As plants use fibrous carbohydrates to provide structure, animals use a variety of fibrous proteins. These proteins are usually made up primarily of highly charged (positive or negative) amino acids. The ionic interactions between the chains make very strong connections applicable to holding things together such as bones, ligaments, and tendons. Usually they have little nutritional value unless highly processed (usually boiled or steamed) to denature them, that is, to break down the quaternary, tertiary, and secondary structure to form just the simple peptide chain. The simple peptide chain can then be more easily broken down by *proteases.* They are basically insoluble in water unless heated for long periods of time, and are generally resistant to most digestive enzymes. However, when cooked to break down the tough structure they can be a good source of amino acids. This is routinely done to make soups and in some wet dog and cat foods. These proteins or by-products from them are also used in many medical and nutritional situations as wound healers, gelatins, and the like.

Collagen is the protein that makes up most connective tissue, ligaments, tendons, muscles, basement membranes, and the "base" structural membranes upon which many organs build their connective tissue and give cells a structure to attach to. As nutritional proteins, if broken down to be digestible, they can be a good source of amino acids and are somewhat high in lysine.

Elastin is another special protein found in connective tissue, ligaments, tendons, muscles, and basement membranes. They have special properties that give them elasticity. You've all done the hair test: You pull out a hair and pull on it. It stretches and finally breaks. The amino acid sequence is such that there is some give in the ionic bonds before eventually it stretches too far and you actually break one of the tough peptide bonds. Hair is not an elastin, but it illustrates the point. Elastins are structured to do this all the time, giving flexibility to skin, ligaments, and tendons. If not for elastin, we would always have pulled tendons.

FUN FACT

Elastase

Elastase is an enzyme that breaks down elastin. When females are close to giving birth, a hormone named *relaxin* increases the activity of *elastase* to partially break down the elastin in the pelvis. This provides more flexibility to move and expand as the young proceed through the birth canal. So if you see someone or some animal near term, walking like they are falling apart, they are. This is also a reason why sometimes near birth or after, animals may have a hard time walking around (there are other reasons too, but this is one).

Keratin is a specialized hard protein that makes up hair, nails, hooves, exoskeletons of insects, and so on. This is a very insoluble and tough protein, only animal in origin. It really does not have any nutritional value without extensive acid digestion far beyond the capacity of digestive tracts. Sharks might be an exception, but basically this is an animal protein, not a food protein.

Conjugated Proteins

So far we have learned about proteins that are pure protein, proteins not usually combined with any other type of chemical. Yet throughout the animal and plant worlds there are thousands of conjugated proteins. *Conjugated protein* simply means that they have attached to the peptide chain some amount of carbohydrate, fat, vitamin, or mineral. These are extremely important functional proteins in the body. Many conjugated proteins transport fats, minerals, or vitamins to be used in various tissues. Many enzymes (Chapter 3) are actually conjugated proteins, with minerals or vitamins that allow the enzyme to function. Finally, many hormones and their receptors are conjugated proteins. The true importance is what they actually do in the body once the animal makes them from the proteins that it does eat. So let us become familiar with them.

Conjugated Proteins

Glycoproteins (protein with sugars attached)—many hormones and membrane proteins.

Mucopolysaccharides (polysaccharides with peptides attached)—lubricants in intestine, saliva, mucus, and joints.

Phosphoproteins—regulatory molecules in cells and membranes.

Lipoproteins—proteins complexed with fats, usually to move fats in the blood.

Metalloproteins—proteins and metals such as iodine (thyroid hormone); selenoproteins for antioxidation; hemoglobin (Fe).

Phosphoproteins are conjugated with phosphorous. This is an important class of proteins and they have many regulatory functions in the body. They include enzymes and other molecules that control metabolism in cells. Protein phosphorylation is a major method by which cells respond to hormones to change their metabolism. In the flight or fight response, it is the hormone epinephrine from the adrenal gland that tells the adipose tissue to release fat for energy for the muscles to use to fight or flee. And it is phosphorylation of the amino acids serine and tyrosine in the proteins that actually activates the hormone-sensitive enzyme lipase (HSL) to break down fat. This is just one of many examples in which the activity of proteins changes by adding or removing one or more phosphorous molecules. Many metabolic pathways of glucose and fatty acid metabolism are controlled by phosphorylation of protein enzymes.

Glycoproteins are also protein and carbohydrate, but these are usually mostly protein with just a little "sugar" attached. The protein chain is made and during construction or after it is complete, various molecules of carbohydrate are added. This allows the protein to be unique and identifiable when it comes near the receptors on the cells. There are many glycoproteins in blood and along the linings of arteries and veins to provide structures and specific functions for the interaction of blood and organs. Many receptors for hormones on cells are glycoproteins, as are several protein hormones. Some proteins involved in the immune response, as well as several hormones that can cause allergic reactions, are glycoproteins. The addition of sugar molecules to the proteins can provide extra uniqueness to the protein, making it easy to recognize. This is used (as hormones and antigens, for example) for extremely specific functions in the body. Figure 4.5 shows some functions of conjugated proteins.

Mucopolysaccharides are complexes that usually contain more carbohydrate than protein. These generally act as lubricants in nature, including several mucins, which make up the covering for mucous membranes that we know as *mucus.* These would include many proteins secreted by the salivary glands, intestinal cells, vaginal secretions, and joint lubricants. They really are amazing when you think of all the different animal functions requiring protection and lubrication that are carried out by mucopolysaccharides.

Lipoproteins are protein complexes with various lipids, or fat molecules. You might remember that fats are not soluble in water, so how do they get around in the blood? Well, because of lipoproteins. These *hydrophobic* proteins

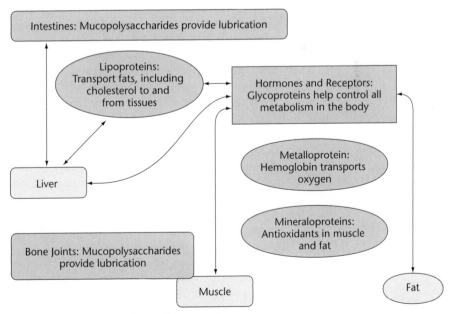

Figure 4.5   Conjugated proteins at work.

help to transport different fats from place to place. They are to keep fats in solutions in the blood and tissues, to transport cholesterol into and out of tissues, and as membrane proteins, some enzymes, and receptors. One major function of lipoproteins is the movement of cholesterol throughout the body. Mammals have evolved a complex and effective way to move cholesterol to where it is supposed to be, including out of the body. Recall the discussion about bile acids, and remember that it is lipoproteins that deliver cholesterol to the liver to be broken down or made into bile acids.

One more type of conjugated protein is the **metalloproteins,** which are proteins with one or more minerals added. These make up another wide variety of proteins, often enzymatic proteins, hormones, or transport proteins. Key examples included hemoglobin (with iron), ceruloplasmin (copper), cyanocobalamin (B12 and cobalt), and selenoproteins (selenium). It is usually the mineral that gives the protein its function: It is the iron in hemoglobin that allows oxygen to bind; selenium in selenoproteins acts as a natural chemical antioxidant to maintain cell health.

Flavoproteins are proteins coupled with flavinoids such as riboflavin (which, in the next chapter, you will learn is vitamin B₂). *Flavinoids* are a class of heterocyclic compounds very good at accepting hydrogen atoms and handing them off to other molecules—this makes up a large part of the electron transport chain to generate energy (Chapter 7). Finally, *nucleoproteins* are those proteins involved in the structure, transcription, translation, and repair of cellular DNA and RNA, another amazingly interesting adaptation to life.

Nonprotein Nitrogen

We have learned about many different proteins. There is a class of nitrogen-containing compounds that are *not* proteins. They are usually breakdown products of amino acids, but also include the nucleic acids in genetic material. These compounds can be ingested by animals as parts of plants, so sometimes we need to account for this nitrogen in feeds. We can measure this in feeds and correct for the nonprotein portion when we make rations. Two major take-home ideas are that for most normal feeds, *nonprotein nitrogen* is usually only a very small percentage of the total N, and it is usually completely worthless to most companion animals nutritionally, because the animals cannot make amino acids out of the nitrogen. This is not true in the case of ruminants, in which the rumen microbes can use the nonprotein nitrogen and carbohydrates to make amino acids.

A major nonprotein nitrogen compound is *urea.* You have already been introduced to this molecule as the normal waste product of amino acid metabolism in mammals. It is made in some other chemical reactions in plants and bacteria. It can be synthesized as fertilizer or a feed additive for ruminants. It can also be formed during spoilage of wet foods. At sufficient levels, it will reduce food intake and it can be toxic to animals, so it is important to process foods correctly and minimize bacterial fermentation. The chances of urea contamination are greater in foods for horses (Chapter 12).

Ammonia is a nitrogen metabolism intermediate in plant and animal life. It is removed from the body as urea in mammals, uric acid in avians, and as ammonia in fish. Ammonia can build up in raw foods under improper warming, with bacterial or fungal action, or in breakdown of animal proteins in a wet and warm environment. So it is important when feeding some raw materials or using them in food preparation to prevent this with cooling and quick handling. Sometimes, ammonia can be important in ruminant diets, because bacteria can use the nitrogen group to make amino acids. But for most other animals it is at best wasteful, makes the food taste lousy, and in worse cases can be toxic.

Nucleic acids are the parts of DNA and RNA that contain nitrogen in what are called purines (adenine and cytosine) and pyrimidines (thymine, guanine, and uracil). They have little practical nutritional value, but can be present in some foods in significant amounts, especially some specialty products containing a large amount of yeast or bacterial cells. Sometimes the amount of nitrogen can be measured as total food nitrogen and if it is assumed this is all part of protein, an overestimation of protein content can occur. This is easily corrected for in our feed analyses.

Practical Application

Because of their importance to life, and because they tend to be the most expensive ingredient in rations, proteins are critical to good nutrition. With too little total amino acids, or with not enough of one or a few, animals will not do well. If there are too many, they will be converted to fat. And always, the animal must make urea, uric acid (birds, reptiles), or ammonia (fish) to get rid of the excess nitrogen from the wasted amino acids. Thus a practical goal is to provide close to the right amount and balance of proteins to supply the amino acid requirement. Changing the mix of different plant and animal protein sources, we can meet the amino acid needs of a wide variety of companion animal species in many stages of life. This is a continually changing and improving aspect of nutrition, as we want to reduce waste and prolong the life of the animal by minimizing the production of potentially harmful nitrogenous wastes.

Words to Know

albumins	*elastin*	*glutelins*
amino acid	*enzyme*	*glycoproteins*
ammonia	*essential amino acid*	*keratin*
catalyst	*fibrins*	*lipoproteins*
collagen	*flavoproteins*	*metalloproteins*
conjugated protein	*globulins*	*mucopolysaccharides*

mucus	*peptide*	*protease*
nonessential amino	*peptide bond*	*protein*
acid	*primary, secondary,*	*side chain*
nonprotein nitrogen	*tertiary, or*	*urea*
nucleic acids	*quaternary*	
nucleoproteins	*structure*	

Study Questions

1. What is an alpha amino acid, and how do chains of them make proteins?
2. What are some major functions of protein in the body?
3. Does the form (type, structure) of a protein in the diet really make a difference to the nutritional value? If so, why? If not why not?
4. Why are enzymes so important?
5. Describe several complex proteins and their functions.
6. What is the difference between a dietary essential amino acid and a dietary nonessential amino acid? Does this mean that all twenty amino acids are not essential to cells in the body?

Further Reading

Case, L. P., D. P. Carey, D. A. Hirakawa, and L. Daristotle. 2000. *Canine and feline nutrition.* St. Louis, MO: Mosby, Inc.

Hand, M., C. D. Thatcher, R. L. Remillard, and P. Roudebush, eds. 2000. *Small animal clinical nutrition,* fourth edition. Topeka, KS: Mark Morris Institute.

McClure, William. 2003. Biochemistry 1. Fall Term 2003 Website. *http://www.bio.cmu.edu/courses/BiochemMols/BCMolecules.html.* Images of sugar, fatty acid, amino acid and protein molecules courtesy of Dr. William McClure, Carnegie Mellon University, with permission.

McMurray, John, and Robert C. Fay. 2004. *Chemistry,* fourth edition. Englewood Cliffs, NJ: Pearson Prentice Hall.

Morris, J. G., and Q. R. Rogers. 1982. Metabolic basis for some of the nutritional peculiarities of the cat. *J. Small Anim. Practice,* 23: 599–613.

National Research Council. 2004. *Nutrient requirements of dogs and cats.* Washington, D.C.: National Academy Press.

Petrucci, Ralph H., William S. Harwood, and F. Geoffrey Herring. 2002. *General chemistry: Principles and modern applications,* eighth edition. Englewood Cliffs, NJ: Pearson Prentice Hall.

Schaeffer, M. C., Q. R. Rogers, and J. G. Morriss. 1989. Protein in the nutrition of dogs and cats. In I. H. Burger, and J. P. W. Rivers, eds., *Nutrition of the dog and cat.* New York: Cambridge University Press.

Chapter 5

Vitamins: Cofactors for Nutrient Metabolism

Take Home and Summary

Vitamins are essential for most metabolic reactions, such that they are commonly referred to as *metabolic cofactors.* The importance of vitamin nutrition was slowly discovered over centuries as humans noted that humans and animals did not do well if they did not consume certain types of foods, in small amounts, even when sufficient total food was consumed. Also, vitamins usually need to be consumed fairly regularly, as all but a few cannot be stored in the body. Thus, throughout history, deficiency symptoms occurred and disappeared as the types of foods consumed changed. Vitamins are essential nutrients, but are not used directly in body structures, are generally not stored, and are not oxidized to provide energy. Therefore their requirement is usually measured by how well they prevent some deficiency symptom. The requirement might be noted as the prevention of a disease, or improved growth or reproduction. Each vitamin has a different role so no general rule for requirements or deficiencies applies, other than reduced performance and general health and in severe cases, illness and death. Animals may survive their entire normal life span with marginal deficiencies of some vitamins. Modern nutritional practice includes formulating rations with a balance of naturally occurring vitamin sources supplemented with additional vitamins to ensure a proper and consistent supply.

Role and Function of Vitamins

It was noted long ago, even in ancient China and Greece, that it wasn't just "enough food" that was needed for optimal health, but specific types of foods. Diets without fruit and vegetables resulted in malaise, sickness, and sometimes even early death. Eating only one type of food usually meant a person did not do as well as when a mix of foods was consumed. Historically, a disease called **night blindness** could be cured by eating yellow vegetables such as carrots. Growth might be increased by providing meat into a diet of only rice, barley, or wheat. Bringing fruits on long sea voyages could prevent scurvy. As humans progressed, and scientific investigation began in earnest, more observations were made about the role of many food types in optimal health of humans and animals.

> The study of vitamin nutrition is a rich history of observation, hypotheses, testing, scientific debate, and practical application for the betterment of humans and animals.

Late in the 1800s and into the 1900s, researchers finally started to make specific hypotheses and conduct experiments to determine exactly what it was in different foods that prevented wasting, deformities, and disease and allowed for improved health and longevity. Fields of biology, chemistry, physiology, botany, and microbiology all came together to solve some of the world's scientific mysteries, and resulted in the eventual removal of a major scourge of humanity (and their animals). Animals and humans live so much easier, healthier, and longer today in part because we know about the role of vitamins and have easy access to them. Figure 5.1 summarizes the role of vitamins throughout history.

The history of vitamin discovery, identification, synthesis and practical application embodies all the best in human inquisitiveness and intelligence. Whole new fields of science were started, and the longevity and quality of human and animal life were dramatically improved. Unfortunately, vitamin deficiencies still exist in large parts of the world.

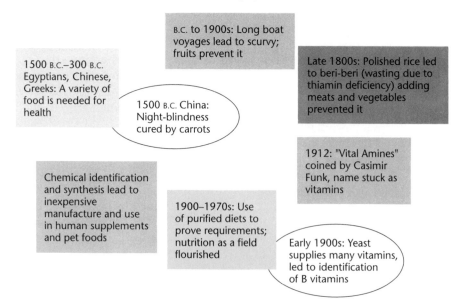

Figure 5.1 Vitamins in human development.

Table 5.1	Vitamins Involved in Energy and Amino Acid Metabolism		
Vitamin Name	**Sources**	**Metabolic Role**	**Deficiencies**
Thiamin—B₁	muscle and organ meats many plants	carbohydrate and fat metabolism	general wasting, slow growth beri-beri
in horses, fern or bracken poisoning from thiamin inhibitors in wet feeds such as fish meal there is an enzyme thiaminase that breaks down thiamin			
Riboflavin—B₂	muscle, many plants yeast, fungi good in legumes poor in cereals milk, eggs, liver, organs, muscle are good	flavin adenine dinucleotide in TCA cycle, electron transport chain	decreased growth, general weakness, and lethargy
Niacin	nicotinic acid or nicotinamide widespread in food supply	major role in energy metabolism part of NAD (see Chapter 7) some niacin can be made from tryptophan protein and amino acid metabolism	pellagra (muscle wasting, weakness) anemia (loss of protein synthesis)
Pyridoxine— B₆	widely distributed in foods	amino acid metabolism	high protein intake may increase requirement deficiencies include slow growth, microcytic anemia (small red blood cells), nerve degeneration, skin lesions (rats), and convulsions

We now know that the primary role for most vitamins is as *cofactors* in metabolic reactions. They aid the enzymes in converting nutrients into compounds the body needs, or aid in the breakdown of nutrients to provide energy. Vitamins act as donor or acceptor groups for metabolic intermediates, helping the interchange of different nutrients into animal structures. Other roles include antioxidant factors, regulatory agents, roles in nutrient absorption and transport, and structural functions in membranes.

Because of their central role as cofactors in metabolic reactions, deficiencies of vitamins are generally noted as reduced performance (slow growth, impaired reproduction), general malaise, skin lesions, hair loss, and sometimes specific disease states characteristic to the type of reactions involved. In the Western world today, because of the normal wide variety of fresh foods available to us, the availability of vitamin supplements, and the regular inclusion of vitamins in diets for our animals, actual vitamin deficiencies and

Table 5.2	Vitamins Involved in Metabolism and Cell Synthesis		
Vitamin Name	**Sources**	**Metabolic Role**	**Deficiencies**
Pantothenic Acid	widely distributed in foods	cofactor in coenzyme A in all metabolism, including membrane synthesis important in all carbohydrate, lipid, and protein metabolism	deficiencies include poor growth, hair loss, skin lesions, reproductive failure, nervous system and intestinal lesions
Biotin	widely distributed in foods	cofactor in transfer of single carbons from one molecule to another to make amino acids and other compounds	poor growth, dermatitis, loss of hair avidin (raw egg whites) binds biotin and prevents absorption (it takes a LOT of raw egg whites to cause biotin deficiency) cooking destroys avidin
Cyanocobalamin— Vitamin B$_{12}$	not found in plant world, only from bacterial synthesis and found in meats	functions in single-carbon (CO_2, CH_4) reactions	secondary folic acid deficiency by blocking folate utilization important in synthesis of genetic material (DNA and RNA)
Folic Acid— Folacin	widely distributed in foods	important in single-carbon metabolism, synthesis of serine, glycine, purines (in RNA and DNA)	important in pregnancy spinal cord malformations in young

related disease states are almost never seen. In some parts of the world, unfortunately, they still exist. Figure 5.2 depicts the role of vitamins in nutrient metabolism.

> Vitamins are metabolic cofactors, usually essential for an enzyme to catalyse a reaction.

In practical application, a vitamin deficiency should only be suspected in extreme nutritional situations (for example, in a situation of profound neglect), or as secondary effects of diseases when intake of food might be reduced. The more inquisitive student should peruse the reading list for many fine publications with historical pictures of vitamin-deficiency symptoms.

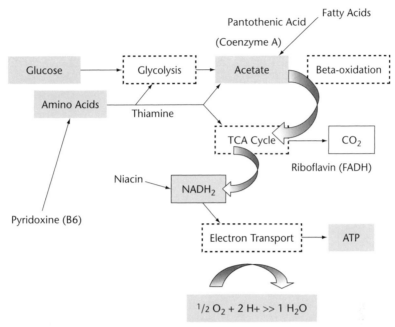

Figure 5.2   The role of some vitamins in nutrient metabolism.

My purpose is for the reader to have an initial knowledge of the vitamins; their role in normal growth, health, and life; an introduction to potential deficiency problems; and finally, a deep appreciation and respect for the century of scientific dedication that has now made life so much more pleasant, healthy, and long for us and our animal friends.

Tables 5.1 through 5.4 provide a quick list to aid in associating each vitamin with its basic function and major deficiency. Each one is described briefly. Teachers and students in various classes may wish to use some of the references to find detail as desired.

Fat-Soluble Vitamins

The major difference between *fat-soluble* and *water-soluble vitamins* is their chemical nature. Fat-soluble vitamins are *lipophilic* (lipo = fat; philic = loving) and *hydrophobic* (hydro = water; phobic = hating) and thus "love" and stay with fat. They can be stored in the body, usually in the liver, unlike water-soluble vitamins. Vitamins A and D, because of their chemical nature and because they can be stored in the liver, may be toxic in high amounts, whereas water-soluble vitamin toxicity is not a problem. They also tend to have very specific roles so that their deficiencies can seem much more devastating, as they affect one or a few systems very dramatically.

Table 5.3	Vitamins with Specialized Functions in Cell Synthesis and Organ Functions		
Vitamin Name	**Sources**	**Metabolic Role**	**Deficiencies**
Ascorbic Acid— Vitamin C	found in many fruits (apples, oranges) found in milk but destroyed in pasteurization green leafy vegetables, potatoes animals can make some vitamin C	functions in formation and maintenance of connective tissue acts as an antioxidant, hypothetical basis for protection against colds	known early on as "anti-scurvy" factor in citrus fruits deficiency is scurvy, loss of connective tissue integrity
Choline	in fats, oils, organs	part of phosphatidyl choline not truly a required vitamin but sometimes a response is seen	essential in metabolism of fat functions in fat transfer within cells prevents fatty liver, but not shown to be therapeutic functions in acetylcholine transfer supplements not always found to be beneficial

Figure 5.3. shows the incredible action of vitamin D. The role of vitamin D is an example of a biological system to help ensure survival and health. Calcium is extremely important to nervous and muscular activity and an acute loss of calcium from the blood can cause paralysis and death. Vitamin D action has evolved to ensure that this only happens in extreme situations. Vitamin D precursors are converted to active forms, which then control calcium coming into the body from the gut and release from the bones, as well as what is lost in the urine. If they receive sufficient UV rays from normal sunlight, animals can survive without vitamin D. Hector DeLuca at the University of Wisconsin is an example of a tenacious scientist who kept working to find "The Final Answer."

Moderate deficiencies of vitamin A and vitamin E have been associated with decreased reproductive function, including lower fertility and reduced fetal growth or numbers in litter-bearing species (primarily agricultural animals). This is still an area of scientific debate, and specific roles and mechanisms of either vitamin A or E in reproduction are not fully understood. Severe deficiencies can affect reproduction, but this may be a generalized effect of loss of membrane integrity. Because the effects to date have been small, it will take large-scale experiments to define any specific roles. However, the

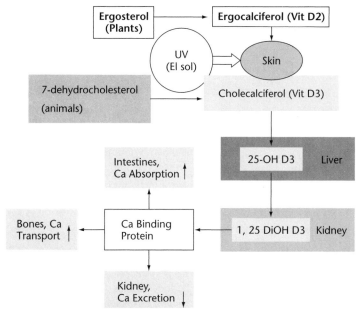

Figure 5.3   Vitamin D: the vitamin that is a hormone.

cost of adding a small supplement of these vitamins is much less than the cost of large-scale experiments. Newer techniques that are evolving as we define the function of the genome may aid in these matters. At present, there is no proof that larger doses (for example, more than twice the normal level) would improve reproduction.

A final note must be clearly remembered here. Naturally occurring deficiencies of most vitamins have rarely been reported for dogs or cats, unless they have been in a situation of actual starvation. The wide variety of dietary ingredients, and the fact that naturally living wild dogs and felids consume animal products, including intestinal contents, practically ensures that vitamin and mineral deficiencies are very rare. Most of the information on deficiencies in this chapter and the next has been gathered (necessarily) from research or observations on rodents or in research situations on dogs and cats, not from natural occurrences. It was only through experiments that withheld vitamins from experimental animals that actually allowed discovery of the functions of many vitamins and minerals. In return, deficiencies are almost unknown because of the proper formulation of rations with a large variety of feedstuffs and the normal inclusion of vitamin and mineral supplements as needed.

Table 5.4	Fat-Soluble Vitamins, Their Sources, Functions, and Deficiencies		
Name	**Sources**	**Metabolic Role**	**Deficiencies**

Vitamin A	beta-carotene
	many plants
	muscle and organs
		functions in rhodopsin
		  synthesis for proper
		  eyesight
		reproductive functions—
		  ovarian tissue
		skin functions—poor coat
		general membrane
		  function
			night blindness, eye
			  damage in severe cases
			bone abnormalities,
			  organ damage, skin
			  lesions
			toxic in large doses—
			  hypervitaminosis A

Vitamin D	vitamin D_2: ergosterol,
	  ergocalciferol:
	the plant form,
	  widely distributed
	vitamin D_3:
	  7-dehydrocholesterol,
	  cholecalciferol
	in animal tissues,
	  this form is required
	  by avians, reptiles
	formed by UV activation
	  of ergosterol (plants) or
	  7-dehydrocholesterol
	  (animals only)
	final active form is 1,
	  25 dihydroxycholecalciferol
		major function is in
		  calcium metabolism
		acts through regulating
		  synthesis of Ca-binding
		  protein
		required in intestinal
		  cells, bone cells,
		  kidney tubules
		can help out when Ca/P
		  nutrition is poor, but
		  cannot replace Ca/P
			deficiency similar to
			  calcium—rickets in
			  young, bone
			  deformations
			is toxic in large doses,
			  toxicity causes
			  increased Ca absorption,
			  hypercalcemia, and
			  symptoms similar to
			  deficiency: decalcification
			  of bone, stunted
			  cartilage growth,
			  calcinosis of soft tissues

Vitamin E—Alpha-Tocopherol	distributed in plants
	  and animal tissues
		acts as intracellular
		  antioxidant
		functions in muscular
		  health
			deficiencies seen as
			  generalized myopathy—
			  muscle weakness:
			  "white-muscle disease"
			note this is not genetic
			  muscular dystrophy
			can act as shelf-life
			  preserver and feed
			  preservative due to
			  antioxidant activity
			generally not toxic in
			  large doses
			very high doses can
			  interfere with normal
			  intestinal function
			interacts with selenium,
			  also an antioxidant

Table 5.4	Fat-Soluble Vitamins, Their Sources, Functions, and Deficiencies *continued*		
Name	Sources	Metabolic Role	Deficiencies
Vitamin K— Phylloquinone	synthesized by anaerobic bacteria (rumen, large intestine) nonruminants may get absorption from lower gut or must practice coprophagy phylloquinone (active compound), menaquinone, menadione (synthetic)	functions in protecting against hemorrhages required for synthesis of prothrombin dietary need most important for avians	deficiencies almost never seen in natural state megadoses not generally toxic, but menadione can cause hemolysis

Practical Applications

Vitamins are critical to the health and well-being of animals. They are present in varied amounts in many different feedstuffs. Most rations are made of a mix of ingredients and this helps provide most vitamins. It is now regular practice to supplement rations with the proper amounts of vitamins to complement that present in the food. It is not necessary to supplement complete pet foods with additional vitamins, however homemade diets should include a vitamin supplement. Vitamins consumed in excess over the normal requirement have not been shown to provide any extra benefits. When intake is low, as in situations of sickness, postsurgery decreased feed intake, or in aging, careful use of vitamin supplements might be indicated.

Words to Know

ascorbic acid	*lipophilic*	*thiamine*
biotin	*metabolic cofactors*	*Vitamin A*
choline	*niacin*	*Vitamin D$_2$*
cyanocobalamin	*night blindness*	*Vitamin D$_3$*
fat-soluble vitamin	*pantothenic acid*	*Vitamin E*
folic acid	*pyridoxine*	*Vitamin K*
hydrophobic	*riboflavin*	*water-soluble vitamin*

Study Questions

The study questions for this chapter are easy. You should be able to give the major function, dietary sources, and a deficiency symptom for each:

1. Vitamin _____ has the function of _____ and deficiency signs that include _____.
2. What pathways of carbohydrate, fat, and amino acid metabolism are the different vitamins involved in?
3. What are the metabolic differences and practical nutritional application differences between fat-soluble and water-soluble vitamins?
4. Describe the metabolism and special functions of vitamin D.

Further Reading

Blomhoff, R., K. R. Norum and M. H. Green. 1992. Vitamin A. Physiological and biochemical processing. *Annual Review of Nutrition* 12:37-57.

Butterworth Jr., C. E. and A. Bendich. 1996. Folic acid and the prevention of birth defects. *Annual Review of Nutrition* 16:73-97.

Case, L. P., D. P. Carey, D. A. Hirakawa, and L. Daristotle. 2000. *Canine and feline nutrition* (Chapter 5). St. Louis, MO: Mosby, Inc.

Castenmiller, J. J. M. and C. E. West. 1998. Bioavailability and bioconversion of caroteniods. *Annual Review of Nutrition* 18:19-38.

Dowd, P., S-W. Ham, S. Naganathan and R. Hershline. 1995. The mechanism of action of Vitamin K. *Annual Review of Nutrition* 15:419-440.

Gross, K. L., K. J. Wedekind, C. S. Cowell, W. D. Schoenherr, D. E. Jewelll, S. C. Zicker, J. Debraekeleer, and R. A. Frey. 2000. Nutrients. In M. Hand, C. D. Thatcher, R. L. Remillard, and P. Roudebush, eds., *Small Animal Clinical Nutrition*, fourth edition (Chapter 2). Topeka, KS: Mark Morris Institute.

Halliwell, B. Antioxidants in Human Health and Disease. *Annual Review of Nutrition* 16:33-50.

Krinsky, N. I. 1993. Actions of carotenoids in biological systems. *Annual Review of Nutrition* 13:561-87.

McDowell, L. 2000. *Vitamins in Animal and Human Nutrition,* 2nd edition. Ames, IA: Blackwell Publishing.

National Research Council. 2004. *Nutrient requirements of dogs and cats* (Chapter 8). Washington, D.C.: National Academy Press.

Omdahl, J. L., H. A. Morris, B. K. May. 2002. Hydroxylase enzymes of the vitamin D pathway: expression, function and regulation. *Annual Review of Nutrition* 22: 139-66.

Pfahl, M., F. Chytil. 1996. Regulation of metabolism by retinoic acid and its nuclear receptors. *Annual Review of Nutrition* 16:257-83.

Chapter 6

Minerals: Providing Many Functions

Take Home and Summary

The body contains minerals that provide structure and function. Although minerals only make up a small percentage of body weight, they are essential for normal processes. Historical observations showed that absence of some foods caused various deficiencies that could be reversed if those foods were included in the diet. Eventually, specific minerals were identified as the critical factor. Calcium and phosphorous are needed every day to grow and maintain bone mass and ensure proper nervous and muscle activity. Potassium, sodium, and chloride are needed every day to maintain ionic balance. Specialized functions, such as transport of oxygen, uses iron; and thyroid hormone uses iodine. Other minerals are only needed in tiny (microgram) amounts for specific functions in metabolism and reproduction. Minerals are found in all feedstuffs but content varies tremendously. For several minerals, true deficiencies rarely occur naturally and a requirement was proven only in experiments removing all sources of the mineral from the diet. A mineral deficiency today would usually be suspected only in cases of low food intake, disease, extremely poor rations such as "all meat" diets, or in horses grazing in certain types of pastures. The majority of mineral needs will come from feed ingredients, but safe nutritional practice includes mineral supplements to complement the natural amounts in foods. There is no biological evidence that minerals consumed in excess of basic requirements can do any good; and some such as iron, selenium, and copper can be toxic in large amounts.

Major Functions of Minerals

Minerals for Bone, Muscle, and Nerve Function

Minerals are essential for composition of animal tissue and the functioning of animal life.

The minerals needed in the largest amounts are *calcium* and *phosphorous* for growth and maintenance of bone mass. They are also critically important in nervous and muscle signaling and contraction. This function is actually

> Minerals needed in several hundred milligrams or several grams per day are *macrominerals.* Those needed in smaller quantities (mg or ug/d) for metabolic functions are termed *microminerals*. These terms mean nothing for the function (or size) of the mineral; they only refer to amount in the diet.

more important for survival, but a calcium deficiency is often noted first in bone problems because the body takes the calcium from the bone to use in nerve and muscle action. This is an example of homeostasis in action: because calcium is so important for nervous and muscle function, several hormones and vitamin systems (see vitamin D, Chapter 5) regulate the supply of calcium to these tissues. The bone is directed to give up calcium and then bone strength is decreased. Muscular cramps and ataxia (impaired movement) are signs of an acute calcium deficiency; those are often noted during late pregnancy and lactation.

> Regulation of blood calcium concentration is an example of physiological regulation called *homeostasis*; maintenance of a steady state. Hormones and vitamin D regulate intestinal absorption, resorption from bone, and excretion into the urine to maintain blood calcium.

Magnesium is also important in bone formation as well as normal cell ionic balance. These minerals are often required in amounts more than one hundred milligrams (for dogs), several grams per day during growth and lactation (up to 1 percent of the total dry-matter intake). In adult animals, calcium, phosphorous, and magnesium are still needed to maintain bone mass. Bone, just like other organs such as muscle or the liver, is always being broken down and rebuilt. Therefore, a constant supply of these minerals is needed; decreased intake will cause early loss of bone mass, often leading to weakness and breakage. Metabolic symptoms of deficiencies may be seen in several days, and long-term affects (stunted growth, decreased milk production) can be noted in just a few weeks.

Minerals for Ionic Balance for Cell Life

The next group of minerals includes sodium, potassium, and chloride (Table 6.1). These minerals help create the ionic balance that keeps all cells alive. *Potassium* is found in greater concentrations within the cells and *sodium* is found outside, in the blood and extracellular fluids. *Chloride* is a negatively charged ion that counterbalances the role of the positively charged potassium and sodium. Their concentration in the blood and cells is closely regulated by proteins called *transportors* that use ATP to keep the ionic balance. Ingestion of a large amount of these minerals in a short time can cause toxic effects because they will quickly overcome normal cell function. Usually this will only happen "by accident," as in the case of an animal getting into something it shouldn't. In situations of a

Table 6.1	Minerals Involved in Bone and Cellular Ionic Balance		

The requirements for these macrominerals are usually measured in a percentage of the diet or several hundred milligrams to grams per day.

Mineral	Function	Sources	Problems
Calcium	bone, teeth, cell ionic balance muscle contraction	bones, calcium carbonate milk products dark green vegetables	rickets: bone malformation improper muscle contractions, cramps interacts with P, Vit D, Zn inhibits absorption absorption inhibited by oxalates in spinach
Phosphorous	bone, teeth, ionic balance, buffer, nervous transmission absorption and excretion primarily in intestine, more in urine in carnivores, about half/half in humans	bones, bone meal muscle meats many plants	interacts with Ca, Vit D, Zn nutritional secondary hyperparathyroidism (osteodistrophia fibrosa) reduced bone mass, causes calcification of soft tissues
Magnesium	bone, ionic balance, metabolism absorbed and excreted in intestines	widely distributed in plants and animals	bone weakness general weakness from poor metabolism
Sodium	ion gradient, buffer extracellular fluid secreted in urine and sweat	widely distributed in many plants and animals can be marginal in seeds sodium chloride	weakness, nervous problems death quickly in severe cases due to acid/base imbalance
Potassium and Chloride	ion gradient, buffer intracellular fluid secreted in urine and sweat chloride: part of hydrochloric acid in stomach	widely distributed in nature, deficiencies very rare	weakness, nervous problems, death quickly in severe cases due to acid/base imbalance
Sulfur	sulfur amino acids wool, collagen, chondroitin sulfate (connective tissue) part of insulin structure and function	primarily comes from sulfur amino acids, but sulfate also utilized a nitrogen:S ratio of 10:1 in the diet is average	poor growth due to cysteine deficiency, poor hair coat interacts with taurine in cats

slight deficiency, the regulatory systems can increase absorption from the gut or decrease loss in urine to help maintain concentration. However, long-term or more severe deficiencies can negatively affect growth, metabolism, and normal health. Although the body has many regulatory systems to control the concentration of these minerals, they cannot be stored and must be replenished regularly—usually daily. These minerals are widely present in feedstuffs and a deficiency is seldom seen in normal situations. As with other minerals, small amounts are added to diets by pet food manufacturers to even out the inconsistencies in feed supply and ensure the proper amount. Great attention to detail should be paid for home-made diets, especially for microminerals.

Minerals in Metabolism

Minerals act as cofactors in many metabolic reactions (Figure 6.1). Magnesium is used in many reactions of DNA and RNA metabolism. Magnesium is also needed for normal synthesis and use of ATP for energy. Calcium is needed for normal muscle contraction. *Selenium* is a critical part of the enzyme glutathione peroxidase, which helps to remove oxidized breakdown products from cells to maintain long cell life. *Cobalt* is part of the protein cobalamin, which works as a part of vitamin B_{12} to help move carbons from one compound to another as the body builds and rebuilds itself. *Copper* is part of enzymes that build connective tissue, and also is part of a protein *ceruloplasmin,* which helps to transport *iron,* which is used to bind oxygen in hemoglobin. Chromium is part of a molecule sometimes called the glucose tolerance factor, and is required for normal glucose entry into cells. It is supplemented into the diets of young pigs because they tend to have a

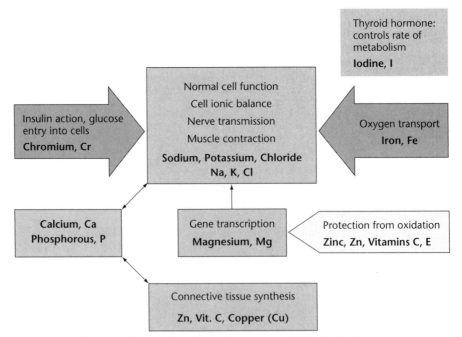

Figure 6.1 General schematic of use of some minerals in the body.

Selenium, cobalt, iron, magnesium, manganese, sulfur, and copper often are *enzymatic cofactors.* Similar to vitamins, they are required for metabolic reactions to take place.

difficult time using glucose. One may read in several publications about chromium supplements, usually chromium picolinate (picolinic acid is a digestible carbon compound). In humans, theories have been touted that chromium supplements will help build performance and muscle mass. The basic theory is that "if chromium is important, more is better, and will cause muscle growth and fat loss because more glucose will be burned up." Controlled studies have shown no benefit of additional chromium for performance. Additional chromium is not usually added to pet foods and will probably not give any benefit in exercising dogs or horses. Figure 6.2 depicts the role of chromium in the body.

Interactions of Minerals in Metabolism

There are many metabolic interactions between minerals in the body (Table 6.2). This means that the presence of one mineral may negatively or positively affect the use of another. Or, one mineral may not be used as well in the absence of another. Copper is necessary for normal iron transport, so a copper deficiency often has the same symptoms as an iron deficiency. Too much *zinc* can decrease calcium absorption, so a horse grazing a pasture of grass (which is naturally low in calcium) and eating a mineral supplement high in zinc, might show a pronounced calcium deficiency.

When any nutrient affects or regulates the use of another, such as a vitamin regulating mineral use (or vice versa), we call this a nutritional interaction.

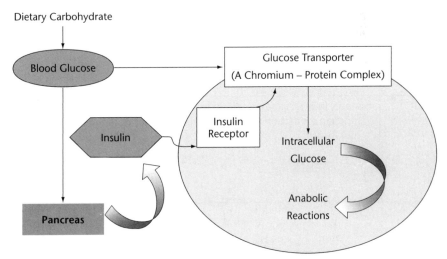

Figure 6.2 The role of chromium in glucose used by the body.

Table 6.2	Minerals Involved in Nutrient Metabolism		

Requirements for these microminerals are usually measured in micrograms (ug) per kilogram of diet (parts per million, ppm) or milligrams (mg) per kg (parts per thousand, ppt) of diet; or in just a few milligrams or micrograms per day. This does not make them any less important than the macrominerals; it just refers to the amount needed.

Mineral	Function	Source	Problems and Deficiencies
Iron	hemoglobin protein to carry oxygen	animal organs, muscles dark green vegetables dirt	deficiency anemia is hypocromic, microcytic anemia, lack of oxygen in cells toxic in large amounts it is poorly transferred through placenta or milk all neonates are born low in iron it is poorly absorbed in inorganic states; organic forms are better absorbed, ferrous (+2) better than ferric (+3) absorption reduced by Cu, Co, Mn, Zn, Cd, P
Copper	metalloproteins function in coat color necessary for hemoglobin synthesis used in formation of connective tissue active in cytochrome oxidase and other antioxidant enzymes	organ meats, muscles variable in plants	deficiency or excess reduces Fe antagonized by excess Zn, Fe, and Mo secondary anemia caused by deficiency problem in tropical areas, low amounts in plants deficiency includes depigmentation, stringy, limp fur or wool can be toxic, normal amount is 4 to 8 mg/kg, toxic at 100 to 200 mg/kg
Cobalt	essential in B 12 vitamin function for carbon metabolism	widely distributed in plants and animals	required at 0.05 to 0.08 mg/kg; toxic at 60 to 120 deficiency seen as wasting, general malnutrition many small areas of deficiency in environment in eastern U.S.

Table 6.2	Minerals Involved in Nutrient Metabolism *continued*		
Mineral	**Function**	**Source**	**Problems and Deficiencies**
Iodine	part of thyroxine hormone, which regulates many metabolic reactions	found in many plants, animal tissues	deficiency is hypothyroidism and goiter. It is often fortified into NaCl. environment (soils) low in northeast U.S., Montana, Great Lakes
Selenium	functions in glutathione peroxidase as antioxidant may have secondary functions in reproduction required at .1 mg/kg, toxic at 3-40 mg/kg	interacts with Vit E, prevents white muscle disease (dystrophy)	extremely variable in plants many areas of deficiency— Pacific northwest coast, northern Rockies, northeast U.S. and southeast U.S. variable in most of western and mid-southern U.S. toxicity areas all across Rockies and Great Plains use of selenium in feeds is highly regulated because of narrow range of benefit
Zinc	essential for normal connective tissue and epithelium functions in carbonic anhydrase, (antioxidant)	variable in plants, marginal in many seeds general adequate in animal tissues	deficiency is parakeratosis, drying and lesions of the skin action of Zn is antagonized (diminished) by excess Ca, Fe, Mo because they interfere with absorption
Fluorine	essential in enamel of teeth	extremely variable in plants and animals	narrow range of requirements too much F can lead to mottled teeth but strength of enamel is not affected
Molybdenum	functions in xanthine oxidase in antioxidant reactions	variable in plants and animals, but very little needed	competes with Cu, Zn, Fe: inhibits their absorption toxicity seen as severe diarrhea and wasting, also causes Cu deficiency
Manganese	cofactor in energy metabolism important in reproduction	variable in plants	
Chromium	assists in insulin action glucose uptake by cells	widely distributed	reduced glucose use

Vitamin D and calcium is a major nutritional interaction. They are so closely related that in practical terms, a calcium deficiency and a vitamin D deficiency have the same signs. Vitamin D is necessary for normal calcium absorption from the gut, and entry into and removal from the bone (see Figure 5.3). Cobalt is part of the molecule *cobalamin,* which works in conjunction with vitamin B_{12} to make *cyanocobalamin,* the factor that moves single carbons from one compound to another to make specific structures and metabolites. There are dozens of these types of interactions that are known, and well-schooled nutritionists spend a lot of time formulating mineral "premixes" (supplements designed to mix into a final ration) that contain the right amount and balance of minerals.

Digestion and Use of Minerals

Changing the digestibility is a normal, but usually slight, adaptation to try to maintain the proper intake by the body. If the diet is low enough, minerals will become deficient, no matter how hard the intestine tries. Thus, several foods may have vitamin and mineral levels that seem high enough, but because of poor absorption, other sources must be added. Some minerals can be toxic or at least detrimental if too much is consumed. Balance and proper amount is the key, and we have determined most of the proper amounts and balances for minerals in the diet through many well-designed experiments.

> The percentage of mineral absorbed by the digestive tract usually declines with increasing amounts of intake of that mineral and the absorption percentage usually increases when intake decreases.

As body growth rate or performance (lactation, exercise) increases, the amounts of minerals needed also usually increases. This generality is followed in growth, as the total metabolic rate is increased and more calcium, potassium, and magnesium are needed for bone. But for lactation, the increase in requirement for minerals that are removed in large quantities in milk (Ca, P, Na, K, Cl) increase more than the increased requirement for minerals such as iron or zinc that are not lost in milk in large quantities.

In most disease states, water flux and ion balance are changed and the mineral requirement can change. Veterinarians will usually provide *electrolytes* when an animal is sick; this refers to a solution of water with sodium, potassium, magnesium, chloride, calcium, and other minerals and buffers to help maintain cellular function.

During tissue repair after surgery or injury, the mineral requirement can increase. Zinc is used in connective tissue synthesis, which is one reason why we use zinc ointments on some wounds, to deliver zinc to the area needing new connective tissue, including skin.

The feeding management regimen also affects requirement of minerals. Horses grazing grass pastures need more calcium in their mineral supplement than horses on alfalfa. Alfalfa and other legumes have a lot of calcium while grasses such as brave or orchardgrass are low. The housing structure can affect requirements, as most animals can receive sufficient iron if they have access to dirt. But if they are kept in a nice clean apartment or concrete stall, we need to add it to the diet.

The specific environment or geographical area also affects mineral nutrition. The trace mineral selenium is deficient in many soils across the United States, while it is adequate in others. In some small areas, it can be found in toxic or deficient amounts. Horses grazing pastures in these areas may be deficient or toxic in selenium. Iodine is known to be low in many areas of the eastern United States. The observation that many people in those areas developed goiter (the hypertrophy of the thyroid gland while it tries to make more thyroid hormone because of the iodine deficiency) led to the iodination of table salt. We use iodized salt routinely in animal diets to supply iodine in a safe and efficient way. For most companion animals other than horses and perhaps rabbits grazing primarily grass, this will not be a major practical problem.

Chronic low calcium with normal amounts of phosphorous can decrease calcium absorption from the food and stimulates *parathyroid hormone* to increase resorption (removal) of calcium from bone, leading to bone degeneration. This is referred to as nutritional secondary hyperparathyroidism in birds (Chapter 15) and metabolic bone disease in reptiles (Chapter 18); these are the same problem, and we will discuss them in detail in these chapters. That is one reason we recommend maintaining calcium to phosphorous ratio in the range of 1:1 to 2:1. It is sometimes worse to have low calcium and normal phosphorous than to have them both too low, although both situations are undesirable. This is because the low calcium affects the function of parathyroid hormone and vitamin D, and, along with the relative excess of phosphorous, makes improper mineralization of bone. This is compared with simply a lower amount of bone formation in the situation when both are low. In practice, two major situations can exist that bring about a low calcium intake when phosphorous is adequate. The first is consumption of large amounts of muscle meats without including calcium sources, as can happen if people think that dogs and cats need lots of meat and forget that they also need the calcium and phosphorous from bones. The second, more common in herbivores such as horses, is when we feed large amounts of cereal grains and plants (corn, wheat, rice, grass hays) without addition of calcium sources such as legumes like alfalfa or clover.

In many plants phosphorous is complexed with a chemical called phytate to make phytin, a complex of inositol, potassium, calcium, and magnesium. Sometimes 20 to 40 percent of the phosphorous in many seeds, stems, and leaves is tied up in phytin and this is less digestible than free phosphorous. The enzyme phytase (found in many bacteria) can break this down, but most animal's intestinal tracts do not have enough phytase. New genetic technologies (growing the enzyme in bacteria) have increased the availability of phytase, which can then be used to treat feedstuffs to improve the digestibility of phytin. This helps not only the animal but also the environment, because we are more efficient in using the mineral (which has only a limited supply on the planet) and lose less to the environment.

It is important to discuss a few general observations people make and perceptions they may develop compared to scientific principles regarding mineral nutrition before we go into more specific functions.

> Sometimes people make observations and then make conclusions from them without fully researching or testing their hypotheses. One example is in mineral nutrition.

People may read or hear that one animal in the wild eats many seeds; therefore, they may think that the mineral content of the seeds is what they need. But they may not observe that the birds eat a large variety of types of seeds, and people may not know that some seeds are high in calcium, and some are low. It is important to recognize the total diet and the variety of possible diets. Another example is the "pure meat" diet touted by some for dogs and cats. The trouble is, muscle and organ meats alone, though high in protein, fat, and many vitamins and minerals, are absolutely worthless sources of calcium. Wild felids and canids do not eat just meat; they also consume bone, connective tissue, and even digestive tract contents. In that way they can get a balanced diet. One hears and reads many theories about the best diets for dogs and cats based on incomplete observations, and that applies directly to misperceptions about minerals. Just because a mineral is present in a feedstuff does not mean it is required by animals in that amount: alfalfa is very high in calcium, more than horses require. Also, absence from a feed does not mean animals do not need it. Most animals in nature do eat a variety of plants or animals and thus receive what they need.

"Orphan Minerals"

There have been experiments and observations over the years that indicate that some other minerals might be required in very small amounts. These would include boron, tin, vanadium, lead, and nickel. It is really an open scientific question as many of these experiments were never repeated and deficiencies in practical situations have rarely if ever been reported. The naturally occurring concentrations of several minerals in plant and animal tissues exceeds any amount thought to be required. The practical reality is that the cost of experiments to verify their requirement almost certainly outweighs the potential benefit of the knowledge gained, and it is unlikely that much more research into the nutritional role of these minerals will be conducted.

Practical Application

Mineral nutrition has in practical application become fairly simple at the consumer level. Because so many excellent experiments have been done to determine mineral needs across a variety of species and life stages, expert nutritionists include the proper balance and amounts of minerals in pet food formulations. Animals receiving rations certified to be "complete and balanced for all life stages" will not benefit from supplementation, and in some cases could be harmed. Because we feed these complete rations, it is easier to have a slight to moderate excess of many minerals than a deficiency. However, for some, such as phosphorous, sodium, and potassium, this may lead over years to an increased rate of kidney function loss. This is why we have moved more into "life cycle feeding," for example, providing a senior diet with reduced amounts of many minerals compared to an "all stages diet." Mineral deficiencies and toxicities should only be suspected in very specific cases such as severe disease, refusal to eat, or in some isolated pastures for horses. A competent veterinarian will be able to determine whether or not a mineral deficiency may be involved. Self-diagnoses and natural remedies will seldom help and will often hurt in these situations.

Words to Know

calcium	enzymatic cofactors	parathyroid hormone
ceruloplasmin	homeostasis	phosphorous
chloride	iodine	potassium
cobalamin	iron	selenium
cobalt	macrominerals	sodium
copper	magnesium	transporters
cyanocobalamin	microminerals	zinc
electrolytes	minerals	

Study Questions

The study questions for this chapter are easy. You should be able to give the major functions, dietary sources, and deficiency symptoms for each.

1. The mineral _____ has the function of _____ and deficiency signs that include _____.
2. What pathways of carbohydrate, fat, and amino acid metabolism are the different minerals involved in?
3. What are the basic differences between and the practical nutritional applications of macrominerals and microminerals?

Further Reading

Case, L. P., D. P. Carey, D. A. Hirakawa, and L. Daristotle. 2000. *Canine and feline nutrition*. St. Louis, MO: Mosby, Inc., chapters 6 and 13.

Goyer, R. A. 1997. Toxic and essential metal interactions. *Annual Review of Nutrition* 17:37-50.

Gross, K. L., K. J. Wedekind, C. S. Cowell, W. D. Schoenherr, D. E. Jewelll, S. C. Zicker, J. Debraekeleer, and R. A. Frey. 2000. *Nutrients.* In M. Hand, C. D. Thatcher, R. L. Remillard, and P. Roudebush, eds., *Small animal clinical nutrition*, fourth edition. Topeka, KS: Mark Morris Institute.

Lei, K.Y. 1991. Dietary copper: cholesterol and lipoprotein metabolism. *Annual Review of Nutrition* 11:265-283.

Lukaski, H.C. 1999. Chromium as a supplement. *Annual Review of Nutrition* 19: 279-302.

National Research Council. 2004. *Nutrient requirements of dogs and cats.* Washington, D.C.: National Academy Press. (lists approximately 260 references on mineral nutrition)

National Research Council. 1997. *The role of chromium in animal nutrition.* Washington, D.C.: National Academy Press.

Underwood, E. J. and N. F. Suttle. 1999. *Mineral nutrition of livestock,* third edition. Oxfordshire, U. K.: CABI Publishing.

Chapter 7

The Basics of Nutrient Requirements: Water, Energy, and Protein

Take Home and Summary

Water is a nutrient used to transport other nutrients and chemicals from organ to organ to remove waste and maintain life functions. It can come from fresh water, water in the food, or from metabolism. It is our responsibility to provide sufficient water for normal life and to maintain good oxygenation with a regular exercise program. Energy is a concept used to keep track of nutrient use. We measure changes in *energy*, defined as *the ability to do work*. The driving force of nutrient metabolism in the body is the reduction of oxygen to water with hydrogen. This reaction supplies cells with the energy to maintain cell function and to make the structures and products of life, growth, and reproduction. Glucose, amino acids, and fatty acids absorbed by the digestive tract enter the cells and are broken down; the new chemicals formed are used to make cell compounds. Through energy-balance experiments we determine how well a food supplies energy in different life stages. Protein in the diet supplies the proper amount and proportion of amino acids for animals to make their own proteins. Protein quality is determined by the amino acid content and the digestibility. We measure the nitrogen content of feedstuffs to determine protein requirements. It is not efficient for protein to supply energy; it is too expensive, and the nitrogen waste can harm the environment. Modern energetics now includes the role of genetics in obesity, and ongoing nutrition includes study into the roles water, specific amino acids, and proteins play in health and longevity.

Functions of Water

Water is the most important nutrient. The proper balance of water, food intake, and activity is critically important to maintain animal health, which really means keeping the cells in all the organs functioning normally. Severe dehydration has been recognized as a life-threatening problem for ages, but in the last few years we have also considered the role of mild dehydration in health. One observation to drive home this point is the presence of water bottles everywhere one goes, not just in the gym or on the hiking trail. Many people carry water bottles to stay hydrated. In fact, hydration has received so much attention that now medical folk are warning people that too much water may not be beneficial and (obviously) the same amount of water is not needed for everyone.

Functions of water include ion balance across cellular membranes, transport of nutrients within and among cells, elimination of waste products through the urine, removal of heat (evaporative cooling), and lubrication in joints and internal organs. Animals adapt to a shortage of water by reducing filtration through the kidneys, and can survive for a while in a dehydrated state. However, if dehydration becomes too severe (usually a loss of about 5 to 10 percent of body water), cell functions diminish to the point at which the activities that make up life, primarily respiration, cannot proceed sufficiently to maintain life. Respiration, or the reduction of oxygen to water, is the primary driving force of the living cell. If the water content of the cell decreases by just a small amount, the structure and function of the cell changes. The different concentrations of mineral ions affect enzyme function and the cell cannot reduce oxygen as fast. Oxygen transport to the cell decreases because of the decrease in blood volume, and these two problems conspire to "finish off" the cell.

Water Balance in the Body

Sources of Water

We often do not think of water as coming from multiple sources. Yet animals do not have only one source of water, free or *drinking water.* They also obtain water from food *(food-borne water)* and from *metabolic water.* The hydrogen from the nutrients combines with oxygen in the electron transport chain to trap energy as ATP, making metabolic water. This happens in the breakdown of carbohydrates, fats, and proteins. This water will function just the same as any other water from the diet or drinking.

Water is derived from three sources:

Free water
Water in food
Water produced in metabolism

Because water is so important to respiration, and respiration is a function of metabolic rate, it is not surprising that we can estimate the water requirement of an animal based on its metabolic rate. For every kilocalorie (kcal) of energy (that amount roughly contained in 0.25 grams of starch or protein or 0.11 grams of fat), an animal requires about 1 milliliter of water. In a practical example, a 30-kg (66-lb) dog requires about 1,800 kilocalories (kcal) of energy

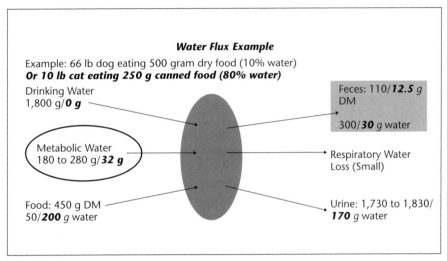

Figure 7.1   Water flux in animals.

a day for normal maintenance, and about 1800 ml (1.8 L, just under 2 quarts) of water. A 4-kg (9-lb) cat would require about 320 kcal of energy per day, and about 320 ml of water (about 10 ounces, a little more than a cup).

> The total water requirement of an animal is related to energy used: About 1 milliliter of water is required for each kcal of energy used.

Almost all food has some water in it. Something without any water at all is very dry indeed—try eating dry flour. Most commercial food products range between 90 percent dry matter (10 percent water) to 80 percent water (20 percent dry matter). A dog consuming about 500 g per day (about 1.1 lbs) of dry food would consume about 50 g, or 50 milliliters, of water with the food or only about 50/1,700ths (see previous paragraph and Figure 7.1) or 3 percent of the total water need. However, that mythical 4-kg cat, consuming a diet of canned food at 80 percent water, would consume about 87 grams of dry matter in the food and about 345 grams (ml) of water from that food—basically all the water the cat needs. This has important implications to the overall nutrition, health, and feeding management of cats on dry versus wet diets.

With average dog and cat foods, about 10 to 16 grams of metabolic water are made for each 100 kcal of energy consumed (more metabolic water on higher-fat diets). Thus, the 66-lb dog using 500 g of food at 3.5 kcal/g (about 1,800 kcal/d) is going to make about 180 to 280 grams of metabolic water, about 10 percent of what it drinks. The 9-lb cat eating 500 grams of wet cat food (100 grams actual dry matter) will obtain 400 grams of water from the food and make 32 grams (10 g water/100 kcal of energy) of metabolic water.

This is why animals, especially cats, consuming wet foods need very little free drinking water.

Figure 7.1 shows water flux in a 66-lb (30 kg) dog consuming 500 grams of dry food and drinking 1,800 g (1.8 L) of water per day, and a 10-lb (4 kg) cat consuming 250 grams of wet canned food per day. Note that animals consuming canned food do not often need a significant amount of other water. This may reduce their thirst drive to the point at which they may actually not receive enough water for optimal performance.

Losses of Water

The body loses water through urine, feces, respiration (breathing), sweat, and in milk during lactation. The loss of water is important in preserving proper concentrations of ions and nutrients in the blood for optimal cell function. The *kidney* is the organ that makes urine by filtering blood. Urine is the largest loss of water through the body, usually accounting for 75 to 85 percent of total water lost. Waste products, excess mineral and acidic or basic ions, pass from blood into urine through the kidney.

> Filtering and excretion of wastes falls under a number of chemical, neural, and endocrine (hormonal) control systems.

It is extremely important for the body to maintain the proper ionic and acid-base (or pH, the "pressure of hydrogen") balance, and proper water flux through the kidney is critical to doing this. Too little water and the wastes and excess ions are not removed sufficiently, leading to short- or long-term metabolic problems such as acidosis (too many acid by-products) or alkalosis (too many basic by-products). These variations in blood pH (acid content) can diminish cell function and damage cells. Over time, there can be an increased loss of kidney function.

Feces are the largest source of water lost from the body next to urine. On a dry diet, food will enter the digestive tract at about 90 percent dry matter (10 percent water), and undigested feces is excreted from most dogs and cats at about 30 to 40 percent dry matter (DM), or 60 to 70 percent water. As an example: 100 grams of food contains 90 grams DM and 10 g water. On an average diet, 75 percent of the DM is digested and absorbed by the body and 25 percent is lost in the feces, so there is about (90×0.25), or 22.5 g DM in the feces. At a water content of 60 percent in the feces (40 percent DM), this is 22.5 g dry feces/40 percent dry matter = about 56 grams of total feces. It is easy to see, then, that on dry diets, intake of fresh water must be at least greater than another 46 grams of water per 100 grams of food consumed, or the animal will quickly lose water and become dehydrated. Figure 7.1 shows an example of water balance for a dog consuming a dry diet and a cat consuming a canned diet. Cats and dogs can obtain almost all of their total water needs from the food when fed diets greater than about 70 percent water, as is the case in canned food.

Respiration through the lungs is also a route of water loss from the body. This is not the same as the reduction of oxygen with the hydrogen from nutrient metabolism. The term can be used to mean either breathing, the taking in of oxygen in the lungs and by the cells, and taking in oxygen to make water and energy as ATP. Water vapor is lost from the lungs in exhaled air. This can be a major loss from companion animals, especially during activity and in the heat.

Sweat is not usually a large loss of water in the animals we discuss in this book, but can be under many circumstances. Evaporation from the tongue can be important in dogs and cats. Horses, however, can lose many grams of water a day in sweat and several minerals such as sodium (Na), potassium (K), and chlorine (Cl) are lost in the sweat and must be replaced.

If a female has had young and is nursing them, she is *lactating*. **Lactation** is the process of making milk. Lactation is a large loss of water, and the increase in water loss signals an increase in thirst and drinking behavior. The milk of most species runs from about 10 to 15 percent solids, or about 85 to 90 percent water. Marine mammals (Cetaceans) such as whales have milks that are about 50 percent fat by weight, and only 30 to 40 percent water. Rodent mothers can easily secrete up to 30 or 40 percent of their body weight per day in milk. A horse nursing twin foals might make 15 kg of milk, or 3 to 5 percent of its body weight.

The amount of water in the body is under direct regulation by the neural and endocrine systems. We know *thirst* as the hunger or desire for water. If an animal is in the initial stages of dehydration, the hypothalamus secretes a hormone (antidiuretic hormone) that reduces water loss from the kidney. In addition, the brain through neural signals increases the desire for water, also to reduce water removal from the kidneys and into the digestive tract. However, please note that by the time noticeable behavioral changes occur, the body is already short on water.

Dehydration is a lack of water in body tissue. A minor loss of about 1 percent of body water will lead to the adaptations described earlier; this loss is often "unnoticed" by the animal. If more than 1 percent of body water is lost, these adaptations are no longer sufficient to maintain normal oxygen and ion concentration, and the "thirst drive" kicks in. If water continues to be lost, cell function is diminished, and there may be problems with nervous coordination and sight and hearing. If body-water loss exceeds 5 to 10 percent of body weight, the loss of ion and oxygen balance can cause cessation of nervous and heart activity, and death. For managing companion animals, these situations would only occur if the owner does not supply water correctly.

Increased activity increases the need for oxygen by cells, and water is removed first from the digestive tract to increase blood volume. Increased sweating and respiratory water loss further decreases body water, which must be replaced by increased intake; otherwise, cells will start to lose water. An increase in environmental temperature large enough to increase respiration for body cooling (for example, panting) will also increase water loss. An excessive amount of intake of ions, such as sodium, potassium, and chlorine, can increase water loss through the kidney, as the kidney must release some water along with the ions. This may happen with the feeding of too many salty snacks, or if the animal "gets into something" with a high ionic content (remember that table salt is NaCl, but there are many other salts and ions which affect water balance).

> Dehydration will affect the cellular function of all organs, and the deficiency must be quickly corrected.

Although not as common as the lack of water, in some cases there can be too much water in the body. Severe cases can lead to edema, referred to as *hydropsy,* the buildup of water in tissues of the body. This affects cell function because of changing ionic strength, and is a painful situation. This can be a life-threatening situation in some cases because of the excess pressure it puts on the heart to pump blood through the saturated tissues. In aging animals, the normal loss of kidney function and heart strength can reduce the ability to remove the excess water from the body. This may also occur during the last several days of gestation and first days of lactation as the ionic and water balance of the body rapidly changes. This situation can be treated to a certain extent with drugs, such as *diuretics* (water-removers), to increase the removal of water through the kidney.

Another imbalance of water use, which is rare but can happen, is known as *hydremia,* or water intoxication. This can occur in young animals (primarily puppies) that consume a large amount of water in a short amount of time. Control of water balance is not fully developed in the young animal. The symptoms are similar to a lack of sodium ions—loss of appetite (anorexia), lethargy (depression), muscular weakness, and lack of coordination. In any situation in which these are suspected, a veterinarian should be consulted immediately.

Oxygen, Respiration, and Energy

All life on Earth requires some form of energy transformation to live. Animals require oxygen supplied by respiration. Respiration supplies the potential energy of oxygen to transfer energy from the chemical nutrients consumed into chemical structures that the cell needs to sustain all life functions. Respiration is the reduction of oxygen with two hydrogen to water that supplies the energy. The major chemical formed in this reaction is *adenosine triphosphate,* with the well-known acronym of *ATP.*

> The two major chemical forms of energy in animals are adenosine triphosphate (ATP) and nicotinamide adenine dinucleotide dihydride ($NADH_2$).

ATP contains a large amount of potential energy used to make body components. The other key chemical containing tremendous potential energy is *nicotinamide adenine dinucleotide dihydride,* or *$NADH_2$.* This chemical co-

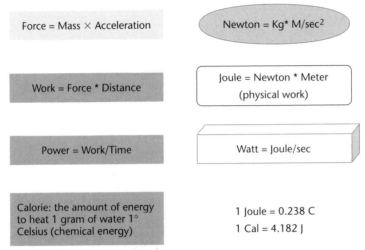

Figure 7.2   Energy: the ability to do work.

factor is formed in many catabolic pathways, and is used directly in other energy-using reactions, or enters the *electron transport chain,* a complex metabolic pathway, to supply the two hydrogen to reduce oxygen to water, making more ATP along the way. Figure 7.2 shows how chemicals are used to supply the body with energy for work.

Energy: The Ability to Do Work

The discovery of energy—in all forms, physical, chemical, and biological— is a beautiful tapestry of human curiosity, intelligence, industriousness, determination, and willingness to fight dogmatic thinking. Early observers such as Lavoisier simply looked at things, asked questions, and did experiments. One observation led to another, many more people became involved, the economies and governments of the world expanded and became enlightened (slowly), and more and more scientific discoveries were made. From 1750 to 1850, much of the chemistry we still use today was discovered using rudimentary techniques. From 1850 to the early 1900s, the laws of physics (energy) and chemistry (chemical transformation) allowed biologists to explain things they had been observing for decades. By 1900, we completely understood that animals obeyed the laws of physics, were comprised of physical structures that ran complex chemical reactions, and were far more complex than any machine or chemical reaction dreamed of up to that time. The first half of the twentieth century saw an explosion of knowledge into biology and chemistry and created the fields of biochemistry, nutrition, and physiology. Thousand of scientists made millions of small discoveries, and little by little these were brought together to understand more of how the complex systems of the body work. The pathways of carbohydrate, fat, and amino acid metabolism were discovered. Many metabolic diseases such as diabetes, milk fever, pregnancy toxemia, phenylketonuria, and less common but deadly diseases were identified and treatments or preventions developed.

In the last fifty years, work on nutrition and metabolism continued and still continues today. However, the final understanding of the structure of the chromosomal genetic material, deoxyribonucleic acid and ribonucleic acid (DNA and RNA, respectively), "blew away" all other understanding in that now we knew what controlled the complex metabolism, nutrition, physiology, reproduction, and health. Just twenty years (a tiny fraction of time on a scientific scale) passed from the discovery of DNA as the genetic code to the actual altering of the genes of plants and animals. Obviously, society is still reeling with the power and ramification of this knowledge.

We now have tools and understanding to identify the genetic basis of nutrient use and requirements. In many cases, we know what traits have high *heritability:* a large potential to be passed from one generation to another. We understand far beyond hair coat and eye color. We know that muscle growth, potential for obesity, and kidney and heart problems all have genetic components. Proper nutritional management can help prevent many problems, but not all. We are now studying the genetics of nutrient metabolism, immunity, and longevity. Developmental orthopedic diseases in dogs and horses are a great example of the interaction of genetics and nutrition (see Chapter 12). This knowledge came from just regular people who were curious and did something about their curiosity. It took a lot of work, a lot of time, many people, and a lot of wealth to find these things out. There really are no easy answers. As we continue discussion of metabolism and nutrient requirements, remember that it is hard-working people who have the support of wealthy people and willing and wealthy societies who understand that knowledge equals power and knowledge for everyone.

Nutrients enter the cell through the cell membrane and immediately begin to be used in various catabolic or anabolic processes. Figure 7.3 shows energy use in the body, or the *net energy system.* The breakdown of glucose begins in the metabolic pathway and is known as *glycolysis.* This pathway was discovered in yeast in the early part of the last century, and was one of the first metabolic pathways discovered that told us how nutrients actually made carbon dioxide (CO_2) and supplied other forms of chemical energy such as ATP and $NADH_2$. This pathway is actually *anaerobic;* it does not require oxygen. This is how yeast convert sugars to CO_2 and alcohol to make bread or malt beverages or wine, through the process of *anaerobic fermentation.* Anaerobic fermentation also occurs in the rumen of cattle and the large intestine of many animals including humans, horses, rabbits, and rodents. But to really get the full energy out of glucose, the breakdown products of glucose require more catabolism in the metabolic pathway—the *tricarboxylic acid cycle (TCA)*—and transfer of hydrogen through $NADH_2$ to the electron transport chain to make ATP, and that requires oxygen.

Fact Box	**A Short and Biased History of Chemistry, Biology, and Nutrition**

1500 B.C. to the 1700s:	Early philosophers and "doctors" observed effects of eating different types or amounts of food: gluttony, starvation, deficiencies. In the writings of Aristotle is the beginning of "science" as it is understood today. Different behaviors in eating, drinking alcohol, and exercise led to different ends in health, weight, and longevity. Unfortunately, after the fall of the Greek Empire, humans were fairly distracted by other matters to pay too much attention to their own bodies. It truly wasn't until the 1600s that society recovered enough to have the leisure time and wealth to ponder weightier questions. After major discoveries in math, physics, and astronomy by folks such as Galileo, Copernicus, Newton, Pascal, and others, people turned their attention to the air, plants, and animals.
1700s:	Lavoisier contended that life itself requires oxygen and gives off carbon dioxide.
Early 1800s:	"German chemists" Fischer and others discovered the structure of glucose, fats, and amino acids.
Later 1800s:	Atwater and Voit prove that animals obey the laws of physics, chemistry, and thermodynamics. They take in oxygen and nutrients; change mass; and give off heat, carbon dioxide, and urea (nitrogen waste) in amounts that are consistent with the basic chemical reactions. Inputs always equal the total outputs and change in animal energy. Physiology begins with observations on digestion system and blood flow, and nervous system function. Sherrington, Cannon, and others painstakingly observe and experiment to show activity of what we now know as the endocrine and nervous system in control of body functions and nutrient use.
1920s to 1950s:	Hans Krebs identified many pathways of amino acid, glucose, and fat metabolism, including the tricarboxylic acid (Krebs) cycle. He and others discovered the actual chemical reactions that accounted for the changes in energy.
1900s to 1980s:	Brody, Kleiber, Blaxter, Van Es, Thorbek, Flatt, Tyrell, Lofgren, Garret, and dozens of others conducted many specific experiments with respiration calorimetry to determine specific chemical interactions in the body and animal requirements. This was the "heyday" of nutrition. Much of what we understand as nutrient requirements, especially for energy, was determined worldwide during this period.
1930s to 1950s:	Mitchell and Rose determined amino acid requirements, by means of using semipurified diets in humans and rats. These still form the basis of many of our requirements.
1950s to 2000s:	Quinton Rogers and Jim Morris explored the amino acid requirements of dogs and cats, and discovered the unique metabolic nature of cats.
1960s to 2000s:	Ransom Leland (Lee) Baldwin (V) and others formulated quantitative metabolic models that used computers and dynamic integration over time to demonstrate the actual chemical reactions that accounted for requirements in the net energy system. Peter Van Soest developed, refined, and expanded the detergent system of fiber (cellulose, hemicellulose, and lignin) in feeds in use globally today. Dale Bauman researched control of nutrient use in lactation to define the concept of homeorhesis, control of metabolism to support a dominant physiological state. George Fahey and Dave Baker study use of fiber and amino acids in dogs. Boon Chew helps discover the role of specific fatty acids in immune functions.

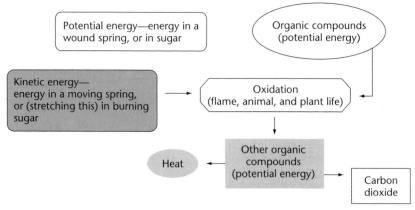

Figure 7.3   The ability to do work from chemicals.

Tricarboxylic Acid Cycle

As nutrients are broken down to supply energy or precursors for making other compounds, the two-carbon compound acetate enters the tricarboxylic acid cycle. There are several metabolic reactions in this cycle, and each progression through the cycle uses one acetic acid molecule and releases two carbon dioxides. The pathway supplies energy, and also converts many nutrients into *metabolic intermediates* that can be used to make dozens of different carbon compounds. In this way, glucose, fatty acids, and amino acids are converted to exactly the compounds the cells need. In the cycle, carbon dioxide is produced, along with some ATP and some $NADH_2$. The ATP produced is used immediately for any reaction that needs energy, but the $NADH_2$ must go through one more pathway to make ATP.

Use of Amino Acids for Energy

Amino acids are oxidized through many complex pathways; most of them end up as just a few chemical intermediates. Amino acids' primary purpose is to make protein. If the total amount of each amino acid available is not great enough to make the needed proteins, then the animal either must eat some more or obtain more amino acids by converting some amino acids into other ones. If there is not a sufficient amount of total amino acids entering the body, then the animal will take them from proteins in the body to survive in the short term. Amino acid metabolism is always taking place as proteins are broken down and resynthesized (turnover, see Chapter 2). In some classes, the instructor will want to go into more detail here.

All fatty acids and glucose, and many amino acids, are eventually oxidized to the two-carbon fatty acid we call acetic acid, which you may know as vinegar. If you have a bottle nearby take a break and go smell it. Just think, every cell in your body is making vinegar all the time. It is one of those "mysteries of nature" why such a simple compound is a major cornerstone of all

> The primary purpose of amino acids is to make protein. Excess amino acids are oxidized for energy or converted to fat, and the excess nitrogen must be removed as waste. This is inefficient for the animal, inefficient for plant and animal production to make the protein, and damaging to the local environment that must absorb the excess nitrogen.

metabolism. It is also a major preservative and flavoring agent in high concentrations (by the way, acetate is so critical it is even made in anaerobic metabolism, such as that carried out by yeast and fungi in fermentation for bread, beer, or the breakdown of biomass in our forests and throughout the environment). Once acetate is formed, it is used in many different metabolic reactions, some to generate energy and some to make various compounds including amino acids, fats, cholesterol, and some vitamins.

Electron Transport

The $NADH_2$ formed in the tricarboxylic acid cycle is the primary source of energy for the formation of ATP. The $NADH_2$ enters into a complex set of reactions called the electron transport chain. This is a complicated array of many enzymes with the singular purpose of making ATP. From this pathway, for each molecule of $NADH_2$ entering into the chain, three molecules of ATP are made. This is a tremendous amount of energy for use by the cell to maintain the ionic pressure that keeps the cell alive, and to make all the structural and functional proteins and other molecules for cell function.

Energy Nutrition

Now that the basics of biology and nutritional chemistry are understood, we can begin to discuss actual nutrient requirements. Nutrient requirements are the interface of chemistry, animal biology, and the environment. A body requires certain chemicals to maintain life and the environment supplies them, but not always in the exact form needed. As nutritionists, we usually define nutrient requirements at the biological level of the animal. For example, we may say: "Growing dogs require 22 percent crude protein in their diet"; or, "An average 8-lb cat requires 400 kilocalories of metabolizable energy per day." In reality, as we have learned in the previous chapters, each individual cell in every organ has a requirement for specific nutrients. Each food does not deliver exactly the nutrients that are in it because of inefficiencies in digestion, and the animal's requirements change with phases of the life cycle. It is difficult and expensive to measure all chemical reactions in the body. However, we can measure the total energy going into and leaving an animal, and we know from many previous experiments the chemical relationship between that and the amount of glucose and fatty acid. The measurement of energy transformations is an efficient and proven way to track practical nutrition.

> The energy requirement of an animal is a function of age, physiological state, activity level, genetics, and environment. All of these together make up a specific *energy requirement*.

During the 1800s it was determined that animals did best on foods that were a mix of proteins, fats, and carbohydrates, although they could survive and even do well when the diet was made up of a majority of protein. Animals did not do well if the diet was almost all carbohydrate or fat, the first real proof that animals needed nitrogen to live. Several experiments determined the exact amounts of carbon-containing compounds, protein, and oxygen animals needed, really defining the chemical bases of nutrient requirements. Data and equations developed over 100 years ago are still in use today to estimate some nutrient requirements because the researchers applied rigorous chemical analysis to the inputs and outputs of animal nutrition. We continue to refine and expand our knowledge base with research today.

The Net Energy System

In order to determine the value of a food for growth, milk production, or maintenance we must have a system to measure and compare among feeds and animals. Over several decades, chemists, biologists, and then nutritionists conducted several experiments and measurements of the amount of energy that animals needed to maintain their body weight, to grow, or to reproduce. Eventually, these coalesced and evolved into what we call the *net energy system*. This system describes a *measurable* and *repeatable* way to evaluate the value of feeds for livestock production. However, as the laws of thermodynamics apply equally to cattle raised for beef, Labrador puppies, calico cats, and you and me, the same basic system is in use for all species. When you look at your label from that snack you bought out of the machine this morning, the sentence "nutrient amounts based on a 2,000 kcal per day requirement" is referring to the application of the net energy system to determine the average daily requirement for energy. Table 7.1 shows examples of the major chemical composition of common feeds.

> The net energy system tracks the uses of food energy by the body and is used worldwide to determine the efficiency of diets and animals in all situations.

The system determines the actual, or *net*, amount of energy that is available to the body from the digestion and metabolism of various feeds of different physical and chemical characteristics. The system also allows for the determination of the *net energy* that the body requires for its different needs during the life cycle, such as growth of muscle, bone, and fat; maintenance of body weight as an adult; or reproduction of the next generation. Using the principles of thermodynamics and the chemistry that we already know, we can define how

Table 7.1	Energy and Protein Contents for Some Foods and Ingredients			
	ME, kc/g DM	% CP	% fat	% CHO
Starch	4	0	0	100
Cellulose, average, for horse	2.0	0	0	100
Fat	9	0	100	0
Protein	4	100	0	0
Sugar (sucrose)	4	0	0	100
Beef muscle—average fat content	4.8	78	17	0
Alfalfa hay, mid-bloom (horse)	2.0	17	2	5–15
Beef liver	5–6	68	25	<1
Chicken by-product meal	4–6	79	16	0
Ground corn	3.5	9	3	60–70
Dog milk per kg DM	5.3	36	43	15
Cat milk per kg DM	5.2	44	28	37
Dog food, dry, all life stages	3.5	22–26	15–18	30–40
Cat food, dry, all life stages	3.7	25–33	15–18	25–35
Required, minimum, by 30-kg dog		10	3.3	none
Required, minimum, by 4-kg cat		15	5	none
Required by 500-kg horse*		8	1	60–70

Content of by-product meals will vary and needs to be tested regularly. The student should practice making combinations of animal and plant products that would meet the contents of protein, fat, and carbohydrate in the dog and cat foods shown.

*There is not a strict requirement for CHO but in practice NDF and some starch must supply glucose.

much ingested energy is used by each of the various energy-consuming or energy-storing functions. The net energy system is a set of functions based on strict quantitative measures. It is quite useful for demonstrating requirements of animals in different life stages or carrying out different biological functions and for identifying how well specific feeds supply these requirements.

Determining Energy Requirements

The energy requirement is measured by an *energy balance experiment.* We measure the input of energy into the body from the food and all the known outputs of energy from the body in feces, urine, carbon dioxide, heat, hair, sloughed skin cells, and milk if it is a lactating female.

A *calorie* of energy is the amount of energy required to increase the temperature of 1 gram of water 1° Centigrade. (The actual measurement of a calorie also includes the starting temperature, for the nutritional calorie this is 14.5 C). It is not a large unit. When we speak of calories in nutrition, we usually

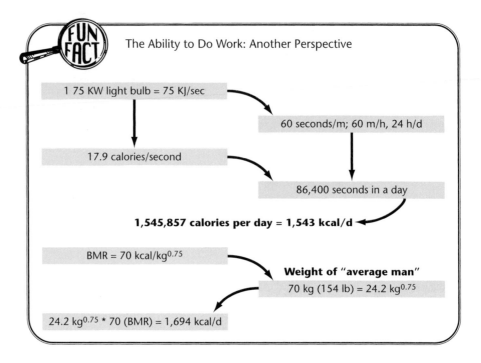

FUN FACT

The Ability to Do Work: Another Perspective

1 75 KW light bulb = 75 KJ/sec

60 seconds/m; 60 m/h, 24 h/d

17.9 calories/second

86,400 seconds in a day

1,545,857 calories per day = 1,543 kcal/d

BMR = 70 kcal/kg$^{0.75}$

Weight of "average man"
70 kg (154 lb) = 24.2 kg$^{0.75}$

24.2 kg$^{0.75}$ * 70 (BMR) = 1,694 kcal/d

refer to kilocalories, or one thousand calories. For example, a 30-kg (66-lb) dog requires about 1,600 kilocalories of metabolizable energy (actual energy used) per day. We measure calories by *calorimetry,* the measurement of heat, either directly or indirectly. All forms of energy have the *potential* to do work, whether in a spring holding something up (like the poor springs in my chair before I sit in it again), or in glucose and fat that are consumed and then oxidized (*kinetic energy:* energy actually in transition) to make the chemical energy that allows muscles to contract. The heat is a by-product of the reactions leading to work. Even a light bulb does work, converting electrical energy into light and heat.

Direct calorimetry was used in the late 1800s and is still used today to measure heat output by small animals. The heat that the animal generates from all of the chemical reactions taking place in the body is measured as an increase in temperature of a known quantity of air (or in an increase in temperature of a known amount of water surrounding a chamber of a known amount of air). You might be thinking that such a contraption is expensive to build and would be difficult to maintain and obtain exact measurements. You would be right; this is why we do not use direct calorimetry much anymore.

Indirect calorimetry uses our exact knowledge of chemical reactions to calculate the heat produced based on the amount of oxygen entering the body and the amount of carbon dioxide and urea leaving the body. For example, glucose has six carbons, twelve hydrogen, and six oxygen. If it is eaten and

> ### Moles and Avogadro's Number
>
> Are we talking about little furry blind creatures and some strange mathematician? No. A mole is a chemical definition of a certain number of atoms, 6.02×10^{23}, to be exact, of any chemical. It is a way to keep track of the chemical characteristics such as atomic weight or energy content. Why? We *must* have a system to keep track of how the chemicals change and interact. The weight of (more technically, the mass of the molecule) carbon or oxygen, or even glucose or protein, is based on the mass contained in one "Avogadro's Number" of atoms of that compound. This name follows the rule of giving chemical laws silly-sounding names. Thus the *molecular weight* of carbon is 12 grams per mole, nitrogen is 14 grams per mole, oxygen is 16, water is 18, glucose is 180, and some large proteins can have molecular weights of 300,000 grams per mole (obviously you never get a whole mole of this stuff together). If you divided the molecular weight by 6.02×10^{-23} you would set the mass of one molecule. This system is essential for tracking chemical changes in biology on a consistent basis. Even a few molecules of a chemical can affect an animal for better or worse. In practical situations, we track food intake and body weight, but research nutritionists have to keep track of every molecule.

converted 100 percent to energy to support life functions, then six CO_2 are made along with six H_2O. We know based on direct calorimetry measures that one glucose has 673 kcal/mole of total (gross) energy (1 mole of glucose is 180 grams, 0.395 lb). Regardless of how it is oxidized (with acid, burned, or metabolized in animals or plants) it gives off the same number of calories, the same amount of heat. According to the laws of thermodynamics, energy release from oxidation of any matter is the same, whether it is blown up in a bomb, burned up in a fire, or *oxidized* in metabolic pathways in bugs, plants, or animals. And that is why we can measure heat by only measuring oxygen and carbon dioxide.

Based on the known chemical reactions, and the heat given off in each one, we can indirectly determine *heat production* by animals if we measure the oxygen going in or the carbon dioxide going out. If we have measured the total energy going into the animal from the food, and we have measured the total energy coming out in urine, feces, and heat, then the difference is the energy that was retained or lost from the animal body (net energy system, Figure 7.4).

> Indirect calorimetry is an effective scientific tool only because we know the exact chemical reactions that form the basis for the system.

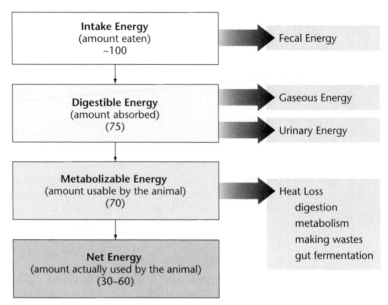

Figure 7.4 The net energy system.

You might be wondering how we measure the oxygen and carbon dioxide. Well, you need what is called a *respiration chamber*. This is simply a box big enough to hold a rat, dog, bird, large cow, or human, with tubes carrying known amounts of oxygen in and devices to measure the amount of carbon dioxide going out. Respiration chambers can be prohibitively expensive, which is why there are not many respiration chambers left working in the world.

Figure 7.5 shows examples of direct and indirect calorimetry. Direct calorimetry is the measurement of the change in heat, usually of water, surrounding some heat-generating body such as a small dog. Indirect calorimetry is the calculation of heat from the amount of energy in, the oxygen in, and the carbon dioxide out. Because we know exactly how much heat is given off, oxygen used, and carbon dioxide given off in each chemical reaction, such as the oxidation of glucose, we can calculate the heat if we know the reactant in, oxygen in, and carbon dioxide out.

Most of the values that we have in tables for requirements for dogs and cats were derived from a few respiration chamber studies made directly with dogs or cats, or extrapolated from values taken from other species such as pigs or rats. We can trust those numbers based on two known things. First, the energetics of nutrient use in rats is the same as it is in dogs, cats, birds, fish, and so on, based on the known chemical reactions taking place and the laws of thermodynamics. Second, based on a number of direct experiments, we can compare the nutrient use by different species and decide whether and how we can extrapolate from a study done in one species to another. We know from some experiments that there are similarities as well as differences between animals. However, we cannot easily extrapolate from an experiment done on a pig to the nutrient requirements of a horse because the chemical pathways in

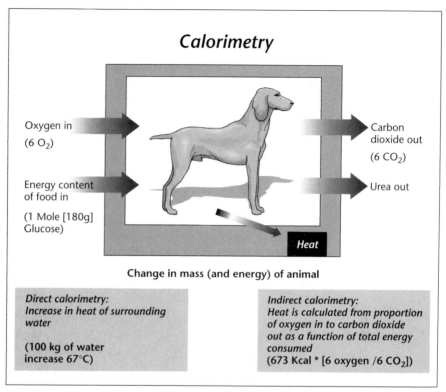

Figure 7.5 Respiration calorimetry: direct and indirect methods

their digestive tracts and organs are more different than similar. In general, rats, pigs, dogs, and cats follow the same basic pathways.

It is a simple economic fact that to replicate every nutritional experiment in every species would involve an expense that would require the cost of your simple dog food to rise tremendously. It is also a chemical fact that because we know the chemistry, biochemistry, and energetics involved in animal nutrition, we do not need to do feeding trials on every food, and we can carefully extrapolate from one species to another when appropriate.

Energy and nutrients are used by the animal for two broadly defined functions: *maintenance* and *production*. We use the term *maintenance* to mean the simple continuation of normal life with no change in body weight or composition. This is the energy used just to obtain food and stay alive, without growing, exercising, or making little animals or milk.

Maintenance energy varies with three functions:

processes to keep cells alive and maintain organ function

normal voluntary activity (food-obtaining activity)

maintenance of normal body temperature

Energy use is categorized as closely as possible to the chemical and biological functions we know:

GE = **gross energy:** total energy value of a given amount of a feedstuff

IE = **intake energy:** total energy value of feedstuffs eaten

FE = **fecal energy:** energy lost to the body in excreted feces

DE = **digestible energy,** or IE − FE: energy actually digested by the animal

UE = **urinary energy:** energy lost to the body in chemicals of urine, primarily urea

GPD = **gaseous products of digestion or gaseous energy:** including carbon dioxide and methane lost in respiration or digestive tract.

ME = DE − UE − GPD is **metabolizable energy:** energy remaining after digestion and excretion of waste products. This is the energy actually usable for the animal.

NE = **net energy:** energy used by the animal for breathing, nervous transmission, blood pumping, or the actual energy contained in muscle, fat, glycogen, milk, eggs, hair, wool, bone, and fetuses. It is the energy remaining in the animal, the net after all the losses are accounted for.

The minimal amount of energy needed to sustain life has been given many names over the years, but the two major ones in use in energy research are *basal metabolic rate* (BMR) or *fasting heat production* (FHP). Fasting heat production is the amount of metabolism that is minimal to keep the animal alive, with no change in body composition, and running all basic bodily functions of involuntary muscular and neural activity is the basal metabolic rate. We also use the term fasting heat production because the food energy used to drive all of these functions is all converted to heat, and thus when the animal is fasting (minimal life), the heat produced is fasting heat production. The measurement of basal metabolic rate must be made under consistent conditions.

The four basic conditions which must be met to measure BMR are:

1. The animal must be at rest.
2. The animal must be awake.
3. There must be no nutrients being absorbed from the intestines (fasting).
4. Animals must be in their zone of thermoneutrality, not expending any energy to keep warm or to stay cool.

Let us go through these conditions in more detail: If the animal is using energy to move muscles (other than just those needed to keep the organs alive) then it is not basal metabolism. The third condition is also obvious: The body cannot be receiving any nutrients and cannot be using energy to digest or transport nutrients or to make new compounds. During sleep, hormonal and neural activity changes the rate of energy use in unpredictable ways, and energy use during sleep is different from during waking. In practical terms, most animals are awake for more time than they are asleep (note *most*), so we use *waking metabolism* as the norm.

Finally, **thermoneutrality** is defined as that range of temperature in which the animal neither expends energy to keep warm nor to keep cool. This will vary for different species, size of animals, hair coat, whether they are wet or dry, the temperature, or the wind speed. Most animals that are companions are **warm-blooded**; they must use energy to stay at their normal body temperature. If the outside temperature is too cold, below what is called the *critical temperature*, animals will generate heat by *shivering thermogenesis* (involuntary twitching of muscles to generate heat) or *nonshivering thermogenesis*, which is heat generated from the energy in fatty acids that are broken down in specialized cells called brown adipose tissue.

Brown adipose tissue cells have a large number of mitochondria that are specialized to generate heat instead of ATP in the electron transport chain. There are usually increased amounts of this tissue in young animals in order to stay warm and alive. Animals that have adapted to a wide range of temperature swings, such as desert-dwelling rodents and reptiles, may also have significant amounts of brown adipose tissue. It is not a major heat-producing organ in adult dogs, cats, or horses. However, basal metabolic rate must be measured in that range of temperatures and situations in which the body is *not* expending energy to stay warm or cool.

The early researchers realized that the amount of heat given off by animals in basal metabolism was a function of body weight. That is, the bigger the body, the more heat. But it was not a "linear equation": It did not follow a straight line. Smaller animals gave off more heat in relation to their body weight than did larger animals. Animals that were bigger because they had a lot more muscle gave off more heat than animals that were bigger because they had a lot of fat. From this observation came the well-known concept of "metabolic rate increases with decreasing body size." A common example is that a small mouse has a much faster rate of metabolism then does a large elephant.

However, historical scientists also thought (or thought something close to this): "Wait a minute—isn't the energy cost of maintaining ionic gradients across a cell, or pumping blood, or breathing, or muscle protein, or fat synthesis the same for the same amount of work done? Shouldn't the basic metabolic rate (BMR) per unit body weight be the same?" Well, almost. What the scientists discovered was that the rate of calorie use for BMR was linear across many species of varying size if it was graphed on a basis of *metabolic body weight*.

The concept of *metabolic body weight* takes into account that in all bodies, there are organs that generate more heat than other organs, so that weight of the body alone is not always the best indicator of energy use.

Organs such as bone, connective tissue, and regular adipose tissue are not highly active and do not generate as much heat per unit mass as more metabolically active organs such as the liver, intestines, heart, and skeletal muscle. Most of the early researchers knew, from physics, that bodies generated heat in proportion to their surface area, and the surface area of a sphere was a function of volume to the power of 3/4 (or 0.75). After lots of experiments and arguments, most scientists agreed that the simplest way to indicate the metabolic rate of animals was to raise the body weight to the 0.75 power, and to this day, metabolic body weight (MBW) is $BW^{0.75}$. If the heat produced (BMR) is then plotted against the metabolic body weight (MBW) then a linear line is obtained. Doing this, several researchers discovered that the BMR in many species was very close to 70 kcal per kg $BW^{0.75}$. They then spent the next thirty years arguing about whether or not 0.75, 0.72, or 0.73 was a more exact exponent to use, but that's science. There is a more in-depth introduction to this concept in the recent NRC book (*Nutrient Requirements of Dogs and Cats*) and Linda Case's book; both are listed in the Further Reading list. Entire books have been written on this subject and nutritionists still argue the point today, but the basic principles we have covered have been proven out in many experiments over seven decades.

> The concept of MBW is important because it demonstrates that the underlying chemistry and biology is similar among different animals, and provides a basis for determining the important differences.

The practical reason for continuing to study energy nutrition is that it allows us to estimate, from the body size, the basal energy needed by that animal and we do not need to do an experiment on every animal, breed, or species over and over again. Basal metabolic rate must be known to have a basis, a frame of reference, to determine the additional costs of things such as exercise, making muscle, or digesting food. This concept is used in the determination of energy requirements of animals all the time, because of the excellent research and concepts developed by the early scientists in this area. Although you don't need to understand this to buy a bag of dog food, it must be understood by the people you trust to put good stuff in that bag and tell you how much to feed your animal.

We have covered BMR, the first component of maintenance. The second component of maintenance is *voluntary activity*, the energy required for normal movement, usually associated with gathering and eating food. Obviously, this is going to be different for the rich cat who has her food brought on a silver platter and the busy farm dog who roams all over the place.

The last component of maintenance is *thermal regulation*: the amount of energy required to either keep the body warm or cool. Below the critical temperature, an animal must either shiver or increase metabolic rate to keep warm. If an animal is at a temperature at which it needs to generate some heat to keep warm, this would, over time, increase the requirement of energy to maintain the body composition—because more is going just to provide heat, not for any other function.

Do Smaller Animals Have More Fun?
(And Live Longer?)

Historically, there have been as many different theories about aging and normal limits to lifespan as there have been generations. Eat less, eat more, don't smoke, sleep upside down, don't eat meat, God doesn't really like small animals, try aromatherapy, and so on. Scientists have tried and tried, and are still trying, to come up with the answer to: "Why do we die?" One theory actually sounds just as good as many and much better than most. This theory states that there is a finite limit to how many chemical reactions can take place before they start to fail. Examples include: There is a finite (and as yet unknown) number of times a cell can reproduce before it starts making too many mistakes in copying the genetic material, and the "not quite so perfect as the earlier" cells cannot do things as well. Or, with more metabolism, either resting or by activity or eating, there is more oxidative damage to cells. Alternatively, there are only so many times nerves can give signals before they start to deteriorate. Or, only so much respiration can take place before the chemical wastes cannot be removed as well and start to break down the normal structures of the cell. Actually, the second two examples are really extensions of the first. Therefore, a general observation is that because smaller animals tend to replicate so much faster, they may go through several billion or so cell divisions in just a few weeks or months and thus the genetic material of their cells breaks down faster and they die sooner, whereas a large whale or elephant has a much slower rate of cell division and can thus live a longer time before the "error rate" catches up with them. (Cancer is an example of a genetic error, and we do know that incidence of cancer increases with age). There is a well-done recent study on this topic by Speakman, Selman, McLaren, and Harper (2002) which, as is usual in science, suggests that a little bit of all of the previous information is true. It is also "true" in a statistical sense that animals and humans who live just below their normal average weight and food intake (calorie-restricted diets) tend to live longer and healthier lives. So, we come back to "all things in moderation" and "eat (a little) less and stay in shape." Perhaps those early observers had it right in the first place.

We call the temperature at which an animal will begin to overheat the *point of hyperthermal rise.* This will also vary with temperature, wind speed, hair coat, exercise, and the like. If an animal starts to overheat, respiration increases to try to remove heat from the body and, in turn, generates a little more heat. Thus, we should strive to avoid situations in which our companions may overheat, as this is often a more life-threatening situation for most animals than being too cold.

Maintenance energy (ME) = BMR + Voluntary activity + Thermal regulation. Maintenance is the condition in which most adult animals spend the majority of their lives.

Now let us turn to the energetic functions when an animal consumes food over and above the amount needed for maintenance, whether that is because this is required for growth, reproduction, or exercise, or just because they are eating too much. If an animal is growing, it is accumulating two major chemicals: protein and fat. We call these *energy retained as protein* and *energy retained as fat*, respectively. If the animal is making little animals, that is the *energy of conceptus*. If the bird is making eggs, we call that *ovum energy*. If the mammal is making milk, we call that *lactation energy*. All of these are part of net energy, the energy that is *kept* or *lost* by the animal, above that needed for maintenance and what is lost as heat.

Remember, maintenance requirements are always met first by the animal, and then any additional energy consumed can be used for growth, lactation, exercise, or reproduction. So if we have an animal that is supposed to be growing, but we are not feeding it enough, only that energy left over after the maintenance needs are met can be used as net energy for growth. Alternatively, if we are feeding the poor little dear more than is needed for normal protein and fat growth, it will primarily just be used to make more fat, increasing the retained energy as fat.

Finally, remember that none of the processes we have described runs at 100 percent efficiency. Each chemical reaction loses some heat, whether it be pumping blood or making muscle or milk. **Heat increment** is the energy given off as heat by the body during normal digestion and metabolism of food over and above the energy retained by the body. This represents the *inefficiency* of food use, and is often the largest percentage of energy consumed by the body. There are many other terms that have been used to describe this same function, such as *specific dynamic action*, *thermic effect of food*, and *dietary thermogenesis*. All of these terms refer to the amount of heat given off in the metabolism of ingested nutrients above that amount needed for maintenance.

> ### Heat increment contains four sources (see Figure 7.4):
>
> *Heat of digestion:* given off by chemical reactions in the digestive tract.
>
> *Heat of fermentation:* given off by chemical reactions in the bacteria in the digestive tract.
>
> *Heat of waste product formation:* given off in making waste products, primarily urea or uric acid.
>
> *Heat of nutrient metabolism:* total heat from all other metabolic reactions.

So what are some basic figures for the energy content of foods? A practical rule of thumb was determined over 100 years ago by Dr. Atwater. The usable (metabolizable) energy value of the three main feed ingredients—carbohydrate, protein, and fat— are known as "Atwater's Physiological Fuel Values" and are still in use today. These values are 4.0 kcal of ME (metabolizable energy) per gram for carbohydrate and protein, and 9 kcal/gram for fat, or 2.25 times as much energy as carbohydrate. These values can be applied to any species. They are, however, based on experiments and foods with high digestibility (purified fats, starchs, and proteins). For most mixed prepared dog and cat foods, these factors are adjusted somewhat to account for different digestibility, and factors of 3.5, 3.5, and 8.5 kc/g are used. In addition, the digestibility of fat by cats is a little lower than other animals, so we use 8.5 kc/g. The basics are the same; however, we can easily estimate the metabolizable energy content of foods because we know the chemistry.

Energy, Food Intake, and Obesity

Now that we have discussed the basic measurements of energy use, let us introduce a major problem of excess energy: obesity. Remember, energy cannot be created or destroyed. If energy is consumed in excess of what the animal needs for normal life, growth, or exercise, storage as fat increases. A 10 percent increase in energy intake over the actual costs of maintenance will add up to a 100 percent increase in just ten days! Before you know it, that extra energy is adipose tissue (body fat). We will discuss the problems, prevention, and treatment of obesity in more detail in Chapter 11 on dogs. Figure 7.6 shows the general pattern of what happens to blood glucose, the hormone insulin, and body-fat synthesis during the course of a day.

Obesity is not caused by one factor. The genetic makeup of the animal dictates its basal metabolic rate, controlled by many hormones and neural pathways. We control the environment with the amount of food offered and the chemical content (energy density) of the food. We provide different potential rates of exercise. Energy intake, amount of activity, and genetics will account for the majority of cases of obesity. The pattern of food intake can also affect the energy use by the animal over time. In general, at the same energy intake, animals that consume fewer and larger meals are more efficient in energy use and will store more fat. This has been shown in pigs, chickens, rats, and humans. We do not know if this occurs in dogs and cats.

In Figure 7.5, we look at the basis for this difference in metabolic efficiency. The heavy dashed line shows that normal glucose needs do not vary that much during the day (this increases with exercise). For every meal, there is a period in which blood glucose concentration exceeds what the animal needs. When glucose supply exceeds the need by the brain and nervous system, the pancreas releases insulin, a hormone that increases glucose uptake by muscle and adipose tissue and increases conversion to glycogen in the liver. The heat increment is the indicator that excess glucose (as well as fatty acids and amino acids) is oxidized for energy. Animals that eat larger, fewer meals have a greater heat increment, thus at the same energy intake will, over time, use less energy and store more as fat (see Chapter 3 in NRC, 2004).

Insulin increases body-fat synthesis (to store the energy from glucose). Large meals cause a large increase in glucose, then insulin, which increases fat synthesis. Smaller meals (even when the animal consumes the same

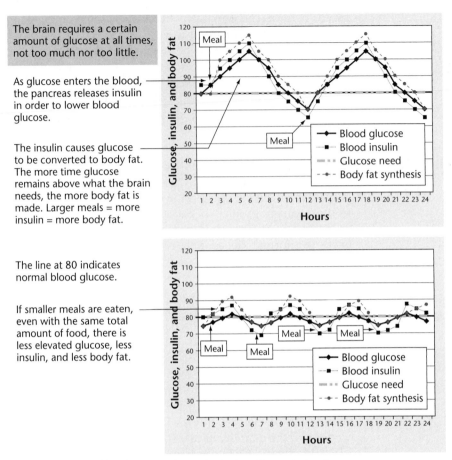

The brain requires a certain amount of glucose at all times, not too much nor too little.

As glucose enters the blood, the pancreas releases insulin in order to lower blood glucose.

The insulin causes glucose to be converted to body fat. The more time glucose remains above what the brain needs, the more body fat is made. Larger meals = more insulin = more body fat.

The line at 80 indicates normal blood glucose.

If smaller meals are eaten, even with the same total amount of food, there is less elevated glucose, less insulin, and less body fat.

Figure 7.6 Food intake control, nutrients, insulin, and obesity.

amount in twenty-four hours), do not cause the glucose to increase as high, so insulin does not increase as high, and the body does not store as much fat. The student should look at Figure 7.5 and imagine, then, what would happen if the meals in the example were either larger or smaller (what would happen to body-fat synthesis if more or less food was consumed?). In practice, how much this contributes to obesity in dogs and cats is not known. There is a lot of variation among animals and species. Also, total energy intake and activity level are more quantitatively important. But remember from the discussion that every calorie counts, and feeding smaller and more frequent meals, making certain that the total intake is correct for the animal, is a proven and long-practiced method to reduce the incidence of obesity. It also helps prevent other problems in feeding behavior and stress on metabolic organs. In general, the closer we match the actual rate of metabolic needs during the day, the more efficient the animal will be, and the lower the disease problems.

Determining Nitrogen Balance: Defining Growth and Production

The other major nutrient requirement in addition to energy is for amino acids, supplied by protein. As you have learned, because of the complexity of amino acid and protein chemistry, and because proteins contain nitrogen while other major nutrients do not, we actually measure protein requirements as nitrogen requirements. On average, most feed proteins contain about 16 percent nitrogen, whereas many animal proteins contain slightly more. We can measure the N content of feeds and animal tissues relatively easily by chemical digestion to the basic elements, capturing the nitrogen released, reducing it to ammonia produced, and titrating with acid. We then multiply by (100/16) or 6.25 to determine "crude protein." In fact, even though nowadays this might sound crude indeed, it is actually a useful measure, and has certainly improved protein nutrition of humans and animals by providing a sensitive and consistent way to estimate the protein content of feeds.

In a simple but real way, we can take the entire previous discussion on energy and substitute *nitrogen* for *energy* and we will understand most of the quantitative requirements for nitrogen; that is, protein. The exception is that nitrogen itself cannot be broken down with energy lost as heat, but amino acids can, and we measure the nitrogen given off in their breakdown. Protein is consumed, broken down to amino acids in the digestive tract, and the amino acids are absorbed (review Figure 4.5). The amounts that the animal needs for muscle and organ protein synthesis are used for those purposes. If the animal is female and pregnant or lactating, then a certain amount of amino acids are used to make the little animals or milk protein.

> ### Utilization of Nitrogen in the Body
>
> IN = intake of nitrogen
>
> FN = nitrogen in the feces
>
> DN = digestible nitrogen (total amount absorbed by the animal)
>
> UN = urinary nitrogen (urea in mammals, uric acid in birds, ammonia in fish)
>
> MN = metabolizable nitrogen (available to the animal to use for protein synthesis)
>
> RN = retained nitrogen as protein in body tissues (muscle, nerves, organs, bone, skin)

Although we use *nitrogen* as an estimate of *protein*, we have to remember that it is the amino acids that are required by the cells of the animals. Just as no metabolic pathway proceeds with 100 percent efficiency, protein synthesis does not either. The amino acids consumed in excess of what is needed for protein synthesis will undergo metabolic degradation. The nitrogen portion is converted first to ammonia, and then to urea in mammals (uric acid in birds and reptiles) and excreted in the urine. The carbons, however, will be converted to carbon dioxide, glucose, or fat, depending on the needs of the animal at the time. If the animal just needs a little extra energy (or if protein was fed just slightly above what the animal needs) then glucose can be formed and used for energy, or stored as glycogen. If a greater amount of amino acids is consumed above what the animal needs, the carbons will be converted to fat and stored in the adipose tissue. Some amino acid carbon chains are oxidized by the TCA cycle to generate $NADH_2$ to make ATP in the electron transport chain.

Nutritional Measures of Effectiveness of Diets to Meet Protein Requirements

There are several practical measures to evaluate different feeds for their worth as protein sources. We use the term ***protein quality*** to describe how well a food meets the exact amino acid needs of the animal. If a horse is eating primarily grass (either pasture or grains) then the protein quality will be lower, because grasses contain less of the amino acid lysine. Thus, the animal needs to eat more of that food to get enough of the total amount of each amino acid it needs. The other amino acids consumed in excess will be metabolized to urea, glucose, fat, and carbon dioxide. The two measures of protein quality used the most in practical nutrition are *biological value* and *net protein value* (NPV). The first value tells us about the amino acid balance, because inefficiencies of digestion are already accounted for. The NPV combines the inefficiencies of digestion with the imbalance of amino acids for a net use value. Figures 7.8 and 7.9 show how the body determines the biological value and net protein value of different proteins.

> Biological value is N retained in the body divided by nitrogen absorbed (digested). Net protein value is N retained divided by the N consumed.

The real and practical problem with protein nutrition is that it is almost impossible for an animal to consume exactly the amount and balance of the twenty or so amino acids it needs every day. Each animal has a unique genetic pattern (genotype) that dictates which proteins need to be made. Then, the environment affects the expression of the genotype so that a few different proteins are made (hormones, different muscle proteins, changes in liver or kidney metabolism). An animal would never consume *exactly* the same pattern of amino acids that it needs (although in some situations it can come pretty close). For example, in a horse grazing in a pasture, the horse is eating plant proteins, which have a completely different amino acid pattern than animal proteins. So the horse must take the essential and nonessential amino acids from the plants and reorganize them into the pattern of amino acids it requires. In the process, the excess nitrogen from the amine group of amino acids is lost in the urine. This may result in a biological value of about

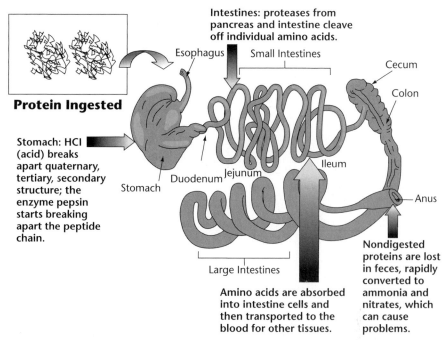

Protein Ingested

Intestines: proteases from pancreas and intestine cleave off individual amino acids.

Esophagus

Small Intestines

Cecum

Colon

Stomach: HCl (acid) breaks apart quaternary, tertiary, secondary structure; the enzyme pepsin starts breaking apart the peptide chain.

Stomach

Duodenum Jejunum

Ileum

Anus

Large Intestines

Amino acids are absorbed into intestine cells and then transported to the blood for other tissues.

Nondigested proteins are lost in feces, rapidly converted to ammonia and nitrates, which can cause problems.

Figure 7.7   Protein digestion.
Source: Protein graphic used with the kind permission of Dr. William McClure. Digestive track redrawn from *Small Animal Care and Management,* second edition by Warren. © 2002. Reprinted with permission of Delmar Learning, a division of Thomson Learning: www.thomsonrights.com. Fax 800 730-2215.

75 percent and a net protein value of about 65 percent. However, a suckling dog consuming only mother's milk will have casein as the sole protein source. Casein is highly digestible (over 95 percent) and the amino acid pattern is very close to exactly what the growing pup needs (greater than 95 percent amino acid match). So in this situation the biological value (BV) is over 90 percent and so is the net protein value (Figures 7.7 and 7.8). The BV and NPV are usually lowest in cereal grains, because of their low lysine content, in the range 50 to 60 percent. Legume plants have a better amino acid balance, so the BV and NPV can be 60 to 70 percent. For animal muscle, the amino acid content is much closer to what the animal needs, although digestibility is a little on the low side, so the BV and NPV of muscle meats is usually in the 75 to 80 percent range. The highest-value proteins are those supplied by the mother to the young, as their need is the greatest. Therefore, milk proteins and egg proteins have BV and NPV over 90 percent. Not surprisingly, quality is a direct driver of price. It is true that plant proteins are of lower quality than animal proteins, but this does not mean that they cannot and should not be fed. As with all nutrition, there is always a valid balance between quality and cost. A diet higher in plant proteins can be completely fine for dogs and cats, and the overall efficiency of natural resource use will be higher, and cost will be lower, than a high-meat diet.

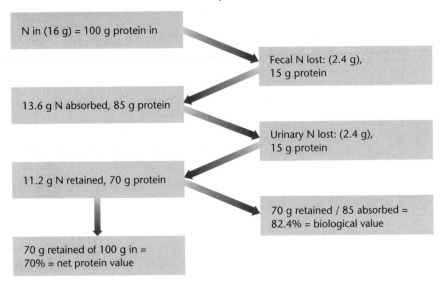

We measure the nitrogen as it passes through and into the body to determine biological value and net protein value of different proteins.

N in (16 g) = 100 g protein in

Fecal N lost: (2.4 g), 15 g protein

13.6 g N absorbed, 85 g protein

Urinary N lost: (2.4 g), 15 g protein

11.2 g N retained, 70 g protein

70 g retained / 85 absorbed = 82.4% = biological value

70 g retained of 100 g in = 70% = net protein value

Figure 7.8 Nitrogen and protein nutrition.

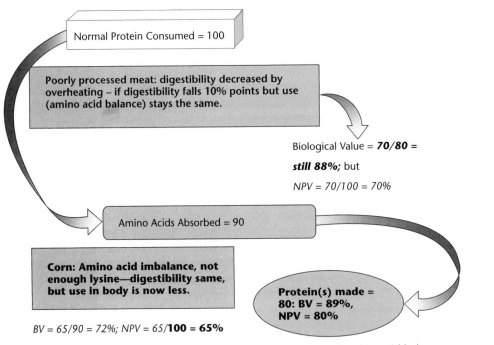

Normal Protein Consumed = 100

Poorly processed meat: digestibility decreased by overheating – if digestibility falls 10% points but use (amino acid balance) stays the same.

Biological Value = ***70/80 =***

still 88%; but

NPV = 70/100 = 70%

Amino Acids Absorbed = 90

Corn: Amino acid imbalance, not enough lysine—digestibility same, but use in body is now less.

Protein(s) made = 80: BV = 89%, NPV = 80%

*BV = 65/90 = 72%; NPV = 65/**100** = **65%***

Figure 7.9 Nutritional effects on protein quality: digestibility and amino acid balance.

Proteins vary in quality—both digestibility and amino acid balance.

	Biological Value	
Egg albumen	90–97	highly digestible, excellent amino acid balance
Milk protein	85	highly digestible, excellent amino acid balance
Beef muscle	69–76	good amino acid balance, digestibility varies with processing
Beef liver	77	good amino acid balance, digestibility varies with processing
Soybean meal	65–70	fair to good amino acid balance, fair digestibility
Wheat	67	poorer amino acid balance, digestibility increases with decreasing particle size
Corn	60	poor amino acid balance, digestibility increases with decreasing particle size (protein score)

Net Protein Value (body N growth − body N growth on protein-free diet)/Total N Intake

Animal proteins—use by rats

Whole egg	91	
Fish (cod)	83	
Egg whites (albumin)	82.5	
Dried milk	75	
Beef muscle	71.5	
Meat meal	35–48	(chicks)

Plant proteins—use by rats

Wheat germ	67
Corn	55
Rice protein	36

Adapted from Pond et al. (1997).

Amino Acid Metabolism and Protein Use

Now we need to learn about specific situations regarding amino acid nutrition that help us to understand and define protein nutrition. Some of this might be a little beyond the scope of the course you are in; in more advanced courses, the instructor might spend more time on these aspects of amino acid nutrition.

Amino acid imbalance is a general term to indicate that the overall pattern of amino acids does not match what the animal needs. There may be two few or too many of some amino acids. In practice, there is always some level of amino acid imbalance, but this usually is significant only when a sole source of protein (usually plant) is included in a diet. In reality, a minor amino acid imbalance is not usually a problem for companion animals.

The *first limiting amino acid* is the amino acid that the body will run out of first as it is making proteins, if the diet does not supply enough of it.

The ***first limiting amino acid*** is a major concept in animal nutrition describing the efficiency of protein synthesis. If, for example, the body is making insulin, quite an important protein in the body, and diet provides enough of all amino acids except the methionine, then methionine is the first limiting amino acid. When there is not enough of an amino acid, then protein synthesis does not just slow down, it stops (Figure 7.10). The first limiting amino acid is the one that is most out of balance on the negative side; it will run out first. Then, for very high-priority proteins such as hormones, enzymes, and nervous tissue, the animal will rob the blood proteins and muscle proteins of the needed amount. Obviously, this situation cannot go on indefinitely, so we

Amino Acid Deficiency: not enough of one or more amino acids.

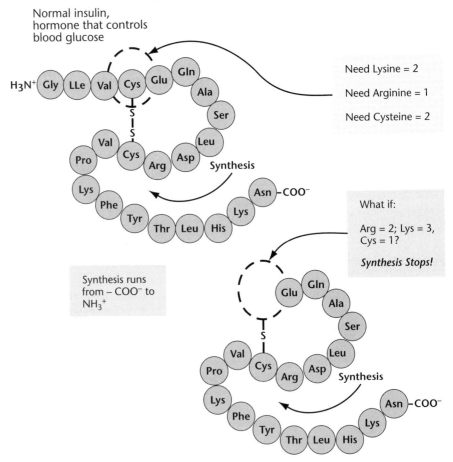

Figure 7.10 Amino acid deficiency.

strive to feed a well-balanced protein source. If the animals are consuming a major proportion of cereal grains, lysine is first limiting. This is actually the most common situation in feeding animals, as animals need proportionally more lysine than is present in most plant products. Thus, a legume plant source, an animal protein, or a lysine supplement must usually be used. Another common situation is when the amino acid methionine is limiting. If we feed a diet with soybean meal making up all of the protein, quite often methionine is the first limiting amino acid, because soybean meal has plenty of lysine but not enough methionine. This is a major reason why nutritionists pay so much attention to providing both sufficient amounts of total protein and the right amino acid balance.

Amino acid competition is the chemical inhibition of the uptake or use of one amino acid by another. Because many amino acids are similar chemically, and use the same cell membrane transporters, they sometimes can inhibit the transport of other similar amino acids across cell membranes. Lysine can inhibit arginine uptake and metabolism, sometimes interfering with the urea cycle. Additionally, they might inhibit each other in their use in protein synthesis. There can be several potential amino acid competitions, but the major ones are between arginine and lysine, and competition between the branched chain (hydrophobic) amino acids. In practical application, it is difficult to actually create an amino acid competition, as most rations are properly balanced to avoid this. Also, it often takes a significantly greater amount of one amino acid than another to create a competition. This situation might only occur with improper use of amino acid supplements.

Amino acid sparing is a situation in which some of one amino acid can be used to make another amino acid, thus *sparing* the amount we need to put in the diet. For example, cysteine requires methionine to be made, thus addition of more cysteine to the diet *spares* methionine, as less methionine needs to be used to make up the deficit in cysteine. Tyrosine is made from phenylalanine; therefore, addition of tyrosine *spares* phenylalanine. In some species (birds, reptiles), more serine can be made from glycine, so additional serine is added to spare glycine. As for amino acid competition, with normal balanced rations this is not a major issue for companion animal nutrition, but you should be aware of it. It is more important for formulating rations for rapidly growing agricultural animals, especially chickens, whose rapid growth rate dictates that the amino acid balance be quite good. This also applies to companion birds.

Practical Application

Animals consume foods containing potential energy, most of which is in fat and carbohydrates and some of which is in protein. Protein provides specific amino acids so that the animal can make its own protein. We run experiments to define energy and protein requirements for different species in different life stages and environmental situations. We strive to supply just the right amount and balance of energy and protein to supply the needs without risking obesity or extra work for the kidneys. Feeding diets designed for the specific life stage is the best way to do this. These diets are formulated from a mix of different types of plants (grasses and legumes), animal protein sources and, if needed, specific amino acids. Young growing animals need extra energy and protein, while animals at maintenance need (and thus should be fed) less. Older animals need a proper balance of amino acids and usually less energy. Further chapters will cover specific foods and food preparation, and then we will go into detail on feeding specific species.

Words to Know

adenosine triphosphate (ATP)
amino acid competition
amino acid imbalance
amino acid sparing
anaerobic
anaerobic fermentation
brown adipose tissue
calorie
calorimetry
dehydration
diuretics
drinking water
electron transport chain

energy balance experiment
first limiting amino acid
food-borne water
glycolysis
heat increment
heritability
hydremia
hydropsy (edema)
kidney
lactation
metabolic intermediates

metabolic water
net energy system
nicotinamide adenine dinucleotide dihyride (NADH$_2$)
protein quality
respiration
sweat
thermoneutrality
tricarboxylic acid cycle (TCA)
warm-blooded

Study Questions

1. Describe in brief the history of discovery in energy use and metabolism in animals.
2. Describe the total flux of water through animals, including the factors that increase requirement or loss of water.
3. What are some of the effects of dehydration?
4. Describe the catabolism of nutrients, transfer of energy, and synthesis of new compounds in cells.
5. Describe the measurements and uses of the net energy system.
6. Describe the uses of measures of protein quality.

Further Reading

Asimov, Isaac, ed. 1979. C is for Celeritas. *Asimov on Physics*, (pp. 128–141). New York, New York: Avon Books.

Baldwin, R. L. 1968. Estimation of theoretical calorific relationships as a teaching technique: A review. *J. Dairy Science*, 51: 104–110.

Burkholter, W. J., and P. W. Toll. 2000. Obesity. In M. Hand, C. D. Thatcher, R. L. Remillard, and P. Roudebush, eds., *Small animal clinical nutrition*, fourth edition (Chapter 13). Topeka, KS: Mark Morris Institute.

Burger, I., ed. 1993. *The Waltham book of companion animal nutrition*. Oxford, UK: Pergamon Press.

Case, L. P., D. P. Carey, D. A. Hirakawa, and L. Daristotle. 2000. *Canine and feline nutrition* (Chapter 9). St. Louis, MO: Mosby, Inc.

Gross, K. L., K. J. Wedekind, C. S. Cowell, W. D. Schoenherr, D. E. Jewelll, S. C. Zicker, J. Debraekeleer, and R. A. Frey. 2000. Nutrients. In M. Hand, C. D. Thatcher, R. L. Remillard, and P. Roudebush, eds., *Small animal clinical nutrition*, fourth edition. Topeka, KS: Mark Morris Institute.

Jurgens, M. H. 1997. *Animal feeding and nutrition*, eighth edition. Ames, NY: Kendall Hunt.

Kleiber, M. 1961. *The fire of life*. New York: John Wiley and Sons.

Lofgreen, G. P., and W. N. Garrett. 1968. A system for expressing net energy requirements and feed values for growing and finishing beef cattle. *J. Animal Science*, 27: 793.

National Research Council. 2004. *Nutrient requirements of dogs and cats*. Washington, D.C.: National Academy Press.

Pond, W. G., Church, D. C., and K. R. Pond. 1988. *Basic animal nutrition and feeding*, fourth edition. New York: John Wiley and Sons.

Pullar, J. D., and A. J. F. Webster. 1985. The energy cost of fat and protein deposition in the rat. *British Journal of Nutrition*, 37: 355–363.

Rogers, P. J., and J. E. Blundell. 1984. Meal patterns and food selection during the development of obesity in rats fed a cafeteria diet. *Neuroscience and Biobehavior Reviews*, 8: 441–453.

Sims, E. A. H. 1976. Experimental obesity, dietary-induced thermogenesis, and their clinical implications. *Clinical Endocrinology and Metabolism*, 5: 377–395.

Speakman, J. R., C. Selman, J. S. McLaren, and E. J. Harper. 2002. Living fast, dying when? The link between age and energetics. *Journal of Nutrition*, 132: 1583S–1597S.

Warren, D. M. 2002. *Small Animal Care and Management*, second edition. Albany, NY: Thomson Delmar Learning.

Chapter 8

Getting Ready to Make Foods: Ingredients, Preparation, and Processing

Take Home and Summary

Animals meet their nutrient requirements by eating a variety of foods, each with a different nutrient content. Modern feed-manufacturing processes can supply the nutrients animals need in one mix of foods, usually referred to as a **complete food.** Ingredients from plant and animal origin contain different amounts of protein, carbohydrates, fats, vitamins, and minerals. When properly formulated, a ration combines ingredients in proportions to provide the correct amounts and balances for a given species or a certain life stage. A variety of processing methods allows preparation of diets of the correct chemical nature as well as the appropriate physical form to improve digestion, palatability, safety, shelf-life, and ease of use. Processing also removes natural contaminants or compounds that inhibit some body functions. Vitamins and minerals are added to complete foods to ensure the proper amounts and balances to complement those found in the ingredients. There has been some attention lately given to raw-food diets as a better, safer, more nutritional alternative to processed, prepared foods. Although a few valid points are raised, and animals can be fed on a carefully formulated, safely stored raw-food diet, years of experience and scientific experiments have proven the overall benefits of high-quality prepared foods for health and longevity.

Introduction

The term *food* usually refers to any item that supplies nutrients. Foods might be in the natural state such as fruits, vegetables, forages, meat, and dairy products, or may be "prey" animals if we are talking about wild predators.

> Foods consist of ingredients, often called *commodities*, such as corn, oats, soybean meal, safflower seeds, meat (structural muscle), animal by-products, mineral supplements such as dicalcium phosphate, and specific vitamins.

Foods may be a mix of many different ingredients to make *prepared foods*. These foods may be, for example, a snack mix of animal tissues and some plant products with sweeteners or other flavors; or a complete food, a mix of different ingredients, vitamins, and minerals that meet all the nutrient requirements of the animal being fed. The term *feed* usually refers to foods fed to agricultural animals such as cattle, pigs, and chickens, or when we refer to specific commodities such as corn, beef, or wheat. We usually reserve the term *food* to apply to ingredients and products consumed by humans and our animal companions, but do note that this is a convention, not a scientific definition.

We will first run through the ingredients used historically and still used to make up foods, then we will move on to discussing the various methods used to combine, process, and complete what we refer to as "pet food." In the next chapter we will discuss at length the types, forms, and analysis of pet foods.

Types of Feedstuffs

To help you start putting foods together with nutrients, the categories described next classify feeds and ingredients by the major function or nutritional component that they supply. However, few feeds other than fats and oils are all energy, protein, or fiber; they usually contain some of all nutrients, but also usually supply one primary nutrient. Also, few plant or animal sources by themselves contain the perfect mix of all nutrients. That is why animals in the wild usually choose a variety of foods (or eat the "whole thing" when they consume prey animals), similar to animals in the past who fended for themselves. That is why we must use feed ingredients to make foods.

Energy Feeds

Energy feeds primarily provide energy-yielding nutrients such as starches and fats. They also provide protein and some vitamins and minerals, but are not necessarily a good source of these other nutrients. Some feeds can be energy and protein feeds at the same time, such as good quality forages, plants, and hays that can provide significant energy and protein for horses and rabbits, some birds, rodents, and reptiles.

Grains usually means cereal grains, which are the seeds of plants, commonly referred to as grasses, technically monocotyledonous plants. These include corn, barley, wheat, oats, and rice. They do not include soybeans, other beans, canola, and cottonseed. The take-home message is that most of these seeds are high in starch, relatively low in protein (although many can supply enough protein for adult animals at maintenance), low in calcium, and variable in other minerals and vitamins. They are used primarily to supply starch for energy. Use of too much grain will result in a ration short on protein, fat (when fat is necessary), and calcium. A little educational note is that the term *cereal* comes from the Roman *Ceres*, or Corn Goddess, also sometimes referred to as the goddess of agriculture and close cousin to *Saturn*, god of seed time and harvest, and

Pomona, goddess of fruit trees and fruit (think of pomegranate, pomme [apple], and "pomidore," or "apple of love," commonly known as the tomato).

Oilseeds are high in oil and protein; they are generally processed before using. Major oilseeds include soybean, canola, cottonseed, sunflower, and safflower. Most are seeds of legumes or dicotyledonous plants. They are often extracted to separate the oil from the protein and starch, in a process designed to provide two or more different nutritional ingredients from one feed. This provides greater flexibility in formulating rations to supply the exact proportions of nutrients needed by specific animals. Some oilseeds contain **nutritional inhibitors** such as **trypsin inhibitor** (inhibits protein digestion, Chapter 4) in soybeans or **gossypol** (thyroid antagonist) in cottonseeds. Soybeans are heated to destroy trypsin inhibitor. Usually only soybean and soybean derivatives are used significantly in pet foods, usually because of lower cost and more acceptable taste.

The greatest source of energy is fat. Fats may be trimmed animal fat: tallow from beef or lamb, lard from pork, or poultry fat. White grease and yellow grease are usually mixes of animal and plant fat used in the restaurant trade and can be used directly as fat sources for pet foods. Fats are more saturated (high in stearic acid and palmitic acid); they are usually solid at room temperature. They are usually heated and sprayed into or onto foods to encourage good mixing. All major fats are good sources of essential fatty acids.

Oils are those fats that are liquid at room temperature. These are almost always vegetable oils, but some fish oils may be liquid at room temperature. Vegetable oils can be "pure": corn, canola, cottonseed, sunflower, soybean, rice, and safflower. They can also be mixed in any proportion. These are usually polyunsaturated (such as oleic acid and linoleic acid). You might see "hydrogenated vegetable oils" on a nutrition label. These oils are hydrogenated to make them solid, usually to be used in cooking (baked goods) or in frying. This process also increases the proportion of trans-double bonds in the fats, referred to on the street as *trans fats*. It is now known that these fats can have some detrimental effects on immunity and arterial health, and more care is taken now in their use. You may have used shortening in baking, which is hydrogenated vegetable oil. You almost certainly have used margarine, which is hydrogenated vegetable oil and contains trans fats, as opposed to good old natural butter, which is of course also all fat and contains cholesterol. So which is better, which is worse? The point is that they are both all fat and should only be used in moderation (see Chapter 3). By definition, if hydrogenated, fats contained increased amounts of saturated fatty acids such as stearic acid or palmitic acid.

Protein Feeds

Protein feeds are, of course, those that supply a large amount of protein. Usually a protein feed contains a minimum of 20 percent crude protein, depending on the source, but may also be a good source of energy (fat or starch). Protein feeds include animal products, such as muscles, organs, and blood, as well as legume seeds such as soybeans and sunflower. Soybeans are the

> **Fact Box**
>
> It is a fact that most plants and many animal products actually contain a greater percentage of protein than is required by animals. For example, a rabbit might only require 10 percent crude protein for maintenance, and many grasses are greater in protein than 10 percent for a large part of the growing season. A cat may require at most 25 percent crude protein (during growth or lactation), and most animal muscles are 50 to 70 percent protein on a dry-matter basis. It makes sense that nature provides more than enough protein—it would not make a lot of sense to have a system that did not provide sufficient nutrients. A question that remains is how much protein is too much? Can diets be too high in protein?

primary example of a food that is high in both protein and fat (more than 20 percent of each for whole soybeans); alfalfa is rich in protein (15 to 25 percent) and fiber (20 to 40 percent); and muscle meats can be greater than 50 percent protein and 20 percent fat (though some can be very low in fat, less than 10 percent). Most commonly, a mix of animal and plant proteins are used in foods to make a balance between amino acid balance, total amount of protein, and cost of the food (plant proteins are usually cheaper than animal proteins).

Protein also provides energy to the animal (amino acids can be broken down and used for energy; see Chapter 4), but protein feeds are generally too expensive to use as a major energy source. It is also a disadvantage for animals to eat significantly more protein than they need for long periods. First, it uses extra energy to metabolize the unnecessary amino acids. It can also put extra stress on the kidneys to remove nitrogenous waste; although this is still a controversial topic in nutrition (we will discuss this more in Chapter 10).

Fiber Feeds

Fiber feeds are usually whole plants or the vegetative parts (nonreproductive and reproductive parts that include flowers or seeds) of plants. These can be the major or sole feedstuff for horses, rabbits, and some rodents and birds. For dogs and cats, processed forages, parts of forages, or the hulls of seeds may occasionally be used to increase fiber for maintenance or weight-loss diets. Fiber feeds are usually those that are greater than 20 percent cellulose and lignin, or Acid Detergent Fiber (ADF). Hay is a general term for any cut and dried forage. Alfalfa, cut and dried as hay, is a high-protein, fiber, energy, and calcium source for herbivores and omnivores. Grass hays (orchardgrass, bromegrass, Kentucky bluegrass) are lower in protein, high in fiber, and low in calcium. *Straws* are generally the grass stem part left over after the seed is harvested. They are usually used as bedding but might be used as a food for horses who like to eat but could stand to lose a little weight. Straws can be from oats, barley, wheat, or any grass; processed forages can be any of these that are usually ground to mix into other foods or to make complete rations with a reduction in bulk. Dehydrated alfalfa and alfalfa meal are examples.

Hays, Forages, Roughages—Confusing Terms

Forage refers to all foods made up of the whole plant or the vegetative part of it.

Roughages are forages that have increased fiber (more than 50 percent) and very little other nutritional value.

Hay is any forage that is cut, dried, and stored for later use.

Meat Meals, Processed

Any part of the animal carcass can be and is used in pet foods (As an aside, in many countries today, any and all parts of the animal carcass are used in human foods, too.) Ingredients are cooked and ground before use. These are used as supplements for protein, and can also be very high in calcium, phosphorous, and other minerals. They are generally a variable source of vitamins, but can have significant amounts of fat-soluble vitamins. The protein content is usually high and the amino acid balance is good, but protein quality will vary with the heat and duration of processing.

Meat meals also include fish meal (many kinds), poultry meal, meat meal, meat and bone meal, bone meal, and blood meal. If it is meat meal, that is usually beef, occasionally pork or lamb. But the species is usually identified. These are all cooked and ground or powdered products to be used as protein supplements. Because they are animal products, their amino acid balance is usually quite good. The real problem with meat meals is in the variability of the processing. When processed properly, without too much heat and with some attention to detail, they are fine products. If they are overheated, the digestibility will decrease, so although their amino acid balance and total protein content is still good, less will be available to the animal. Pet food companies do measurements in the laboratory to determine the digestibility of various sources of meat meals.

There is a lot of information on the Internet and in various publications on the relative quality of animal and plant protein for dogs and cats, including how bad meat meals are (they do contain intestines, which some may find offensive). There are some facts in this argument, as we have discussed in previous chapters and will discuss in later ones. But remember that at the end of the day, it is the actual protein digestibility and amino acid content of the food that is the important characteristic, not solely how much animal or plant protein was used or the organ from whence it came. That friendly little wild dog or felid isn't offended by the prospect of chomping down on some good, ripe intestines. There is no valid scientific reason for not using them in animal foods today.

Fact Box

Bovine Spongiform Encephalopathy, or, Can Dogs Get Mad Cow Disease?

You may have heard of a ban on using animal tissues and products in feeds for ruminants, as a proactive measure to prevent transmission of bovine spongiform encephalopathy (BSE), or as the newspeople say, "mad cow disease." Although this is a debilitating and fatal disease, with much yet to be learned about it, there is no such ban in place on foods for pets. This is because there is no evidence of disease transmission in feeding ruminant animal tissues to most nonruminants. This seems to be a disease that ruminants (cattle, elk, deer, sheep, and goats) tend to be more susceptible to, so feeding ruminant by-products and meals to llamas and alpacas should be avoided. There is also a spontaneously occurring encephalopathy present in felids (FSE). In the early 1990's a new variant of BSE (vBSE) was discovered in cattle. There is evidence that about 150 cats were infected with the variant prion during the outbreak of BSE. It appears that cows in Europe were affected by consuming meat meal from sheep that contained the causative prion. Sheep get a disease called *scrapie* that is similar to BSE. Since the affected cows were killed and the feeding of ruminant by-products to ruminants was stopped, few new incidences have been reported. We should usually err on the side of caution; however, to date there is no indication that any encephalopathic disease has been transmitted to other companion animals.

Humans can spontaneously develop a similar disease, called Creuztfeld-Jacobs (pronounced KROYTS-felt-JAH-cobs) disease (CJD), which forms plaques in the brain, causing lack of brain function and dementia. A new variant of CJD (vCJD) arose in humans about the same time that BSE was found in many cattle in England and Europe. The evidence that CJD may have been caused by the consumption of meat from animals with BSE is impressive. Since the ban on feeding ruminant tissues to ruminants and the destruction of hundreds of thousands of older cattle, the incidence of vCJD has almost gone to zero (one case in the last two years). The pathogenic agent itself is a **prion**, a very simple protein with no genetic material. Normal prions are thought to be involved in memory function, but we truly know little about their function yet. The variant prions (they have one to a few different amino acids) can precipitate out and cause vacuoles in the brain, leading to destruction. The incidence of spontaneous CJD in humans and animals is about one in a million, as is the incidence of spontaneous BSE.

With the discovery of *one* infected animal in the United States in December 2004, the U.S. Department of Agriculture (USDA) moved to restrict the use of ruminant nervous tissue in foods for any animals. It is now known that this animal came from Canada, was older, and probably consumed ruminant meat by-products before the bans were put in place. In the United States, many companies have voluntarily stopped use of some by-products in pet foods. There is no doubt that this has been a serious problem. It is easy to quote statistics and say this is a small problem but we must remember that for the people affected and their families, it is devastating. However, as rational beings, we must keep in perspective that regardless of press and Internet coverage, the fact remains that about 160 people worldwide have been reported infected with vCJD since 1994, and only four or five in the last four years. In just four years, more than a hundred thousand have died of preventable seasonal influenza. Testing for BSE has increased in the United States, and it will likely confirm that the normal incidence of spontaneous BSE is about one in a million. There is no excuse for bad practices, but application of the scientific method must take precedence in keeping each situation in perspective.

Mineral and Vitamin Sources

Minerals and vitamins are found in all plants and animal products in an extreme range of concentrations. Many analyses have already been done and many tables exist that provide laboratory measurements (see Chapters 5 and 6). In general, most plants are good sources of sodium, potassium, phosphorous, magnesium, and most other minerals other than iron and calcium. Grasses are notoriously low in calcium, whereas legume plants are higher in calcium. Seeds tend to be lower in calcium and tend to be lower in most minerals than the whole plant. The fat-soluble vitamins will vary widely as discussed earlier, with seeds tending to have more fat-soluble vitamins than the total plant. Whole seeds (with germ) are good sources of vitamin E and vitamin A. Animal products such as liver can be quite high in iodine, phosphorous, iron, zinc, copper, and in fat-soluble vitamins such as A and D; it is also fair in vitamins E and K.

When we formulate rations for animals, we always make a final balance of vitamins and minerals by supplementing to make up for what might still be lacking in the feeds. Normal mixes of ingredients actually come pretty close to the vitamin and mineral requirements of most animals, with some exceptions. However, it is nutritionally, morally, and ethically accepted and required that minerals and vitamins are added to foods in amounts slightly more than is needed. This is simply to ensure that the vast majority of animals fed these diets will have enough and not too much. This is not a bad idea if you think about it. The pet food industry should never be in a situation where it knowingly makes foods that do not supply sufficient nutrients. However, this extra allowance can sometimes lead to potential problems for animals consuming large amounts of complete foods, as they may be getting too much. This is usually only a problem with minerals such as calcium and magnesium, or in some situations fat-soluble vitamins. There is no research that shows, other than for magnesium in cats and potentially excess vitamin C, that slight excesses of most vitamins or minerals have caused health problems. The presence of deficiency signs in animals consuming poorly formulated homemade diets is a problem much greater than any such incidence for commercial rations.

> Why add in minerals and vitamins in amounts greater than requirements? From any aspect, it is much worse to have a vitamin or mineral shortage than a slight excess.

Mineral Supplements

Mineral *supplements* are mixed into prepared food in the proper balance for that animal. They can be purchased separately in pet food stores, but are not usually necessary for dogs and cats eating commercial foods. Supplements can be concentrated from various sources, extracted, or present as many mineral

Table 8.1	Types and General Composition of Plant Feeds
Legumes	*Whole plants*
	Stems, leaves: for horses, rabbits, some rodents, and birds.
	Alfalfa, clovers, timothy, dehydrated alfalfa (as protein and fiber source).
	Seeds
	Soybean, canola, cottonseed (these are high in protein and fat).
	Whole, they are 20 to 40% oil, with 15 to 30% crude protein, 30 to 50% starch.
	Usually cooked first to destroy trypsin inhibitor, usually extracted to form the separated oil and meal (which is high in protein and used as a protein supplement).
Grasses	*Whole plants*
	Orchardgrass, fescue, bromegrass, bermudagrass, corn, wheat, oats, barley.
	These are usually food for cattle but also for horses, rabbits, and birds.
	Seed grains
	Corn, rice, oats, wheat, barley, rice, sunflower, safflower.
	Generally 50 to 70% starch, 9 to 12% crude protein; 2 to 5% crude fiber, 2 to 4% fat, but sunflower and safflower are higher in fat.
Tubers	*Whole plants*
	Potatoes, sweet potatos.
	Similar nutrient concentration to grains.

forms—many different kinds are available; for example, see Table 8.1 and *Nutrient Requirements of Dogs and Cats* (NRC, 2004). Some major ones used are dicalcium phosphate, oyster shell (mainly calcium carbonate), calcium chloride, magnesium oxide, and zinc oxide. Table 8.2 lists the mineral supplements common in companion animal rations.

Vitamin Supplements

Vitamin supplements are usually the specific vitamin itself, or concentrated sources such as yeast or seed germs. They can be purified from foods or synthesized and purified from yeast or bacteria grown in culture. Then they are usually combined into a *premix* for addition to the ration during mixing. Premixes will be specifically designed for the ration being made and the animal and life stage it is prepared for. Because many vitamins, as many minerals, are only needed in tiny amounts per ton of food, the premixes allow the manufacturer to be more precise in mixing the rations (it is not really easy to mix

Table 8.2	General Types and Compositions of Animal Products
Muscles	*Beef, chicken, lamb, pork* Fresh: 70% water, 10 to 20% protein, 5 to 20% fat. Usually 50 to 70% protein when dried, 5 to 30% fat depending on the muscle and trimming done, high in iron, basically devoid of calcium.
Organs	*Liver, kidney, spleen* Same general nutrient content as muscles, except usually higher in protein and lower in fat. Can be very high in vitamins A and D, and minerals such as iron and copper, leading to toxicities if fed as a major portion of the diet over several weeks or months.
By-products and meals	Intestines, viscera, trim, cooked, and ground. High in protein, also contain greater amounts of calcium. Quality (digestibility) can vary depending on cooking temperatures.
Bone meal	Ground and cooked bones, primarily a Ca, P, and Mg source, also contains some protein.
Blood meal	Cooked, dried, and powdered whole blood. Very high-quality protein.
Fats and oils	80 to 100% fat, no appreciable carbohydrate or protein. Plant oil: separated (extracted) from grains and oilseeds. Animal fat: trimmed from animal carcasses.

100 ug directly into a ton of food). Vitamin supplements can be purchased by consumers, but are not necessary for dogs and cats eating commercial foods, as there are always enough already in those foods. There can be any variety of vitamin premixes that can be used by feed manufacturers, or they may add specific vitamins in free form to supplement the diet.

Processing and Preparation of Feed Ingredients and Feeds

So, now you understand the variety of ingredients that are available to make diets or food. Plants and animal products usually need to be handled and processed in many ways. Sometimes processing gets a bad rap. A percentage of the vitamin content is destroyed with cooking, and some vitamins are more susceptible to heat destruction than others. We control this loss by heating for no longer or no hotter than necessary to increase digestibility and destroy pathogens, and by the judicious and inexpensive use of vitamin supplements. Processing, although not an absolute necessity, is often the wisest course for many reasons.

Not all nutritionally useful feedstuffs are palatable, desirable, highly digestible, or even safe in their "pure" form. Most are usually processed in some way to improve their nutritional or overall value. You may find many sources of information on "raw foods" or the "BARF" (bones and raw food) diet on Internet sites, in popular press articles, and in scientifically sound books and papers. This is without a doubt a scientific controversy in some circles. The interested student may look very closely at some Web sites and try to sort out what are biological facts and principles and what might only be categorized as "marketing." (Just as in reading a label, which we will discuss in the next chapter.)

Reasons for Processing Food Ingredients

Advantages

1. Increases palatability and acceptability.
2. Increases availability (digestibility).
3. Removes toxins, inhibitors, unwanted fractions.
4. Increases shelf-life and storage time.
5. Improves handling characteristics.
6. Improves texture, taste, or appearance.
7. Increases or decreases nutrient density.
8. Adds flexibility in consumer choices.

Problems

1. Decreases palatability and acceptability.
2. Can decrease digestibility.
3. Can accidentally introduce contaminants.
4. May negatively alter texture, taste, and appearance.

In this text we will not go into detail for the making of homemade diets, especially for dogs and cats. There is a good and sound text out on the topic that some may find useful (Strombeck, 1999). I have also listed a scientific article on this topic from the recent literature (Murray, Patil, Fahey, Merchen, & Hughes, 1998). Factors to consider in using homemade diets must include at least your time, cost, constant availability of ingredients, safe food handling and storage, and the relative benefits to the animal. We will discuss the specific digestive and metabolic differences of different types of diets in the specific chapters on each species.

Types of Processing

There are various time-tested types of processing for the feeding objectives listed earlier. We will touch on the major ones.

Mechanical

Most plants and animal products need to be mechanically processed to reduce particle size. This allows different types of foods to be mixed together into a total food. It also greatly increases digestibility and this is better for the animal and more efficient for use of natural resources. Seeds can be *dehulled* to remove the fibrous outer coat. In *rolling*, seeds are pushed through two rollers with a wide gap between the rollers to knock off the fibrous hull. Parts of the seed include the outer coat, or *hull*; the inner coat, *bran*; the *endosperm*, primarily starch; and the embryo, or *germ*. The various parts can be used to make a wide variety of foods and products with different textures, tastes and nutrient contents.

Seeds and all feeds can be *ground*. Seeds can be ground between two large stones or rollers. More commonly today, the seeds are "hammered," or pushed in a mill against some type of cutting edge until a uniform size is reached. Some mills push the grain through a cutting edge machined with holes of a uniform size, and these plates can be made in any particle size. Thus the feed can be ground to any fineness. Flour is an example of a very finely ground seed, whereas normal coarse-ground corn resembles bread crumbs more than flour. Remember that as for any process, the desired size will be the average size—there will always be a "bell curve" of particle sizes, some larger, some smaller than average.

In general, the smaller the particle size, the more digestible it is, and the easier to handle. However, if too finely ground, the particles will absorb too much moisture and become clumped, reducing the ease of use. In addition, very fine grinding will reduce palatability. When fine particles are mixed into a food, they can break off easily, or separate too quickly in the digestive tract and increase the chances of damage to the stomach or intestinal lining or impaction of food somewhere along the digestive tract. The processing industry thus has grinding and rolling down to a fine science, setting the mills to any average particle size needed, so this avoids either underprocessing or overprocessing. Figure 8.1 provides an overview of the processing of plant and animal parts.

All ground feed contains *fines,* the small, dust-like particles that break off during processing. If you have dumped out a bag of any processed food, be it for your dog, your cereal, or crackers, you have probably seen fines in the bottom. These are generally not easy to eat, and thus become useless. Companies spend a lot of time, money, and effort to reduce fines in order to increase efficiency. Depending on how the ground product will be used, these fines will be filtered out over sieves of various sizes. Fines can also form through normal handling. High-quality finished products will have most particles in a very narrow size range, so that the particles are uniform, and there are little or no fines. Presence of a lot of fines indicates a lower-quality product.

Further rolling can result in a variety of seed components listed above: the hull, to increase dietary fiber; the bran, high in fiber, vitamins, minerals; the endosperm, for starch; and the germ that contains more protein, fat, vitamins and minerals. Seeds can be also be *steam rolled, dry rolled*, or *wet rolled*. Adding the steam or water softens the seed so that it does not shatter as much during rolling, and reduces small particles known as fines. Wetting also starts a process called gelatinization of starch—the starch particles in the seed take up some water and start to swell (you have seen this if you have mixed flour and water or starch and water together, or if you have made gravy or sauces containing flour or starch). The gelatinization process makes the seed more

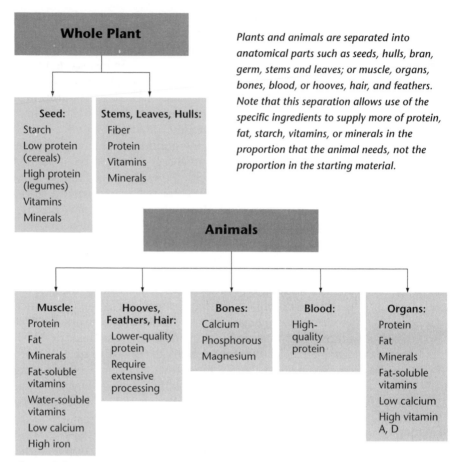

Plants and animals are separated into anatomical parts such as seeds, hulls, bran, germ, stems and leaves; or muscle, organs, bones, blood, or hooves, hair, and feathers. Note that this separation allows use of the specific ingredients to supply more of protein, fat, starch, vitamins, or minerals in the proportion that the animal needs, not the proportion in the starting material.

Figure 8.1 General processing schematic.

digestible once eaten, as the animal does not need to spend time starting this process. Rolling breaks the seed coat and reduces the particle size, but does not grind to a uniform size. This increases digestibility, packing, and handling ability, but does not provide optimum performance, either physically (handling) or nutritionally. Figure 8.2 explains the use of many processing ingredients added to foods.

Whole grains can be steam *flaked*. **Flaking** is a type of rolling process. The grain is wetted previously, or steamed to soften, then put through rollers that are not quite as close together as in normal rolling. This breaks the seed coat and improves digestibility, but keeps the seed intact. This increases the bulkiness of the grain, so there is less grain per unit volume. Breakfast oatmeal is an example of a steam-flaked grain. This process is not usually used for foods for dogs or cats, but can be for horses, rabbits, and some bird feeds and fish feeds. Your morning oatmeal is usually steam-flaked.

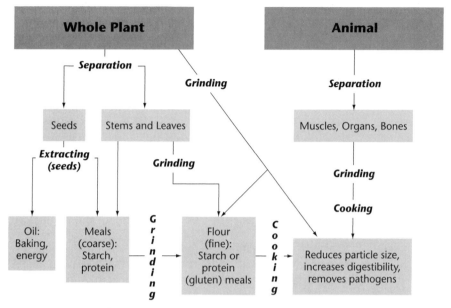

Figure 8.2   Uses of processing to turn ingredients into foods. Separation of anatomical parts supplies purified nutrients to remix into the specfic requirements of many animals. Physical reduction of particle size increases digestibility.

Extruding is the process in which the heated, moist, complete food is pressed rapidly through an orifice of a particular shape and size. Individual pieces are cut off and allowed to dry.

Extruding is generally the final process in mixing and making a feed. The ground feed, usually moist and at a raised temperature, is forced through the holes, which are smaller at the entering end than the exiting end. This sudden change in pressure disrupts the starch granules and makes them more digestible. In addition, it decreases the density of the diet, making it lighter and more palatable. The holes can be of any shape, and this is where the shape of the different feeds can be given. Almost all dry pet foods today are extruded. The various little shapes used (usually in cat foods) are made by the shape of the final die (hole) in the processing machine. Extruding is also used in human cereals and snacks. Because of the machinery and the heat needed, it is somewhat expensive to do, but usually increases the digestibility, handling characteristics, and marketing options.

Changing the density of the feed particles (how much actual food is in a nugget) can provide for different options for different animals.

Pelleted foods are mixed with some type of pellet binder, usually a gelatin-type compound (guar gum, for example), wetted, and pressed into pellets of various diameters. These can then be broken at different lengths. This used to be the most common form of pet foods, before extruding became viable. Pelleting is still widely used in food for agricultural animals to increase the density of feed and reduce the waste caused by fines in ground feed. Some pet foods, usually semi-moist snack foods, are still pelleted. Pellets can be *crumbled*, which involves taking pellets and crushing them through rollers. This process reduces particle size but maintains density, and results in pellets that are low in dust and easier to chew and swallow. Crumbling is not usually used for companion animal food, other than for young rabbits and horses. This is in order to provide two products with the same nutrient content, but one with a larger particle size for the larger growing animals and one with a smaller particle size for the younger animals.

Heating

Heating is an important processing method. Animal products must be cooked to start the breakdown of protein and reduce bacterial growth. Grains are often heated to remove excess moisture that can cause spoilage and to start the gelatinization of starch, which increases digestibility. Drying of grains is usually with forced air at a relatively low temperature (80–120°C). Any feeds or feeds above approximately 15 to 20 percent moisture cannot be stored safely, as they will heat up, due to bacterial action and will spoil. Dryness is the easiest and most common feed preservative method.

Higher-temperature (280 to 320° F; 120 to 160°C) heating is used to actually cook some grains or seeds. Often called *roasting*, this process results in the seed being slightly expanded as the water heats up and starch particles are broken (similar to the gelatinization process when wet). This process decreases density, which increases palatability; it also improves digestibility of the starch. Roasted seeds to be used in pet foods are usually then ground and the final moisture content is 5 to 8 percent.

Cooking of raw animal products is done to eliminate bacterial contamination and improve handling, taste, texture, and digestibility. Usually this is some type of wet process—either steaming or boiling. Products are then ground or chopped to different sizes depending on use. Ground products are usually called a *meal*, such as bone meal, meat meal, by-product poultry meal, or fish meal. These meals are used as protein and mineral sources. Meals may be dried to allow for longer storage and shipping.

> Heating is a time-proven process to improve safety and digestibility. However, the time and temperature are critical to preserving quality. Heating should only be as long as and at the minimal temperature needed to ensure safety and increased digestibility. After that, more heat reduces quality.

Extracting

Extracting is the process of removing one thing from another, and in feed processing refers to the removal of oil from seeds. The process is often mechanical, using high pressure in a press. The seeds are often ground, wetted, and heated to increase the liquidity of the fat and then the fat is pressed out. The remaining low-fat meal can be used directly as another feed ingredient such as in soybean meal, or perhaps further separated to starch and protein, as in the making of corn oil, corn starch, and corn gluten meal (*gluten* is a generic term for the type of protein in corn and other seeds, see Chapter 4). Extraction can also be done with a food-safe solvent, such as very hot water or alcohol to remove the fat; the solvent is then removed by heating or evaporation.

Presence of Inhibitors

Nutritional Inhibitors

Some plants contain chemicals that are unsafe for animals or that interfere with digestion and metabolism. Many of these plants may have developed this as a defense mechanism. They basically just got tired of doing all the work of setting down roots, growing, surviving through drought, and making seeds, just to have some inconsiderate buffalo, deer, or bird come along and eat them down to the nub so that they had to start all over. Some antinutritional factors are part of several plants, and are not highly digestible or they interfere with digestion. Yet even so, these plants can provide safe nutrients to animals, increasing our efficiency of use of natural resources.

In many plants there are complex carbohydrate polymers such as beta-glucans and arabinoxylans. In grains such as oats, rye, and barley these can interfere with digestion. Remember that we eat oats to reduce our cholesterol. Oats don't really suck cholesterol out of the body, but the naturally occurring soluble fibers such as beta-glucans bind fats including cholesterol in the digestive tract and reduce their digestion. This can be a good thing in humans who have plenty to eat and not a lot of work to do, but is not necessarily a good idea for a rapidly growing kitty who needs a lot of fat for energy and membranes. These complex carbohydrates are very viscous when wet. In greater amounts they can interfere with normal digestive processes and irritate the intestinal lining, especially of young animals. Enzyme preparations are now available to partially break down these compounds to smaller molecules, which are not as viscous and can be easily digested. These are primarily used to treat grains meant for agricultural animals and are not widely used in pet foods as yet, but could be. This can open up perfectly good low-cost foods for use by other animals, improving use of natural resources and reducing cost.

Other compounds exist in plants that can more severely affect an animal. Many plants contain actual toxic chemicals that can poison an animal—most of these are relevant here only to horses on pasture. Some otherwise

In Canada several years ago, a bunch of bright young scientists thought it a real waste to grow this stuff just to use in industry when it also contained good nutrients as well. So they simply used basic plant-breeding techniques to remove the genes for erucic acid and glucosinolates out of rapeseed so that the plant could be used for food! Then, the marketers realized that products called rapeseed protein and oil were simply not going to catch on, and decided on a new name for this plant: **canola.** You can now go down to the store and buy canola oil, a perfectly good oil; and also many grain snacks that contain roasted canola seeds, which are high in protein, fat, and vitamins. In so doing, you are buying a highly nutritious product developed with good practical science by the Canadian Oil Association. Canola oil and meal are now widely used as agricultural animal feeds but are not widely used in pet foods, primarily due to cost and because some people and animals do not enjoy the flavor of the oil.

nutritional foods are not used widely in pet foods because of some naturally occurring compounds. As mentioned, gossypol is a toxic product in cottonseed, making cottonseed a good food for ruminants whose rumen bacteria can break some of this down, but limiting its use in nonruminants. Cottonseed is high in protein and fat, so otherwise could be a good source of nutrients.

Erucic acid is a polyunsaturated, long-chain fatty acid that can disturb normal membrane function. It is found in a plant called rapeseed that has been used in the past to make industrial oils for paint and lubrication. Rapeseed is similar to soybean and cottonseed in that it is naturally concentrated in protein and fat, making it a good feed for many animals. Also, the fat is highly unsaturated, which is considered to be more healthful for humans. *Glucosinolates* are a problem with industrial-grade rapeseed. These compounds are *isothiocyanates*, which inhibit thyroid function and can cause goiter.

One of the most widely present naturally occurring nutritional inhibitors in protein feeds is the trypsin inhibitor, which inhibits action of the intestinal protease, trypsin. It is a protein present in soybeans and is destroyed by heating. Almost all soybeans are roasted before use; this destroys trypsin inhibitor, and soybeans are a fine protein source for pets.

The enzyme *thiaminase* (see Chapter 5) is present in uncooked animal products and some plants. Wet feeds containing products such as fish meal or wet brewer's grains are often oversupplemented with thiamin to overcome the enzyme activity.

Food Contaminants

Ergot is a common mold of small grains that contains alkaloid compounds, which can affect the hormonal systems of women and interfere with pregnancy and lactation. In fact, one very common ergot compound is used by the medical profession to inhibit lactation in women who do not want to breastfeed their children. *Aflatoxin* is a common mold that can infect corn and other small grains, usually in an unusually wet year, or if the grain is stored too wet. *Vomitoxin* is another food contaminant. These compounds can and have caused illness and death in animals and humans. However, as is true for some chemical reactions, there are safe levels that can be included (levels that have been shown to have no effect). When increased amounts of these toxins are found, usually because wet weather has delayed harvest or grain has been stored too long, the grain can be diluted out with tons of clean grain to nontoxic amounts. Alternatively, if the concentrations are too high, they are not used in pet foods, but can be used in some foods for ruminants, as the rumen microbes will destroy the toxins.

> Many plants and seeds can be naturally contaminated by bacteria, fungus, or other compounds during growth and storage. Proper practices, especially keeping stored grains dry, should be followed. The lack of quality control by some does not mean that all grain is inappropriate for use in dog and cat foods.

A student may find, on the Internet and in various books, strong words against the use of commercial products containing "cheap grains" and low-quality grains, even referring to cases in which some contaminated grains have been used in pet foods resulting in illness and death of dogs and cats. These occurrences are rare, accidental, preventable, and unacceptable, and in fact have usually happened in smaller (super-premium) companies who cannot buy large quantities of grain at one time. Most pet food companies go to great lengths to inspect, test, and be aware of problems. This argument (potential presence of these toxins) should not be used as a blanket argument against the use of grains in foods for dogs and cats.

Effects of Processing on Palatability

Physical Form

A major reason for processing foods is to make them more palatable. Some feed particles may be too big, too small, too dusty, or just right. Extruding provides the shape of the final product, from X's or O's to stars, little fish, and the like. The actual shape of a processed food in itself is *not at all important to nutrition*. However, animals can and will develop a preference, or habit, for a certain shape, and this will directly affect the acceptability. This is really a neat marketing trick, doesn't hurt the animal, and helps deliver a good product. Not altogether a bad idea, really, but remember that it is not really necessary nor is it bad.

Taste Factors

Processing can remove or alter some tastes that animals do not like, but does not really affect the nutritional value of the food. Nevertheless, if addition of taste factors will encourage the animal to eat a nutritional product, that is usually a good thing. Alfalfa can have a bitter taste in dried form, so we often add molasses as a sweetener to horse or rabbit food. Thus the animal gets good protein and a little fiber, and just enough sweetener to affect taste, not add a lot of calories.

The right fat content can improve flavor and texture. However, overheating or storing too long can lead to oxidized fatty acids, which is one type of rancidity. *Rancidity* usually refers to chemical changes in unsaturated fats caused by heat, light, or bacteria. The double bonds accept an oxygen molecule to form a triangular bond. This has a distinct odor and flavor but does not necessarily change nutrient value. Too many free fatty acids that have been cleaved from triglycerides increases this chemical rancidity. The feed may still be nutritionally acceptable, but can taste acidic, sour, rancid, or even soapy. Processing can prevent this: We add antioxidants to prevent this chemical reaction.

The olfactory system is critical in palatability and we can help this along by processing to enhance or reduce smells. Smell factors such as bitter or rancid, and flavor that the animal is not used to, can be diluted or masked by molasses or other sweeteners. This should not be used to encourage consumption of a nutritionally inadequate formulation, however.

The problem is that the use of too much sweetener such as glucose, corn syrup, high-fructose corn syrup, or other digestible sugar will provide too much energy too fast, leading to an insulin response and over time leading to an increased propensity for diabetes (see Chapters 7 and 10). There are few complete feeds that use significant amounts of these ingredients; their use is usually limited to snacks. Remember that a snack is a snack, and limiting sugar intake is always a good idea. It is also a good idea to read the label.

Additives in Pet Foods

There are few areas of nutrition that are as poorly understood and misinterpreted as the use of additives. Are additives absolutely necessary for feeding animals? The answer is no. Are additives necessarily bad, unsafe, dangerous, corrupt, right-wing-conspiracy-centered terrible things? Again, the answer is no.

> A feed **additive** is defined as any compound that is added to a diet formulation that is not normally occurring in those feeds.

Additives have a place in the safe and efficient production, distribution, and nutritional use of food. A simple example is salt (sodium chloride). Salt is naturally occurring, but by definition, extra salt added to food is an additive. If a compound is added to the diet and was not present in the starting feeds, it is an additive by definition. Additives used in pet foods are regulated by the *FDA* (Food and Drug Administration) and by the *AAFCO* (Association of American Feed Control Officials; a group of state officials responsible for developing guidelines for pet food). Additives must fall into what is called the GRAS (Generally Recognized as Safe) category, otherwise they cannot be used without complete testing under FDA authority. An additive that was not categorized at the start as GRAS, such as a synthetically made sweetener (aspartame) or a food coloring (red dye #2), would have to go through many years of testing to prove it was safe. If at any time new testing shows that the compound is not safe, it can no longer be used. This testing was used to remove several dyes from the market that had been shown to cause cancer (at levels of use that were thousands of times greater than ever used in foods, but nevertheless they were removed from use). The fact of the matter is that we have hundreds of effective and safe food additives, so there is very little development of new food additives. The cost of this testing effectively precludes pet food companies from developing any other compounds, and the ones we already have work quite well.

Additives are used for several purposes:

1. To improve nutritional value.
2. To improve handling and storage and prevent deterioration of food.
3. To alter the texture, color, smell, or taste to improve acceptability.
4. To prevent bacterial or fungal contamination.

There is always a lot of controversy around the use of additives, but I think most reasonable people would agree that these four purposes are good and necessary to a safe food supply. The balance is in what processes and additives are used. Regardless of the many opinions to the contrary, having a poor-looking, poor-handling, easily spoiling, less digestible food (like many foods in their natural state) is not a good way to go.

Additives that meet these purposes fall into different categories. One is *emulsifiers,* which help keep fats in solution, so the food does not separate. These are often phospholipids (Chapter 3) that naturally occur in plants and animals and are purified from plant or animal products. *Lecithin* is a common name for the most common emulsifier: *phosphatidyl choline.* Phosphatidyl choline is a natural fat that occurs in all life forms, often in the membranes. These kinds of agents are found in many foods—salad dressings, sauces, candy, bread, other bakery products, and so on. They have been used naturally for thousands of years and now are used in specific ways. When we put an egg in a cake or pastry, we are adding a natural emulsifier, as egg yolks have significant lecithin. Natural sources of high amounts of emulsifiers are whole eggs, pure oils (not processed or filtered), and some plant hulls and germs. Other emulsifiers include the chemicals known as gums, which are more complex molecules that can get sticky or gummy when wet and in small amounts help hold things together. A cake, cookie, or piece of dog food that is falling apart is not effective or enjoyable, and emulsifiers help hold things together. Some common gums you may see on labels, often for wet foods, include guar gum, xanthan gum, and carageenan. Gums have no nutritional value apart from keeping the food together so that the animal can eat it. Their effective use is only a few tenths of a percent of the total food, so foods are not usually "loaded" with substances that stick up the digestive tract.

Flavors are added to enhance the acceptability of the feed. If you or your pet doesn't like the taste of something, it won't be eaten, no matter how good it is. Small amounts of flavors ensure that the food is accepted. Flavoring includes what is referred to in the industry as "digest." This is a finely ground and solubilized broth of some animal product—chicken, beef, turkey, pork, liver, and so on. It is usually sprayed on the finished product and dried to provide a smell and taste that the animal will like. Thus it is really a food, but in this case it is also an additive. Other flavorings include salt, sugar, molasses, and any natural product that has flavor to it. Generally, artificial flavors are not added (like imitation vanilla, for example), because these are not natural flavors for pets.

> *Colors* and flavors are used to enhance the acceptability of the product by the owner.

If all pet food was a dull gray or brown, and the owner didn't like it, it wouldn't help the pet any either, so coloring is sometimes added. Colors can be various naturally occurring dyes (amaranth is an example, a configuration of nitrogen and carbon atoms that absorbs different wavelengths of light and is the basis for most natural color). In addition, carotenoids such as **beta-carotene** and astaxanthin add color and may act as antioxidants. Coloring truly is unnecessary in foods for dogs, cats, horses, and rabbits. It may have some value in inducing food-seeking behavior in birds, reptiles, and fish. There are sufficient naturally occurring colors that the use of other coloring agents probably can no longer be justified for scientific, nutritional, economical, or safety reasons.

Binding agents and *flowing agents* are used to either hold things together (like gums, as mentioned) or to keep them apart to keep them flowing (like in powdered products). In the pelleting process or to hold extruded foods together, binding agents are needed. We use flour, gelatin, or corn starch as a binder in sauces or bakery products. Similarly, various starchs, gelatins, and gums are used as binding agents to keep the food together. Simple gelatin was used (by a WSU graduate) to make the food sent up with the space shuttle rats, so the feed particles would not float all over. Gums also fall into this category. For powder- or granular-type products, flowing agents are sometimes used—silica is a common example.

Surface-active agents are similar to emulsifiers and are used to hold different forms of feeds (starch, protein, lipids) together in a usable and palatable structure. The simplest surface-active agent is water, but water is not by definition a feed additive. Others include phospholipids and *propylene glycol* (which is *not* the same as ethylene glycol used in automotive engine coolant), which has both ionic and lipophilic properties. These provide for proper binding, holding, and texture of the food.

Antibacterials are used to minimize bacterial growth that spoil the taste or actually cause toxins to build up in the food. A food that is contaminated by bacteria or broken down because of bacteria is not a good diet. Food-making processes are not usually fully sterile, nor could be without exorbitant cost. Normal safe-handling procedures, cooking, and packaging ensure minimal bacterial contamination. To keep this safety level after products are opened, several antibacterial agents are used. These are *not* antibiotic medicines or prescription items. These are naturally occurring, well-known, historically used compounds that have antibacterial ability—such as *ethoxyquin,* which is a close relative of vitamin K. Salt is another antibacterial, and sugar is another (the bases of pickling and preserving—proven techniques to preserve food and slow bacterial growth). Sugar and salt, however, are not added to pet foods for their antibacterial activities, as the concentration required would be too high. Other antibacterials include *benzoic acid,* which is used in human foods but not usually in pet foods, because cats have an unusually high sensitivity to it; *sorbic acid,* similar to but not the same as ascorbic acid and *sodium nitrate,* which is used in most preserved meats and is also used to stabilize color in meat products by reacting with heme pigments. Sodium nitrate is safe for dogs and cats (and humans) at doses used, regardless of the wide negative press nitrates have received. There have been several studies done demonstrating no short- or long-term detrimental effects of these compounds at normal use rates. However, several nitrate containing compounds may increase incidence of some cancers, and research has shown that we can use less nitrates in preserving than were previously used.

Antioxidants and preservatives are used to slow down other chemical reactions that can destroy the flavor or nutritional value of foods. As for bacterial degradation, if normal light and chemical reactions break down the feed and alter its color, smell, texture, taste, or nutritional value, it becomes less useful than what it should be. Oxidative reactions can destroy vitamins, cause fatty-acid rancidity, and create various "off" flavors. Natural antioxidants include *ascorbic acid, vitamin E* and other tocopherols, and chemical agents like *BHA* (butylated hydroxy anisol) or *BHT* (butylated hydroxy toluene), which are on the GRAS list and have been heavily tested and are used as antioxidants in all foods. Concern over the possible carcinogenicity of these latter two compounds prompted renewed testing in the 1970s, and although no carcinogenic effects were found, the effective preservative level was found to be much lower than what was normally used. Now, often these compounds are chemically attached to the package, so that they prevent surface deterioration without actually being consumed.

Practical Application

The animal and plant ingredients that make up foods are of many types with a variety of nutritional contents. By mixing and matching various ingredients, nutritionists make well-balanced complete foods. Processing of many types is used to make ingredients more digestible, safer, more acceptable, increase shelf life, and provide more flexibility in use. Careful use of vitamin and mineral supplements completes the nutrient requirements. Careful use of chemical additives (most of which are naturally occurring) that improve flavor, color, texture, handling, and shelf life complete the construction of outstanding foods. We can now formulate from a wide variety of ingredients and intelligent processing and preserving methods to make useful, nutritionally appropriate, and safe *pet foods*. Homemade foods can be viable, but are not, in general, better or worse than commercially prepared foods.

Words to Know

AAFCO	*erucic acid*	*lecithin*
additive	*ethoxyquin*	*nutritional inhibitor*
antibacterial	*extracting*	*oilseed*
ascorbic acid	*extruding*	*pelleting*
benzoic acid	*FDA*	*prion*
beta-carotene	*fiber feeds*	*protein feeds*
BHA/BHT	*fines*	*rancidity*
binding agent	*flaking*	*rolling*
bran	*flowing agent*	*sodium nitrate*
canola	*germ*	*sorbic acid*
complete food	*glucosinolates*	*supplements*
emulsifier	*gossypol*	*thiaminase*
endosperm	*grains*	*trypsin inhibitor*
energy feeds	*hulls*	*vitamin E*

Study Questions

1. Describe the characteristics and give examples for protein feeds, energy feeds, and fiber feeds. Which ones can actually be both protein and energy, protein and fiber, energy and fiber, or all three?
2. Name at least three each of different protein, energy, and fiber feeds.
3. Which feeds, a few to several in combination, would provide an overall generally balanced ration (protein, energy, and in some cases, fiber)?
4. Give the reasons why we process foods, and the disadvantages of processing foods.
5. Give the reasons why we use food additives, and give some examples of specific food-additive uses.

Further Reading

Cheeke, P. R. 2004. *Contemporary issues in animal agriculture*, third edition. Englewood Cliffs, NJ: Prentice Hall.

Cowell, C. S., N. P. Stout, M. F. Brinkmann, E. A. Moser, and S. W. Crane. 2000. Making commercial pet foods. In M. Hand, C. D. Thatcher, R. L. Remillard, and P. Roudebush, eds., *Small animal clinical nutrition*, fourth edition (Chapter 4). Topeka, KS: Mark Morris Associates.

Crane, S. W., R. W. Griffin, and P. R. Messent. 2000. Introduction to commercial pet foods. In M. Hand, C. D. Thatcher, R. L. Remillard, and P. Roudebush, eds., *Small animal clinical nutrition*, fourth edition (chapter 3). Topeka, KS: Mark Morris Associates.

Hand, M., C. D. Thatcher, R. L. Remillard, and P. Roudebush, eds. 2000. *Small animal clinical nutrition*, fourth edition (chapter 3). Topeka, KS: Mark Morris Associates.

Kellems, R. O. and D. C. Church. 2002. *Livestock feeds and feeding*, fifth edition. Upper Saddle River, NJ: Prentice Hall.

Murray, S. M, A. R. Patil, G. C. Fahey, Jr., N. R. Merchen, and D. M. Hughes. 1998. Raw and rendered animal by-products as ingredients in dog diets. *Journal of Nutrition*, 128: 2112S–2815S.

National Research Council. 2004. *Nutrient requirements of dogs and cats*. Washington, D.C.: National Academy Press.

Patil, A. R., and G. C. Fahey. 1998. Petfood ingredients and ingredient processing affect dietary protein quality. Proceeds of the 1998 Purina Nutrition Forum. *Compendium on Continuing Education for the Practicing Veterinarian*, vol. 21, no. 11(K), Nov. 1999: 38–43.

Remillard, R. L., B. M. Paragon, S. W. Crane, J. Debrackeleer, and C. S. Cowell. 2000. Making pet foods at home. In M. Hand, C. D. Thatcher, R. L. Remillard, and P. Roudebush, eds., *Small animal clinical nutrition*, fourth edition (chapter 6). Topeka, KS: Mark Morris Associates.

Strombeck, D. R. 1999. *Home-prepared dog and cat diets. The healthful alternative*. Ames: Iowa State University Press.

Chapter 9

Formulation, Analysis, and Labeling: Foods to Meet Requirements

Take Home and Summary

Diets are formulated by matching the chemical content of the ingredients with the chemical nutrient requirements of the animal for which the diet is meant. Ration formulation and analysis is very much a science, with a little experience and variation thrown in. We must determine what is contained in the ingredients, make the diets, and test whether foods meet requirements. Testing is a combination of direct chemical analysis of the ingredient or foods and nutritional testing for biological function. Once prepared, foods are analyzed to ensure that they contain the right nutrients. New and specialty products are subjected to animal testing to actually demonstrate that they are safe and effective. The products are finished, labeled, and marketed. The pet food label is a legal document, and it must give specific information: name, species intended, guaranteed analysis, ingredient list, statement of nutritional purpose, and contact information. Some manufacturers voluntarily put more information on the label. The interpretation of labels is truly a crossroads of science, art, and marketing.

Diet Formulation: Basic Process

The formulation of a diet for a given animal, often called *balancing a ration,* is quite simply matching up the chemical constituents of the feed ingredients to supply as closely as possible the actual chemical needs of that animal in that situation.

> Balancing, or formulating, a ration means matching the chemical needs of the animal with the chemical contents of the foods you have available.

So we may say: "I need to balance a ration for a growing, large-breed dog," or "I am formulating a diet that meets or exceeds the needs for a cat in lactation." What we are doing is matching the biochemical needs of the animal with the chemical content of the feeds. By now you should understand that to do this correctly, we must take into account simple things like how much we expect the animal to consume, the average digestibility of the diet, and the metabolic efficiencies of the processes we are supporting in the animal.

Basic steps in formulation of foods:

1. Keep it simple. Complexity by itself offers nothing. Use the simplest effective ration.
2. Determine requirements for the specific animal.
3. Put nutrients and ingredients in proper order:
    ➤ most nutritionally important
    ➤ most expensive ingredients.
4. Build the rest of the ration around the most important.
5. Determine special restrictions or considerations in feedstuffs.
6. Determine actual nutrient concentrations in feeds.
7. Consider expected feed intake.
8. What type of feeding management is expected?
    ➤ one or two meals per day, free choice, or restricted?
9. Determine feed availability and cost.
10. Determine need for any supplements or additives.
11. Balance the ration.

There is no need for complexity that cannot be justified. Simplest is best. First, requirements are identified. Many of the requirements for protein, energy, vitamins, and minerals are already known, or we have close estimates for them, so use of standard reference materials is the first step. If there is some concern that requirements published to date are not adequate, further research should be considered.

When beginning to select feedstuffs, it is prudent to select the most important nutritionally and economically (which in most cases is protein). Then the rest of the ration can be built around that. In this example, we are assuming the basic requirements are known because, in general, they already are. Huge databases of feed analyses of common (and uncommon) ingredients are available, and this is always our first resource. In addition, the present-day assumption is that vitamins and minerals, either in the feedstuffs or as supplements, are readily available. Even though per unit mass they may be the most expensive ingredients, especially the vitamins, when one considers that supplemental vitamins are usually included in the ppm (parts per million) range, the total cost is usually not as much as for protein.

Other questions that should be asked include: Is the amount of animal muscle in the ration going to increase the fat content of the ration above what is needed, or cost too much in relation to other adequate sources of protein? Is the calcium content of the ration going to be too high for normal bone development (in the case of alfalfa for horses)? Can the excess calcium be diluted out with the use of a cereal grain? Is there a legal limit to the amount of a mineral we can add (selenium is one mineral that has a definite legal maximum in diets)? Will there be too much grain, providing too much starch and increasing the chance of obesity?

In addition, if there are any restrictions, either voluntary or by law, in use of feedstuffs, they must be considered. Legal restrictions are very few; they include use of selenium and fluoride. These minerals are required only in a narrow range, so guidelines have been developed for their use. Most other restrictions are of a practical and economic nature. Several choices of feeds may be available, but the need for a practical cost may limit the options.

The nutrient concentration of the feeds is the next step to consider. When adequate data are available, and there is no reason to think that the feeds available to you are significantly different from the average analysis in the database, then this is a perfectly acceptable starting point. If you suspect that your feedstuffs may differ from the average, then it is prudent (and usual) to conduct some level of analysis. The number of samples taken to ensure a statistically meaningful estimate depends upon some previous knowledge of variation in chemical content of the ingredient. Our basic corn will vary between 8 and 10 percent protein in practically 99 percent of cases. Thus, an acceptable testing rate may be 1 in 50 batches. Meat and bone meal can vary more than 50 percent in content of protein, fat, calcium, and phosphorous, so more samples should be analyzed, perhaps 1 in 10.

> We need to test to make sure that what we *think* we are feeding is what we are *actually* feeding.

Some batches are routinely tested; others are assumed to follow the averages. We have an extensive knowledge of the average chemical content of specific feeds. There is always a certain amount of variation, but the reality of life is that it is best to find the balance between chemically testing every batch of every ingredient versus the wise use of average composition from hundreds or thousands of earlier tests. If we continued to test every single batch of ingredients, frankly, we could not afford to feed and keep companion animals.

Now we return our attention back to the animal. Based on available research, we have reasonable estimates for the expected intake. We *must* formulate every ration around the amount we expect the animal to consume. This is one of the most critical and confusing aspects of nutrition. Animals do not have a requirement for a percent of protein in the diet; they have a requirement for a certain amount of protein that will provide the proper amount and balance of amino acids. However, for mathematical and practical reasons, it is much easier to assume a certain food intake (based on past knowledge from research) and calculate the ideal percentage of each ingredient. This is what you see on the label. You often also see expected or recommended intake amounts. One without the other is meaningless.

> Formulating a ration must include estimating the expected intake of the animal for which the food is intended. Animals eat amounts, not percentages.

If the expected intake is 100 grams, and the protein requirement is 16 grams, then the ideal ration would be balanced at 16 percent of protein. However, if the expected intake is 120 grams, and the protein requirement is still 16 grams, then the ideal percentage of protein is now 16 divided by 120, or 13.3 percent, a huge difference.

Remember that, in general, animals eat to meet their energy requirements (Chapter 7). Thus, if we have a low energy diet (say 3.25 Mcal ME/kg ME), and the animal requires 325 kcal ME per day, the animal will on average eat 100 g (3.25 Mcal = 3,250 kcal; 325 kcal required/3,250 kcal/kg food = 0.1 kg or 100 g of food). If we increase the fat content, so that the energy content is now 3.75 Mcal ME/kg, and the animal still requires 325 kcal, the expected consumption is now 325 kcal required/3,750 kcal/kg = 87 grams intake. The protein requirement is still 16 grams, so the required percentage of protein is 16/87 or 18.4 percent. If we stuck with 16 percent, we would only be feeding 0.16 times 87 g, or 13.9 g protein, or 87 percent of the requirement.

Percentages and Intakes Example

Energy Requirement kcal/d	Energy in Food kcal/g	Nutrient, Amount Required	Expected Intake	Percentage Needed in Diet	Actual Nutrient Consumed
1,000	3.25	50 g/d	308 g	16.3	308 × .163 = 50 g
1,000	3.5	50 g/d	286 g	17.5	286 × .175 = 50 g
1,000	3.75	50 g/d	267 g	18.8	267 × .188 = 50 g
If % nutrient remained at the average amount:					
1,000	3.25	same	308	17.5	308 × .175 = 54 g (8% too much)
1,000	3.75	same	267	17.5	267 × .175 = 47 g (6% too little)

This concept cannot be stressed enough. It is core to the science and practice of nutrition, whether you are a highly paid veterinary nutritionist or a normal person trying to feed your companion on a budget. You must *always* consider the intake (measure it) when formulating or deciding upon feeding a certain percentage of nutrients. Figure 9.1 shows a general schematic of creating a food for a given situation. Note that you must take into account all the animal and feed factors to make an efficient and economical ration.

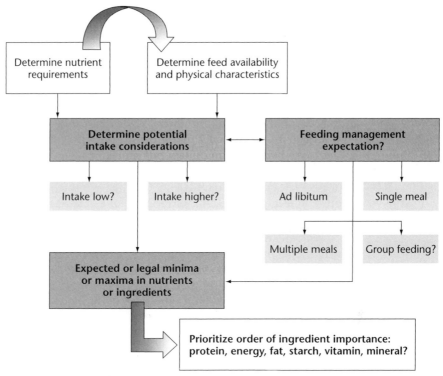

Figure 9.1   General process of ration design.

Related to the intake question is the *feeding-management system*. There are two basic types of feeding: *free choice,* or *ad libitum,* in which the animal always has access to food (the food bowl is never empty); or *meal-fed,* in which the animal may still receive all the food it needs, but it does not always have access to food. *Limit feeding* is a type of meal feeding in which something less than what the animal would freely eat is offered.

Free choice means exactly that: The food bowl is never empty. This may or may not mean the animal is eating too much. If the animal is regulating its intake, this may be fine. But if the feeder is always filling the bowl, no matter what the animal eats, this can lead to problems such as obesity (see Chapters 7 and 10). Meal feeding can mean one meal, two meals, or several during the day. The food is offered for a time, the animal can eat, and then any remaining food is removed. Meal feeding can actually allow less, the same, or more intake than ad libitum feeding, depending on how much is offered and what the animal chooses to eat. Limit feeding restricts the amount of food offered or the amount of time that the food is offered, and is usually used for animals that would otherwise eat more than they need, or during periods of intended weight reduction.

Feeding management is critical:

➤ Free choice—food is always available
Many animals, especially cats, can thrive and not overeat on this system. It is better with dry foods and lower energy foods.
➤ Meal-fed—food is offered and removed (or consumed) once or more per day
The animal may still be able to eat all that it wants to.
One meal a day: all the food the animal would normally eat is provided once; when the animal is done eating the remainder is removed.
Two or more meals a day: the total daily amount expected to be consumed is divided by the number of meals, and fed as for one meal.
Restricted fed: food is offered, regardless of the number of meals, at a lower amount than the animal would freely consume.

With the excellent transportation systems we now have, fresh foods and commodities can be obtained almost anywhere in the world at any time of year, yet cost and quality will vary. Geographical location, season of the year, weather patterns, and truck strikes can all influence true cost and cause a variation in cost. The fact of the matter is that we can derive a good balance and amount of amino acids from plants or from animal sources, and T-bone steaks cost more than soybeans. So we have to decide sooner or later if we actually need or want to spend the extra money for certain ingredients. Sometimes that answer will be yes and sometimes no. Can you think of life-cycle stages for which we might pay more for certain ingredients than we would for other stages?

Availability and cost of ingredients is always a major consideration

A final feeding question remains: Are supplements needed? After we consider the basic ingredients for fat, carbohydrate, and protein, we look at the mineral and vitamin content, the flavor, the texture, and the needed shelf life. Do we need to supplement some ingredient? Should we add fat, amino acids, minerals, vitamins, sweeteners, coloring agents, or flavoring agents? These decisions are made considering the intended user and the chemical composition of the ration (vitamins and minerals for example). We may add a flavor so that the animal eats it and wants to eat it (the food increases the animal's comfort level). Perhaps a ration has a lot of grain, so we need the amino acid lysine. This fine-tuning insures nutrient adequacy and palatability.

Finally, we actually *balance the ration:* We calculate the composition and come up with a final product. During, before, and after this process, there are checks and tests along the way. We will rely on certified trusted tables of analysis, and we will check actual analysis as needed or when we have reason to suspect greater variability. We will know and use the biology of the animal we are feeding to determine the requirements. We will test actual animal performance in experiments. Figure 9.2 reviews the steps of balancing a ration: So now, let us learn about the different types of diets available and how we categorize and analyze them to best meet the needs of our animals.

At the end of this chapter is an example on the actual mathematical balancing of rations. Depending on the course or student, you may spend a little or a lot of time on this process. Certainly, if you intend to be a nutritionist, veterinarian, or well-informed consumer, you should spend some time on the simple examples. Presently we have wonderful computer programs that store the requirements and feed chemistry data of thousands of animals and feedstuffs, and with minimal effort, we can balance a ration. But, this is at the *end* of the application of the knowledge discussed earlier. The bio-math without bio-knowledge is worthless. I have had senior students turn in rations with 20 percent calcium along with the line: "This is what the computer did!" Let us just say that these students quickly learned that they were not done with the assignment.

Types of Pet Foods

There are several different types of pet food, and these types are based often on marketing strategies in addition to actual nutrition. The first criterion is that they are nutritionally adequate, then the question becomes, "How do we get people to buy it?" Although it is my practiced experience and opinion that almost all diets available for sale are nutritionally adequate, there are significant differences in the content and availability (quality) of nutrients among these types of foods.

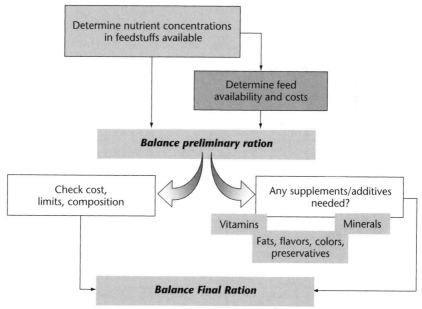

Figure 9.2   Final steps in balancing a ration.

Generic Foods

Generic foods were big during the "back to nature" and "anticorporate" fervor of the 1970s. The term *generic* means legally that the food contains similar chemical content, but has no brand identity. The idea of generics was to do away with the costs of advertising, packaging, and marketing and just pay for the important stuff, the food. Though a good idea, it did not sit well with all the people who actually made a living in advertising, packaging, and the like. But the real reason generics did not catch on was a real lack of quality control which, in fact, does cost money.

Some generics were and still are quite decent foods. However, manufacturers may not necessarily have formulated to meet animal needs, considered nutrient excess or toxicities, considered variability and availability, or tested the product sufficiently. Not all generics are bad; and in fact some are made by big-name companies to salvage value from various feeds. The main problem with generics is variability. They should be used only when you have good knowledge of their content, *palatability,* and *acceptability,* and are aware of any potential problems. Later in the chapter we will cover some simple visual tests that are a start toward determining the quality of the food.

Private Label

A *private label,* or *store brand,* is usually that used by a grocery store or store chain. These are usually made by an actual pet food manufacturer for the store under a contract. You will see on the label: "Made for a store brand by a manufacturer." Look for "Distributed **by** _____"; or "Manufactured **for** _____" as opposed to "Manufactured **by** _____." These are perfectly good foods and nutritionally adequate. There will be variability of ingredients used in the food under the same label. Usually the actual nutrient content is relatively constant, within 10 percent or so (but remember, a 10 percent error can add up).

Popular Brands

Popular brands are those with the name of the manufacturer, such as Purina, Friskies, or Alpo (all trademark names). These are certainly formulated to meet animal needs, adequately tested, and excesses and toxicities considered. These are usually less variable and are perfectly fine pet foods. In every batch of food, the same feedstuffs are used, but the content of each may vary by a few percent to take advantage of changes in cost or to overcome lack of availability. The variability in nutrient content of each batch is not always considered. These are fine for most pets; differences in nutrient content are not usually a problem.

Premium Brands

Premium brands are now fairly widespread, though they used to be uncommon. *Premium brands* in itself has no legal meaning, but is defined as a traditional use of a product that has a more limited distribution, usually more continuous research, and a fixed (or closed) formula. They are usually sold in pet stores or from veterinary clinics. They are manufactured by a company, typically on a smaller scale than popular or private label brands.

> A *fixed,* or *closed, formula* means that the same formula of ingredients will be used in all batches regardless of cost.

A fixed formula means that the variability in nutrient content of feedstuffs in taken into account and the final product always has the same content of nutrients. This definitely increases the cost, as different commodities must be bought when needed, regardless of their cost, and more people must be hired to reformulate and mix rations. In addition, more research is conducted on the foods, and this also increases the cost. The premium brands actually led the way in feeding for different stages of the life cycle or specific situations (disease, allergy, and aging). Major food companies have also developed life-stage formulae and it is common practice to feed this way. Premium brands try to consider optimal nutrition rather than simply minimums and maximums.

Forms of Pet Foods

Dry

Dry foods (6 to 12 percent moisture) now make up the vast majority of dog and cat foods. Types include kibbles, meals, and pellets and they are primarily extruded particles (Chapter 8). The advantages of dry food include ease of storage and feeding; lower cost (one-third to one-half of canned); the free-choice feeding option; and the abrasive effects, which help to reduce tartar buildup on teeth. The disadvantages include less palatability; altered nutrient content resulting from improper heating; restricted amounts of fat and energy; and lower digestibility. All of these disadvantages can be overcome by proper formulation and manufacturing.

Soft-Moist or Semi-Dry Foods

Semi-moist foods (23–40 percent moisture) are exactly that—they are softer in texture than dry food, but not wet to the touch. Their moisture content is such that they are chewier than dry foods, but not sloppy like wet foods. The advantages of semi-moist foods are that they can be sold in bags, do not require refrigeration, can be fed free choice, can be sold in larger amounts to reduce cost, and can be fed as patties. Disadvantages include greater cost of manufacturing and shipping, and the fact that they usually contain acidifiers to lower pH and retard bacterial growth, which can be a minor problem for some animals. They also tend to be higher in sugar to improve taste. There is a lot of bad press on this formulation, for good reasons relating to high sugar and sweetener content. The consumers and industry have taken these concerns to heart, and nowadays such formulations are primarily used as snacks. Yet these products do have some viability for bridging the gap for animals who, for various reasons, cannot handle dry foods and consumers who cannot afford wet foods.

Canned (Wet or Moist)

Canned foods are obviously the highest in moisture (68–82 percent moisture). This was once the most common way pet foods were sold, but the decreased cost and increased flexibility of dry foods have long since made wet foods a much smaller share of the market. Advantages usually include increased palatability (this is because these foods are almost always higher in protein and fat than most dry foods) and the simple fact that wetter foods are easier to eat. This is also a disadvantage, as the protein and fat content can mean too much protein and energy for most animals. Other disadvantages include greater cost and the increased attention to dental care that must usually be taken, as the abrasive effect of dry foods is not in play. Wet foods often are separated into two general types, *rations* or *gourmet*. Ration-type foods are usually nutritionally complete foods, less expensive, and can help prevent food preferences. Gourmet types usually contain a lot of protein and fat, but are not necessarily formulated to be complete feeds (vitamins and minerals may not be adequate). Because of their texture and taste, animals can easily form a food preference. This knowledge is used in marketing these foods to a high degree. The increased protein and fat can lead to degenerative problems if fed over a long period.

> Remember this simple guideline: The quality of a food is not related to its *form*, but to its *formulation*.

Pet Food Labeling

Labels on pet foods are simultaneuosly advertising, educational instruments, and legal documents. Some of the content is required and restricted by law. Some parts are pure advertising. Other parts teach us and help us to better feed the animal we purchased the food for. We will go through each part of the label in detail. Figure 9.3 shows an example of a pet food label.

Guaranteed Analysis

Each label needs to have the ***guaranteed analysis*** of the ration for water, ash, protein, fiber, and fat. The reasons behind this are now fairly historical. You may not believe it, but there was a time when some would try to cut corners and push the limits on what was in their products to save costs. Some early manufacturers of poor-quality pet foods would add water, or cheap minerals or other ingredients like straw, to reduce cost, whereas others would make fine-quality foods. Who could tell the difference before buying? Thus, "big

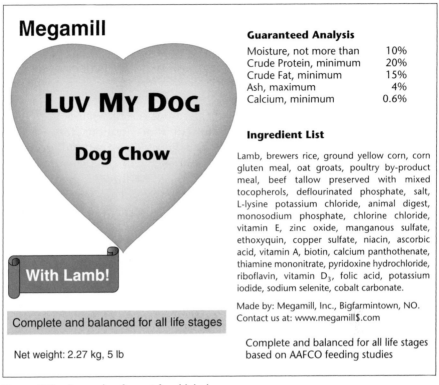

Megamill

Guaranteed Analysis

Moisture, not more than	10%
Crude Protein, minimum	20%
Crude Fat, minimum	15%
Ash, maximum	4%
Calcium, minimum	0.6%

Ingredient List

Lamb, brewers rice, ground yellow corn, corn gluten meal, oat groats, poultry by-product meal, beef tallow preserved with mixed tocopherols, deflourinated phosphate, salt, L-lysine potassium chloride, animal digest, monosodium phosphate, chlorine chloride, vitamin E, zinc oxide, manganous sulfate, ethoxyquin, copper sulfate, niacin, ascorbic acid, vitamin A, biotin, calcium panthothenate, thiamine mononitrate, pyridoxine hydrochloride, riboflavin, vitamin D_3, folic acid, potassium iodide, sodium selenite, cobalt carbonate.

Made by: Megamill, Inc., Bigfarmintown, NO.
Contact us at: www.megamill$.com

LUV MY DOG

Dog Chow

With Lamb!

Complete and balanced for all life stages

Complete and balanced for all life stages based on AAFCO feeding studies

Net weight: 2.27 kg, 5 lb

Figure 9.3   Example of a pet food label.

brother" stepped in and decided pet foods needed to have some limits. This guaranteed analysis section (in the United States and Canada) *only* means that the product is not above a maximum (for ash, fiber, water) or below a minimum (protein, fat, energy). It may mean that one or more of the ingredients are significantly below a maximum or above a minimum. That in itself is not bad, but it does mean that you do not actually always know exactly what is in the product. It is also a fact, for better or worse, that several quality controls and regulations for analysis and labeling on pet foods exceed those for human foods. In Europe, the average analysis must be on the label.

Example of a Guaranteed Analysis Section of a Label

Guaranteed Analysis	Dry Food	Wet food
Moisture, not more than	10%	80%
Crude Protein, minimum	20%	8%
Crude Fat, minimum	15%	6%
Ash, maximum	4%	1%
Calcium, minimum	.6%	.1%

List of Ingredients

Pet food labels provide an *ingredient list* in descending order from most to least. Although this list may not change from batch to batch, the actual chemical ingredient analysis may, as many feedstuffs are not constant in composition. This is why industry is slowly moving more toward a nutrient analysis type of label, in which the actual nutrient analysis is provided. The list of ingredients is helpful for identification of what is in the feed, but only with some knowledge of nutrient concentrations can a person determine the chemical content of the feed. Ingredients that contain greater amounts of water such as meats and meat meals will generally appear first, but the actual protein, fat, and energy supplied by these ingredients may be less than for other ingredients such as soybean meal or corn because these ingredients are usually about 90 percent dry matter. There are thousands of ways to combine just a few ingredients to meet a given chemical content. This is not bad, but until actual percentages of ingredients are on the label, we will not be able to determine the actual makeup of the product.

Ingredient List Example, Wet Food

Water, beef, chicken, soy flour, rice, corn, carrots, whole egg powder, dicalcium phosphate, potassium chloride, caramel color, choline chloride, manganous oxide, cobalt carbonate, calcium iodate, sodium selenite, vitamin A, D-activated animal sterol, vitamin E, thiamine, niacin, calcium pantothenate, pyridoxine hydrochloride, riboflavin, folic acid, biotin, vitamin B_{12}, garlic powder.

Ingredient List Example, Dry Food

Lamb, brewers rice, ground yellow corn, corn gluten meal, oat groats, poultry by-product meal, beef tallow preserved with mixed tocopherols, defluorinated phosphate, salt, L-lysine potassium chloride, animal digest, monosodium phosphate, choline chloride, vitamin E, zinc oxide, manganous sulfate, ethoxyquin, copper sulfate, niacin, ascorbic acid, vitamin A, biotin, calcium pantothenate, thiamine mononitrate, pyridoxine hydrochloride, riboflavin, vitamin D_3, folic acid, potassium iodide, sodium selenite, cobalt carbonate.

At this stage, the student should recognize the various ingredients as energy, protein, fiber, vitamins, and minerals. In some classes, the teacher may want to combine these lists of ingredients with available analyses to estimate the nutrient content. The example labels given are actual labels from products available in the United States.

Nutritional Purpose or Adequacy

Label must state, through an *adequacy statement*, which minimum requirements are met by the product—maintenance, pregnancy, lactation, growth, or special performance. They also may be a complete food, snack, treat, or specialty food. It must be clearly marked as a snack, or *veterinary prescription* product, for example. Many are formulated to meet or exceed the highest requirement and can be fed in all stages if intake is controlled. Others are formulated for maintenance only and should not be used in reproduction unless total diet is adequate. Not every single claim for every single product needs be substantiated by feeding trials. The claims are valid if *either* of these conditions is met: the product passes a feeding test under protocol given by AAFCO (Association of American Feed Control Officials); or contains at least the minimum amount of each nutrient as recommended by the National Research Council (NRC). Nutritional adequacy statements do not address potential nutrient excesses, toxic substances, or palatability.

Quite often, somewhat more than the minimum required is included in the food; you may see statements such as "sufficient in nutrients for all dogs." This may actually mean that it is "oversupplemented" for many dogs. Why is so much often included? Well, there are some good, valid reasons for this, as well as some that are not so good.

> Reasons why complete feeds usually contain somewhat more than the minimums include:
>
> 1. To allow manufacturers to advertise that a product *exceeds requirements*.
> 2. To protect against nutritional deficiencies. If the foods were minimal, some animals would be all right, others would be deficient.
> 3. To protect against deficiencies due to variability of composition of products used. Companies do not and cannot afford to analyze every batch of every feedstuff used.
> 4. Because many feedstuffs are high in nutrients such as calcium, phosphorous, and magnesium.
> 5. Because of the syndrome of *more is better*. Several people still respond to this when purchasing pet foods. In all fairness, this is not the problem it once was and most foods are where they actually should be.

Name, Species, Contact, Weight

These parts of the pet food label should be self-explanatory: the name of the product, what species it is intended for, the total weight (including water), and some means by which to contact the manufacturer or distributor.

Analysis of Diets

We have selected ingredients, analyzed them, formulated a ration, and created a product. How do we now know that it does what it is supposed to do? There are many simple to complex analyses, measures, and experiments that are done to determine the acceptability, safety, and effectiveness of pet food products.

Palatability: How Well the Animal Likes or Eats the Food

This is a function of odor, temperature (also related to odor and mouth feel), texture, size, shape, mouth feel, nutrient concentration, and habit (preference), and is usually determined by a two-pan test. In this test, animals are given the choice of two or more different foods, repeated over several days, and how much of each food is consumed is determined. From this, one can make a general statement that a set of animals prefers one food over another or others.

Acceptability: How Well the Food Is Used and Meets the Requirements

Nutritional Need or "True Appetite"

Because, in general, dogs and cats will eat to maintain their energy requirements if a food is acceptable, the animal will be able to eat it to meet the requirements. If a food is not palatable or not properly formulated so that the nutrient requirements can be met at a realistic rate of intake, then it is not acceptable. For example, a food high in fiber will be limited in intake physically by taking up space in the digestive tract and signaling *physical satiety* (that the animal is full) before the real requirements of amino acids, vitamins, and minerals are met. If a diet is high in fat, the body will think it has sufficient energy, and will signal *chemical satiety* (that it has all the energy it needs), and intake will also stop before the requirements for other nutrients such as amino acids, vitamins, or minerals are met. The only way to ensure that the animal receives the proper amount and balance of nutrients is to feed a food containing the proper balance of nutrients with respect to the caloric density of the diet.

Thus, the terms *nutrient density* or *nutrient per kcal* or *nutrient:calorie ratio* all are used to determine if the food has the right proportion of nutrients in relation to the energy content. If an animal is eating to meet its basic energy requirement, the nutrient needs for protein, fat, vitamins, and minerals must be balanced in relation to the energy content of the diet. Most pet foods are formulated to a specific nutrient:energy ratio, so that the animal will eat the recommended food intake for its weight and activity, and meet the other nutrient requirements in so doing.

Previous Experience, or "Learned Appetite"

If a food, or ingredient, perhaps by providing a smell or other factor, has previously given an animal a "bad" experience in the past, this food may be avoided by that animal. If an animal becomes sick on a certain food, either directly or coincidently, it may avoid similar smells, tastes, and textures in the future. An animal will not always intelligently avoid foods that are bad (either nutritionally or toxicologically), nor necessarily seek out foods that are "good." Taste (palatability) is the major determinant of whether or not it will be eaten.

Determining Food Nutrient Content

Obviously, we cannot make foods and determine if they meet requirements unless we test them. Figure 9.4 shows the general process of food analysis to determine food content.

Analyses may be conducted formally or informally:

1. From chemical (proximate) analysis of the feed—the only way to ensure good knowledge of the feed, which is what companies do.
2. From food label—trust the label to be correct, it usually is.
3. Calculation from the ingredient list—can be done if good numbers are available and average composition of ingredients is known.

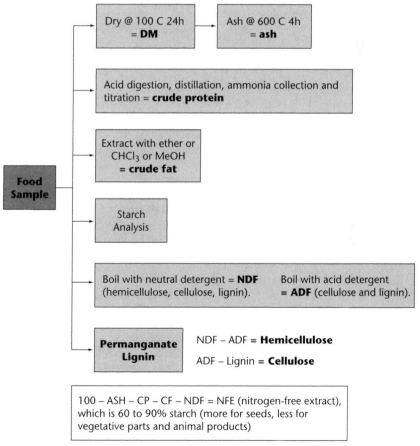

Dry @ 100 C 24h = **DM**

Ash @ 600 C 4h = **ash**

Acid digestion, distillation, ammonia collection and titration = **crude protein**

Extract with ether or $CHCl_3$ or MeOH = **crude fat**

Starch Analysis

Food Sample

Boil with neutral detergent = **NDF** (hemicellulose, cellulose, lignin).

Boil with acid detergent = **ADF** (cellulose and lignin).

Permanganate Lignin

NDF – ADF = **Hemicellulose**

ADF – Lignin = **Cellulose**

100 – ASH – CP – CF – NDF = NFE (nitrogen-free extract), which is 60 to 90% starch (more for seeds, less for vegetative parts and animal products)

Figure 9.4   Analysis of foods.

Feed Chemistry: Proximate Analysis

Feed is made up of protein, starch, fat, cellulose, hemicellulose, other complex carbohydrates, lignin, minerals, and silica. These are determined by simple chemical analyses, which we usually call the *proximate analysis system*. It has been in use, with some improvements, for over 100 years based on simple chemistry. Parts of the results of *proximate analysis* are used directly on food labels, such as *crude protein, crude fat,* and *crude fiber.* Crude protein, crude fat, crude fiber, lignin, and nitrogen-free extract are determined as follows:

> *Crude Protein*—extract feed sample in strong acid, boil off NH_3, and titrate = amount of nitrogen. The amount of nitrogen multiplied by 6.25 is crude protein, because most plant protein is 16 percent nitrogen.
>
> *Crude Fat*—extract feed in ether and measure fat. Chloroform/methanol is now the preferred method, as it is more precise, but the idea is the same: total amount of fat in the feed.
>
> *Crude Fiber*—acid hydrolysis of the feed. The residue left behind and measured is most of the cellulose, lignin, and some hemicellulose and ash.
>
> *Lignin*—permanganate precipitation. Lignin is a complex phenolic compound that makes up cell walls and structures of plants. It is not available as a nutrient and is included in the fiber fraction.
>
> *NFE (nitrogen-free extract)*—what is left over after subtracting crude protein, crude fat, crude fiber, lignin, and ash. The remainder is primarily starch, but also hemicellulose and other soluble molecules. Roughly 60 to 80 percent of NFE for most seeds is starch.

Van Soest System (or Detergent System)

Peter Van Soest, a young animal scientist working in the 1960s, realized that crude fiber was just that and created a better analysis. Using acid or neutral detergents, this separates the major fiber components:

> *NDF* = neutral detergent fiber, which is C, HC, lignin, and ash.
>
> *ADF* = acid detergent fiber, which is cellulose, lignin, and ash.

Thus, from this, the amount of ash, cellulose, lignin, and hemicellulose can be determined.

The *detergent system* is now used in all food analysis. Dr. Van Soest was elected to the National Academy of Sciences for this important work. The combination of the old proximate analysis and the powerful fiber-measuring detergent system is still the standard.

Expressing Nutrient Content

Nutrient content is a major cause of confusion in comparing foods. One must be aware of the basis on which the information is presented in order to compare or evaluate foods. Water, while a worthwhile and necessary ingredient in many foods, does not provide other nutrients. Thus, the actual nutrient concentration or intake must be known, regardless of the water content of the food.

1. **As-fed basis.** This is the nutrient content of the food in the condition in which it is fed. Thus, foods of normally different water contents will have different nutrient concentrations, *as they are fed*, even if the concentration is the same in the dry matter of the food.

2. **Dry-matter basis.** The dry matter is the actual food, without any water. This is determined by a mild heating (for example, 100°C overnight) to vaporize the water. *Please note that this is the only way to directly compare nutrient compositions of different foods.* If one food is normally fed at 50 percent DM, and another at 90 percent DM, then the first food must be 1.8 times denser in nutrients as the second feed to provide the same nutrient intake at the same total intake.

3. **Calorie basis.** As described earlier, as the animal eats in general to meet its energy requirements, the amount of nutrients in relation to the energy content can be a quite useful number. If one knows the energy requirement of the animal in kcal, and the kcal/kg of the food, then one can get a good estimate of the kg intake of the feed. One can then calculate the amount of another nutrient (protein, for example) in the feed per kcal of energy, and determine how much of the nutrient will be consumed in relation to the requirement. This is why the general dog and cat requirements for protein, minerals, and vitamins are often given per kcal of energy.

For example, a 30-kg dog will require 2,000 kcal (2 Mcal)/d of ME, at an energy content of 5,000 kcal/kg DM of food; the dog would be expected to eat (2,000 kcal/d)/(5,000 kcal/kg) = .4 kg, or 400 g of food DM per day. This is about 1.3 percent of body weight (BW), a reasonable intake for maintenance. At a protein:calorie ratio of 30 g protein/1,000 kcal of ME, the protein intake would be 30 g/1,000 kcal * (2,000 kc/d/1,000), or 60 g protein per day. This would be a diet containing 60 g protein/400 g or 15 percent protein on a DM basis. This example is a high energy ration.

Thus, looking at the energy content of the feed, the energy requirement of the animal, the protein content of the food, and the protein requirement, one could judge if the dog would be able to meet the protein requirement with that food. Also, looking at the protein requirement, one could formulate a ration of differing energy density to match different food intakes to the same protein requirement.

Calculation Box *Calculation of Nutrient Concentrations on a Dry-Matter (DM) Basis*

Ok, so what is this dry matter stuff? All foods contain some water. We cannot compare one food to another unless we know what the water content is or, that is, if we know what the dry-matter content is. Let us look at an illuminating example:

	Food A Megamill Dog Chow	Food B Natural Wholesome Dog Food
Moisture	10%	80%
Protein	20%	8%
Fat	15%	7%
Fiber	3%	1%
Ash	8%	4%

So you are trying to figure out what your animal would actually eat. Let us say that if you fed 0.5 lb (227 grams) of food to this animal, this is what it would eat:

	Megamill Dog Chow	Natural Wholesome Dog Food
Moisture	10% × 227 = 22.7 grams (g)	80% × 227 = 182 grams (g)
Protein	20% × 227 = 45.4	8% × 227 = 18.2
Fat	15% × 227 = 34.1	7% × 227 = 15.9
Fiber	3% × 227 = 6.8	1% × 227 = 2.3
Ash	8% × 227 = 18.2	4% × 227 = 9.1

So for the same amount of total food, Megamill dog chow provides more actual nutrients, while Natural WDF provides mostly water. Neither one of these is bad by itself, but let us look at it a different way. Remember that the animal needs the same amount of energy, regardless of where it comes from. So in the actual food itself, without the water, here is the analysis, on a **dry-matter basis**:

	Food A Megamill Dog Chow		Food B Natural Wholesome Dog Food	
	As-Fed Basis	Dry-Matter Basis	As-Fed Basis	Dry-Matter Basis
Dry Matter	(100 − 10) = 90%	100%	(100 − 80) = 20%	100%
Moisture	10%/.9 =	0	80%	0
Protein	20%/.9 =	22.2%	8%/.2 =	40%
Fat	15%/.9 =	16.7%	7%/.2 =	35%
Fiber	3%/.9 =	3.3%	1%/.2 =	5%
Ash	8%/.9 =	8.9%	4%/.2 =	20%

For the actual amount of food eaten, besides for water, the wet food provided 40 percent protein and 35 percent fat compared to the dry food A, providing 22.2 percent protein and 16.7 percent fat. So for each 100 grams of actual *dry matter* eaten, the dog eating food A would get an appropriate 22.3 grams of protein and 16.7 grams of fat, while the little pooch eating food B would receive 40 grams of protein and 35 percent fat. Bottom line: too much. Wet foods are almost always higher in protein and fat than is required. Again, all by itself this is not bad, but often leads to obesity because of the high fat content.

For example, if you wanted that dog to eat 60 grams of protein per day, but also wanted to limit energy density and food intake to avoid obesity, you would have to increase the protein:calorie ratio in the food. If a food contained 3,500 kcal Digestible Energy/kg (by adding fiber sources) so that intake was limited to approximately 400 g DM/d, then the energy intake would be: 3,500 kcal/kg * .4 kg, or 1,400 kc/d. The protein intake would still need to be 60 g/d, so in 400 g that is still 15 percent. But this changes the protein:calorie ratio to 60 g/1,400 kcal or 42.8 g protein/1,000 kcal ME. The bottom line is that this animal receives the same amount of protein while consuming less energy.

Regulation of Pet Foods and Labels

Pet food content and labels fall under regulations and guidelines provided by several agencies. We will briefly describe the makeup and role of each one.

National Research Council

The NRC is a part of the National Academy of Sciences, with advisory power to the president and Congress to set dietary guidelines. This panel is made up of scientists who are elected by members based on their proven worth as a scientist. It is a quasi-government agency, which does not directly receive public money, but does report to the president and Congress. There are many specialties within this agency, including technology, chemistry, physics, medicine, the environment, and agriculture. Within the Board on Agriculture is the Committee on Animal Nutrition, and within this are many subcommittees on each species. Over the last sixty years or so, usually about every ten to twelve years, each subcommittee is asked to review the scientific field and make a new report on the nutrient requirements of each specific species. For most agricultural animals, there are thousands of studies done every year and we have a quite exact understanding of their requirements, although they do need to be occasionally updated. The previous updates for dogs and cats were in 1984 and 1985, and just this year, the two were combined into one volume, released as the *Nutrient Requirements of Dogs and Cats* (NRC, 2004). The purpose of these reports is to help ensure that the animal is adequately fed, and sometimes to set an absolute requirement. Other scientists, especially those at pet food companies, use these reports as a guideline, modified by specific situations and newer data, to formulate pet foods. If things come down to a legal battle, which does happen sometimes, the NRC publications are often considered to be the best scientific information available.

Association of American Feed Control Officials

The AAFCO is an association of state officials, scientists, and industry representatives charged with regulating animal feeds. This body is loosely organized and has limited legal power. It does suggest guidelines on requirements and feed formulations, usually based on the NRC publications. Thus, it does not set the requirements, but is more of an enforcement arm that works with state veterinarians and departments of agriculture to help make common rules for food labels and ensure that labels are factual and foods are what they claim to be. As with any such public agency, its power is determined by what society grants to it through legislatures and laws. Some readers may think that there should be more regulation in the pet food industry, and some may think "big brother" should back off but, in essence, these agencies are an arm of society's desires, nothing more, nothing less. They will do what the majority wants, over time.

The Food and Drug Administration

The FDA regulates the safety and efficacy of drugs and food additives. In that vein, it has regulatory power over what can be used in human and animal foods of all types.

The charge of the FDA includes:

Establishing certain animal food labeling regulations.

Specifying permitted ingredients such as drugs and additives.

Enforcing regulations on chemical and microbiological contamination.

Describing acceptable manufacturing processes.

The Center for Veterinary Medicine in the FDA provides regulation and enforcement for health claims. "A health claim is defined as the assertion or implication that consumption of a food will treat, prevent, or otherwise affect a disease or condition" (Hand et al., 2000).

Physical Evaluation of Food

It is not usually necessary to actually inspect the products you buy. However, it is not a bad idea to look at different foods just to learn. Physical examination of a food can identify some major characteristics of that food, but cannot replace chemical analysis. You cannot define the chemical content, but can recognize various components and ingredients. This section is a summary from more complete information in Cowell, et al., 2000.

Look for the following:

Inspect the package:

1. Torn or open packages—can be contaminated.
2. Dented or swollen cans—especially the latter is a strong indication of bacterial contamination.
3. Putrid, fermented, or sour odors—indicative of bacterial fermentation.
4. Darkened surface on top of canned products is *not* indicative of a problem—this is normal due to the vacuum in the head space of the can; it indicates partial oxidation of the surface carbohydrates.

5. Pour canned product onto plate. Inspect can. Is it lined or not? If not, food may have metallic smell; the can may have a shiny or somewhat dulled surface from reaction with food. This is not necessarily bad, but can be indicative of a poorer-quality product.

Inspect the canned (or semi-moist) food:
1. Cut lengthwise through chunks or loaf of food, spread food out on plate.
2. Inspect contents—mixed foods should be somewhat nonuniform in nature, having small pieces of cereal grains or animal tissues. Foods should be devoid of hair, feathers, large chunks, and seed hulls (unless listed on package for "low-energy" rations). Meat-type foods should have striated muscle evident. Little or no chunks of connective tissue or bone or blood vessels. Foods should be devoid of TVP (textured vegetable protein) unless so labeled, which is soy protein textured and colored to look like meat. Small pieces of charcoal may be evident; this is used to "decharacterize" the food as not used for human consumption; it is fine for the pet food. Semi-moist products should be just that, moist but not wet. If package is broken on semi-moist foods, spots may appear dried out. Same kinds of food particles should be present or absent as for mixed foods.

Inspect dry food:
1. Inspect the package. Look for "grease out," or excessive fat soaking through the container or on the surface of food. Some is normal, but it should not be liquid or grossly evident. Too much can lead to vermin infestation.
2. Take a sample (handful or pour out) from the top and bottom of the bag, spread it on a light plate or piece of paper. Look for fines, the small particles broken off from food. There are always some, but too much indicates lower-quality product, poor formulation, or excessive handling.
3. Look for mold—a musty or stale odor or a white, blue, green, or black dusty coating on the food.

Supplementation

Supplementation of pet food with human foods or protein, vitamins, minerals, or other compounds is often done, and this is where the most problems are encountered with overfeeding, nutrient excesses or deficiencies, toxicities, and poor eating habits. As is true for all of nutrition, several myths abound among the time- and experience-tested simple truths. Human foods can be fed to pets if caution is taken, the total ration of the animal is considered, and certain things are avoided.

1. Meat—good protein source, but by itself is inadequate in energy and several vitamins and minerals. If fed to be adequate in energy, has too much protein and too little calcium.
2. Fish—good protein source, but same problems as meat. Raw fish can be contaminated with several different bacteria, including *neorickettsia helminthoeca* or *neorickettsia elokominica*, leading to "salmon poisoning"—severe depression, fever, enlargement of lymph nodes (inflammation), oculonasal discharge, hematemesis (blood cell damage), and diarrhea, with 90 percent mortality within five to twelve days after ingestion.
3. Fats and oils—can be used to supplement energy. However, if excessive, total intake of the rest of the diet will go down as the animal regulates energy intake. Thus, this is the most common cause of nutrient deficiencies (protein, vitamins, and minerals) in pets. Use only one tablespoon per pound canned food or per 8 oz dry food. Oils and fats are often used to increase taste and intake. Oils in themselves will not improve hair coat over a well-balanced ration, but some specific oils increased in omega-3 fatty acids may help some animal's hair and reduce inflammation (see Chapter 10).
4. Eggs—excellent protein, but should be cooked to avoid avidin binding to biotin. The essential fatty acids in the yolk can improve hair coat in an animal consuming an otherwise low-fat diet. Improvement in coat is not due to the egg protein, but the essential fatty acids in the yolk.
5. Cow's milk—good source of calcium, phosphorous, protein, and many vitamins. However, cow's milk is too dilute and too high in lactose to make up much of the food for dogs and cats. Too much lactose can cause gas and diarrhea in puppies or older animals.
6. Cheese and cottage cheese—good source of protein and fat, low in calcium.
7. Liver—good source of protein, fat, some CHO, and trace minerals and vitamins. However, is deficient in calcium; Ca:P ratio is 1:35. Liver is beneficial when fed raw for sick, weak, or anemic animals. Exclusive liver feeding will cause calcium deficiency and may lead to hypervitaminosis A, depending on the source of the liver. There is no reason not to feed liver because of the myth that it is an excretory organ and contains toxins, unless the source of the liver leads one to believe toxins are present (sick animals, animals feeding off waste ponds, etc.)
8. Vegetables—can be fed as part of a complete food. They can be a good source of CHO, fiber, some protein, and some minerals. However, they are mostly water, and can vary tremendously in content. They are high in moisture and will limit intake. Calcium and protein will need to be supplemented. Vegetarian diets can be formulated using cereal, soy, and vegetable sources, but require great care and attention to detail.

Other Possible Supplements

Vitamin and mineral supplements are absolutely not necessary if a high-quality food makes up 90 percent or more of the pet's ration. Oversupplementation costs extra money, wastes resources, and can be a bigger problem than deficiencies.

1. Bones—hard, long, or knuckle bones can be good chewing exercise for dogs. Small bones and soft (poultry) bones should be avoided at all times due to risk of injury.
2. Table scraps—can occasionally be given, but should never exceed 10 percent of the total ration. Too much of anything, especially fat, at once can lead to digestive upset, diarrhea, and discomfort. For animals that are already obese, stop feeding scraps now.
3. Chocolate and candy should *not* be fed. Chocolate contains *theobromine,* a caffeine relative that is toxic to dogs at 240 to 500 mg per kg body weight, although death at lower doses has been reported. Milk chocolate contains about 1.5 mg theobromine/g; there is 15 mg/g in dark chocolate; thus, a 10-kg dog would be in trouble with only 2.2 oz (60 g) dark chocolate to 670 g (1.5 lb) milk chocolate. Theobromine toxicity symptoms include vomiting, depression, lethargy, diuresis, muscular tremors, diarrhea, and death.
4. Onions and garlic—do not control worms, fleas, or other pests, but can make animal smell like garlic or onions. Too much onion (5 g/kg body weight) can lead to "Heinz bodies" in erythrocytes, hemolytic anemia, fever, dark urine, and death.
5. Grass is sometimes eaten, and is often vomited soon thereafter. No major problem unless occurring often—restrict access to break habit.
6. Vitamin C, brewers yeast, bee pollen—*no* demonstrated benefit in health, disease or pest control, or quality of life. Vitamin C (ascorbic acid) may help dogs prone to joint and cartilage problems and may be given in consultation with a veterinarian. Too much vitamin C might lead to other problems.
7. "Chelated minerals"—can be natural or artificial chelation (binding) of minerals with proteins, CHO, or other chemicals. In and of itself, chelation may increase absorption but does not change use of the mineral. If the compound is more digestible (heme iron vs. elemental iron), the mineral may be absorbed at a higher rate. In the absence of a digestive or clinical problem, chelated minerals are not necessary.

Calculation of Ration Composition

The actual formulation of the ration requires direct matching of the requirements of the animal to the foods mixed together. However, when one takes into account the protein, energy, fat, and all the vitamins and minerals, one can see that the math can get burdensome quickly. In the actual large-scale formulation of rations, computer-assisted programs do all the mundane calculation. Also, few feeds provide only one nutrient, so for several feedstuffs, if they have the proper amount of protein, they also often have the proper amount, or close to it, of other nutrients. However, this is false as often as it is true. Thus, all rations need to be checked, and feeds need to be chosen in an iterative fashion to come up with the best final formulation. You must realize that without using purified ingredients, no food is perfectly balanced, nor need it be; for there is a range of acceptable situations.

The ability to mathematically balance rations will depend on the class you are in. However, you should be able to look at compositions of feeds and nutrient requirements and come up with good qualitative comparisons and pick the feeds best for that situation.

The following section is included so that you can have at least an idea of what is involved. For the actual calculations, we are setting the requirement of any nutrient to a given amount, and then determining algebraically the amounts of two or more feeds that vary in composition of that nutrient. All calculations are on a dry-matter basis.

First, we set our *fixed* ingredients, those which will be added to the ration in a given percentage that do not contain the nutrient we are balancing for, and thus do not need to be included in the calculations. Fixed ingredients usually include vitamin supplements, mineral supplements, sodium chloride, and flavors or preservatives. Of course, if we are balancing for vitamins or minerals, we would not include the supplement in the fixed ingredients, but would include it in the part needing to be calculated.

Algebraic Method of Balancing a Ration

We use simple algebra to match the content of the nutrient in the feed with the requirement. Of course, now we have computer programs to do this for us. Always balance for first-limiting or most expensive nutrient first, then the others, rechecking as you go; for example, balance for protein and amino acids, then check for energy (DE in horses; ME in cats, dogs, fish, birds). The left side of equation is requirement, the right side is feed; thus

$$\text{Requirement } Y\,(\%) = [X * (\text{decimal \% nutrient in } X)] + [(\text{decimal \% nutrient in other feed}) * (100 - X)]$$

Where Y is the percentage required, and we solve for X, which is equal to the percentage of the feed component with the highest concentration of the nutrient, and $(100 - X)$ is the percentage of the feed with the lowest concentration of the nutrient.

Example

Balance a ration for 20 percent crude protein, using soybean meal at 44 percent CP and corn at 10 percent CP. You will have 5 percent of the ration as fixed ingredients (vitamin, mineral, fat).

Requirement = 20%; so set left side (Y) at 20 (% CP)
Percentage of CP in soybean meal = 44;   percentage in corn = 10

Equation = 20 =   .44 * X   +   [.10 * (95 − X)]
(% needed in diet)   (soybean, 44% protein)   (corn, 10% protein; we use 95 because
                                                    other 5% of the matter is already
                                                    designated as fixed ingredients)

$$= .44X + 9.5 − .10X$$
$$= (.44X − .10X) + 9.5$$
$$= .34X + 9.5$$
$$20 − 9.5 = .34\ X$$

Divide through and reorganize:

$$X = (20 − 9.5)/.34$$
$$X = 11.5/.34$$
$$X = 33.8, \text{ or } 33.8\% \text{ SBM}$$
$$\% \text{ of corn} = 95 − 33.8 \text{ or } 61.2\% \text{ corn}$$

Check:

$$SBM = 44\% \text{ CP} * .338 = 14.87\% \text{ of CP from SBM}$$
$$CORN = 10\% \text{ CP} * .612 = 6.12\% \text{ corn}$$
$$14.87 + 6.12 = 19.9, \text{ or } 20\% \text{ CP in feed}$$

Check the balance for the next nutrient; check and adjust; formulate table; and include expected intake. This can also be done on a grams/day basis instead of a percent in the feed basis.

Practical Application

Obviously, purchasing and feeding the proper food in the right amounts for your animal is the ultimate practical application. Using the variety of feeds available to them, pet food companies spend considerable time, effort, and money to formulate, manufacture, test, and distribute good foods. There are a variety of types and forms on the market for a wide variety of foods and consumers. Consumers have the option to buy the type and form of food that meets their needs the best. Basic practices are followed by nutritionists to determine requirements, feed availability and cost, need for supplementation or processing, and eventually to formulate a good product. Supplementation of complete feeds is not necessary. Product labels can tell us quite a bit about the food in the package and should be used to assess the usefulness and use of the product for your animal.

Words to Know

acceptability	crude fiber	meal-fed
ad libitum	crude protein	NDF
adequacy statement	detergent system	palatability
ADF	dry-matter basis	physical satiety
as-fed basis	free choice	popular brands
balancing a ration	generic	premium brands
calorie basis	guaranteed analysis	private label
chemical satiety	ingredient list	proximate analysis
crude fat	limit feeding	veterinary prescription

Study Questions

1. Describe the differences in types of pet foods related to form (wet, dry) or marketing category.
2. Describe proximate analysis and the detergent system and state what number(s) on a given label come from which test.
3. Why do we put some nutrient content on a calorie basis?
4. What are the roles of the NRC and the AAFCO in advisement and regulation of pet foods?

Further Reading

Association of American Feed Control Officials (AAFCO). 2002. *Dog and cat food nutrient profiles* (pp. 126–141). Oxford, IN: AAFCO.

Cowell, C. S., N. P. Stout, M. F. Brinkmann, E. A. Moser, and S. W. Crane. 2000. Making commercial pet foods. In M. Hand, C. D. Thatcher, R. L. Remillard, and P. Roudebush, eds., *Small animal clinical nutrition*, fourth edition (Chapter 4). Topeka, KS: Mark Morris Associates.

Crane, S. W., R. W. Griffin, and P. R. Messent. 2000. Introduction to commercial pet foods. In M. Hand, C. D. Thatcher, R. L. Remillard, and J. Roudebush, eds., *Small animal clinical nutrition*, fourth edition (Chapter 3). Topeka, KS: Mark Morris Associates.

National Research Council. 2004. Diet formulation and feed processing. In National Research Council, *Nutrient requirements of dogs and cats* (Chapter 12). Washington, D.C.: National Academy Press.

Remillard, R. L., B-M. Paragon, S. W. Crane, J. Debraekeleer, and C. S. Cowell. 2000. Making pet foods at home. In M. Hand, C. D. Thatcher, R. L. Remillard, and P. Roudebush, eds., *Small animal clinical nutrition*, fourth edition (Chapter 6). Topeka, KS: Mark Morris Associates.

Roudebush, P., D. A. Dzanis, J. Debraekeleer, and R. G. Brown. 2000. Pet food labels. In M. Hand, C. D. Thatcher, R. L. Remillard, and P. Roudebush, eds., *Small animal clinical nutrition*, fourth edition (Chapter 5). Topeka, KS: Mark Morris Associates.

Chapter 10

Nutrition of Canines through the Life Cycle

Take Home and Summary

Feeding-management goals begin with establishing good eating habits. Owners should feed a properly balanced ration for each life stage, feed only what is necessary for proper growth or maintenance of body weight, and have a regular exercise program for you and the dog. During growth, pregnancy, and lactation, more protein, energy, vitamins, and minerals are needed per unit food fed than during adulthood and aging. Many excellent foods are available on the grocery, pet mart, or veterinary store shelf. Homemade diets are a possibility, but present many challenges. Presently, obesity is the biggest problem in dogs in the United States, but can be prevented. Regular observation of how much the animal eats and its growth and body condition (fat) is essential to long-term health. Feed only what the animal needs, even though it might be significantly more or less than the averages on labels (or in this book). Regular exercise will make sure the nutrients get used properly and not stored as excess fat and will keep the circulatory and muscular systems strong for a long and healthy life. Older dogs should be maintained in a regular, if less vigorous, exercise program. Feeding of maintenance or senior diets to older dogs is preferred, as these provide better protein quality to minimize the stress on metabolic organs. It is the guaranteed analysis on the label, as well as the ingredients, and not the name, that should dictate your choice of food. Situations of increased environmental or physical stress (cold weather, increased exercise, special service) require close observation and adjustments of intake and occasionally require altering nutrient density (feeding a high-protein diet to working dogs or higher fiber for obese dogs). The bottom line is: Know and observe your dog and feed to maintain a good body condition.

Normal Feeding Goals and Management

Many people, including some nutritionists, will often say: "Just feed a well-known dog food and everything will be fine." Well, that has some truth, and for many animals, that might be just about it. However, you should want to know *why* feeding an animal might be this simple. You should want to know the best practical applications of nutritional principles to ensure long and healthy lives, how to prevent or lessen most feeding-related problems, and how to provide an effective solution to problems as they might arise.

The basic structure of the remaining chapters of this book will contain "Goals and Objectives" sections for each stage of the life cycle. The reader may refer back to Chapter 2's discussion of the key points in each stage of life. In the remaining chapters there will be a brief explanation as to why certain practices are used, and I will refer to the specific metabolic background in previous chapters as needed. Throughout, the philosophy is that nutrition can and should be simple; this book can help you understand the reasons behind what we do, as well as show that nutrition should be fun, and so should learning about nutrition. The "Take Home and Summary" section will still appear in each chapter, but from here on, I have dropped the "Practical Application" sections, as everything, I hope, is a practical application. Referring back to Figure 2.1 on the general goals of life-cycle feeding may help you keep things in perspective.

Goals and Objectives for Puppies

The puppy starts out life completely dependent on the mother for survival, both for food and warmth. The first need of the young pup is for energy to stay warm and alive. The other major need is for the antibodies (immune-function proteins) that the colostrum provides. *Colostrum* is the name given to first milk, or the milk produced in the first few hours of the puppy's life. It is nutritionally richer in protein and fat, and contains specialized types of protein called immunoglobulins. These provide immunity against the diseases and infections that the mother has been exposed to. The mammary gland has collected these from the mother's blood for the last several days of pregnancy in order to provide the pups with a large enough dose to provide protection right after birth. The immunoglobulins are consumed by the pups and absorbed as intact proteins through spaces in the intestinal lining. This transfer of immune protection from the dam to the young through the milk is called *active immunity,* and it is critical to the survival of the young. *Passive immunity* is the transfer of antibodies to the developing fetus in the uterus.

The puppy can only take advantage of this protection for the first twenty-four to forty-eight hours of life. The newborn intestine has not yet fully sealed itself off, so to speak, from the body, and intact proteins can pass through the small spaces between the cells of the intestinal lining without being digested. After several hours, intestinal bacteria start to grow and could infect the animal if they entered the body. Thus, the lining matures and closes and intestinal enzymes start to be secreted to break down lactose, fat, and protein. After this time, most immunoglobulins still left in the intestine will be broken down and absorbed as amino acids, and thus will not provide any immune protection.

After the first twenty-four to thirty-six hours of lactation, the mammary gland starts to make "normal" milk to meet the needs of the growing puppies, rather than colostrum. At this time, the mother can make all the milk that the litter demands. As days pass, the litter continues to grow and thus requires even more milk. By a few days to weeks (depending on size of mother and number of pups) the nutrient requirements of the dam can already be two to three times as much as she needed in pregnancy. We will discuss the mother's requirements in depth later. The pups are using lactose for glucose to supply the developing brain and nervous system as well as to provide energy. Milk fat supplies the majority of the energy, and provides essential fatty acids, cholesterol, and other lipids to help make new cell membranes. High-quality milk protein supplies amino acids for the large amount of proteins they need to make themselves. Milk is, of course, primarily water (usually 80 to 88 percent for most land mammals); dog's milk contains about 4 to 6 percent protein, 4.5 to 5.5 percent fat, and lactose is about 4.5 percent (Table 10.1). As a proportion of actual nutrients (not counting water), this is about 34 percent protein, 35 percent fat, and 31 percent lactose, a nutrient-dense and easily digestible ration. This is the right mix for pups, and milk of other species could be too strong or too weak. Note in Table 10.1 that the milk of cows will have too much lactose and not enough fat or protein.

There are many milk replacers available through veterinarians to supply pups that for some reason cannot get milk from the mother. Growth rates should be about 1 to 2 grams per day for every pound of expected adult weight. In perspective, this means that a small dog (10 lbs adult weight) should gain about 140 grams or 1/4 of a pound in a week; a medium dog (60 lbs) would gain about 1 1/3 lb a week and a large breed dog (100 lbs mature weight) would gain about 2 1/3 lbs/week.

Table 10.1	Composition of Milks and the Puppies' Requirement		
Ingredient	Cow's Milk	Bitch's milk	Queen's Milk
Water, %	87.6	77.2	81.5
Protein, %	3.3	4–6	7–8
Fat, %	3.8	4.5–5.5	6–8
Lactose, %	4.7–5.0	4–5	3.5–4.5
Energy, kcal/L	276	700 to 900	800–1200
Ca, %	0.12	0.15–0.2	0.15–0.2
P, %	0.1	0.13–0.18	0.13–0.18
Protein, % of DM	28	34	41
Fat, % of DM	31	35	38
Lactose, % of DM	40	31	22

Note that cow's milk would provide too little fat and protein, and too much milk sugar. This will stunt the growth of the pup or kitten, and usually causes diarrhea as the animal's digestive system cannot break down all that lactose.

Preparing for Weaning

For several weeks, the milk can supply everything the pups need. Some time around week 3 or 4 for many litters, the pups are now too big for the mother to supply everything they need. In the wild, the mother would start to bring food back to the litter to supplement her milk production. For our companions, it is our responsibility to start feeding. Many adaptations must take place in the digestive and hormonal systems of the young during weaning in order to successfully switch from a liquid milk diet to a dry, solid food diet. Review the sections of Chapter 2 covering weaning, as needed.

In practice, we start to present moistened puppy chow to the puppies to introduce them to the concept of eating, which they usually figure out fast. Their natural curiosity, olfactory senses, and taste buds show them that eating food is interesting and fun. It is our responsibility to make sure that food is of the correct texture and nutrient content for what they need, and to avoid allowing them to eat too much. The chow should at first be quite wet, easily mashed to gruel. Over several days to a few weeks, add less water, so that when they stop nursing, they are eating moist, still soft, but not mushy food. Within a couple more weeks, they should be able to have only dry food. This period can be shortened as you observe your animal's ability to eat, but as always a slower change is usually safer.

> Weaning is a time of adaptation from wet, soluble, high-quality milk to drier, variable-quality foods. We can ease this transition with simple practices.

During this time, the mother will still be providing a large amount of nutrients to the young. The basic purpose of supplying the extra food is not so much for the nutritional need, but to start the digestive adaptations occurring. Up to this time, the stomach and intestines have had it easy, only needing a few proteases to break up the casein, lactase enzyme to cleave lactose to glucose and galactose, and lipase to break down fat. In addition, as the food came in wet, there was not a lot of physical grinding to be done in the stomach, and no harsh food particles to rub against the intestinal cells.

Figure 10.1 shows the generalized anatomy of the digestive tract of nonruminants. Note that the stomach is the first major digestive organ, secreting acid and enzymes to start protein breakdown. The majority of digestion occurs in the small intestine, where pancreatic enzymes are added. Most digestion occurs in the duodenum and jejunum, and the absorption of glucose, amino acids, fatty acids, vitamins, and minerals takes place primarily in the jejunum and ileum. Undigested feed, especially plant cells walls (fiber: cellulose and hemicellulose) passes into the large intestine, where it can be broken down and fermented, and the energy absorbed as volatile fatty acids. Note, however, that although dogs and cats do have a large intestine, they have very little capacity to digest and ferment fiber, even when fed significant amounts.

On a dry diet made up of many different ingredients, several proteases are needed. These break the peptide bonds between specific amino acids, so that

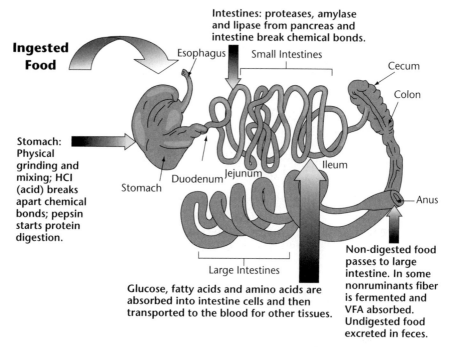

Figure 10.1   Digestion in nonruminants.
Source: Redrawn from *Small Animal Care and Management,* second edition by Warren. © 2002. Reprinted with permission of Delmar Learning, a division of Thomson Learning: www. thomsonrights.com. Fax 800 730-2215.

eventually single amino acids are released to be absorbed into the body. For carbohydrate digestion, enzymes are needed to break down starch (amylase), amylopectin (amylopectinase), and other "-ases" (such as maltase) to release glucose from the various trisaccharides and disaccharides that are the intermediates of starch and amylopectin digestion. Fat still needs lipase, but instead of coming in as already-formed micelles (fat particles) from milk, the fat must be physically broken up into small particles, and bile salts (see Chapter 3) must stabilize these particles (called micelles) to keep them in solution.

The dry food must be physically ground by the action of the stomach muscles to break it down into small particles (this is true also in the wild, as the muscles and organs consumed by wild dogs would need to be physically reduced in size). Some hard particles enter the intestines and can scratch or damage the intestinal lining, causing mild irritation or even damage. Providing wetted food gradually allows the adaptations to take place over a longer period of time. This allows the pancreas, stomach, and intestine more time to make sufficient enzymes, so it is not a shock to the system and undigested food can pass through (it would be a shock if we just dumped dry food down all at once). It also allows the stomach and intestines to slowly adapt to food particles, and this causes secretion of protective mucopolysaccharides (complex lubricant molecules, commonly called mucus) to help coat the intestinal cells.

> There is no need to make up any special food in addition to the puppy chow.

A high-quality puppy chow is formulated to provide the proper mix of protein, fat, carbohydrates, vitamins, and minerals that the young animal needs. Making up your own food is almost always too expensive and time consuming, and it will almost certainly not be as nutritionally balanced, unless you take great care. There is also no need to supplement vitamins and minerals during nursing or weaning. The milk and the food will provide all that the puppies need. Four feedings a day is adequate. Do not wake the pups to feed, they need sleep, too.

> During weaning, meals should be small, three to four per day, with feedings three to four hours apart, as this gives time for the food to begin to be digested in the stomach and pass from the stomach into the small intestines.

Orphaned Puppies

Occasionally, but rarely and for different reasons, a litter of pups will be unable to be fed by the mother. This is a different situation than a litter that simply cannot get enough milk from the mother. That latter situation can be fixed by supplying transitional food as described earlier. The harder part is to note, by daily observation and at least weekly weighing of the litter, whether they are gaining weight normally. However, in an extreme case, orphaned pups younger than five to six weeks of age can be readily cared for. The first thing to do is to go see the veterinarian. Orphans can be supplied with a formula substitute for bitch's milk that will meet all their needs until they can start eating wetted dry food.

First, return to Table 10.1 to compare milk composition, especially comparing cow's milk to bitch's milk. Cow's milk is too dilute for the rapidly growing pup. The calf (and the human infant) grows much more slowly and does not need as rich a source of nutrients. One common mistake is to feed cow's milk to the pups, but that overloads them with water, leading to stunted growth or severe digestive upset. Very young animals will need to be fed with a bottle, available at any pet supply store. You must take care not to feed too much. Young animals do not have a well-developed food intake control system yet, and it is easy for them to eat too much. This can be a problem of potential obesity later on. In the immediate situation, they could eat too much, get digestive upsets, and then in fact do not or cannot eat, and the situation quickly becomes worse. It is better to limit them according to their size and age. Growth rate will be slow. You can start to replace bottle feeding with trained saucer lapping within one to two weeks. As they grow to a few weeks of age, you can start to offer wet food as noted earlier. This is a little bit younger than we would like, but in this case, you have already saved the pups and with proper introduction of wet food supplemented with the bottle, they will do fine. On average, very young pups only need about 10 milliliters per feeding at first; this is only about a tablespoon. Young animals have a stomach capacity about 50 ml/kg body weight. Daily intake should not be greater than 10 percent of body weight. Also, though it might seem obvious, with the mother not there, the main source of heat to the pups is gone. You should provide heat at approximately 70 to 80°F for puppies four to six weeks of age; 80°F from one to three weeks, and 90°F from birth to about one week. Use a thermometer and an approved heat lamp.

Growing Dogs

Once you have your pups on dry food, you need to continue to feed them properly. Because we assume that we are feeding them a complete and balanced ration, the single biggest goal is to *encourage proper eating behavior.* If we do this, we can minimize the chances of an animal eating improperly and avoid most problems associated with digestive upsets and obesity. Do you feel hungry by mid-day if you skip breakfast? Do you wake up feeling hungry and queasy if you skip supper? Well, so do dogs! Start with small meals three to four times per day, but be sure that each meal only provides one-fourth of the total that the animal requires. This can be a downside to feeding often—some people will provide too much. But it is important for efficient digestion, and for setting up good behaviors, that, initially, meals be provided often. After about two weeks, you can gradually increase meal size and decrease the number of meals to three, and then to two, times per day. You should only allow about twenty minutes for the pup to eat all the food; it will normally eat quickly, and when it leaves the food bowl, it is done. Take the food away. This will prevent the pup from eating when it does not really need the food, and will teach the pup that food will be available only at certain times of the day.

The importance of feeding management at this stage of the life cycle must be emphasized because it is so important to the long-term objective: to have our dogs live long and healthy lives. It is a fact that obesity is the single biggest problem in dogs in the United States. It is also a fact that it is easy to avoid if we just pay a little attention. We want to feed our dog to optimize, not maximize, growth and development. The growth curve (Figure 10.2) shows the basic growth curves of average dogs of a few different breeds. This curve, along with the directions on the label of the product you are feeding, is a good starting point on feeding rates and expected growth. *But*, do not forget genetic variation. There will be tremendous variation in the *normal growth rate* of dogs within a breed, and even within a litter. You must observe the individual.

Figure 10.2 shows typical growth curves of dogs. Note that not only is the mature weight different, the rates of growth are different at each month. Larger animals tend to grow very fast early on, and then slow. Smaller breeds tend to grow slowly and mature early. So, a Great Dane at 12 months, though much larger, is still a pup, while that peke is already near its final size and may already be physiologically mature.

Introduction of a regular exercise regimen will help to control body condition and health. Regular exercise puts the proper amount of tension on the

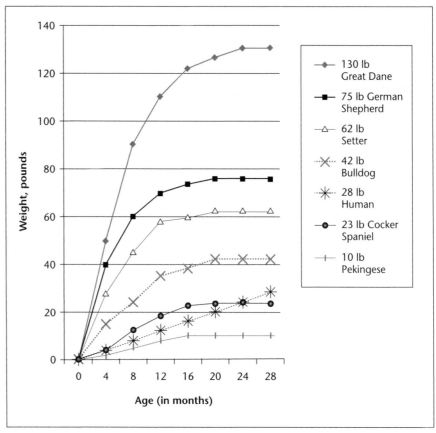

Figure 10.2 Growth curves of dogs.

muscles which, in turn, helps bones to grow strong without excessive growth that can lead to joint problems. Exercise also develops a strong, healthy, and flexible cardiovascular system. Giving the proper amount of food and occasional weighing of the animal can prevent a large part of potential problems. But it is not just the weight that is important, body composition is important as well. We refer to the ***body condition*** of an animal as a practical reference to monitor body composition, primarily of fat (Figure 10.3). An animal with a *high* body condition has too much fat; an animal with a *low* body condition is too thin.

Protein and Energy

From the kind of everyday observation discussed in Chapter 1, plus a little learning at home and school, everybody *knows* that a growing animal needs a lot of protein and energy. But what is "a lot?" What can we learn more specifically here? The animals have to build muscle, and that takes amino acids from dietary protein. The rapid rate of growth requires significant energy from fat and carbohydrate.

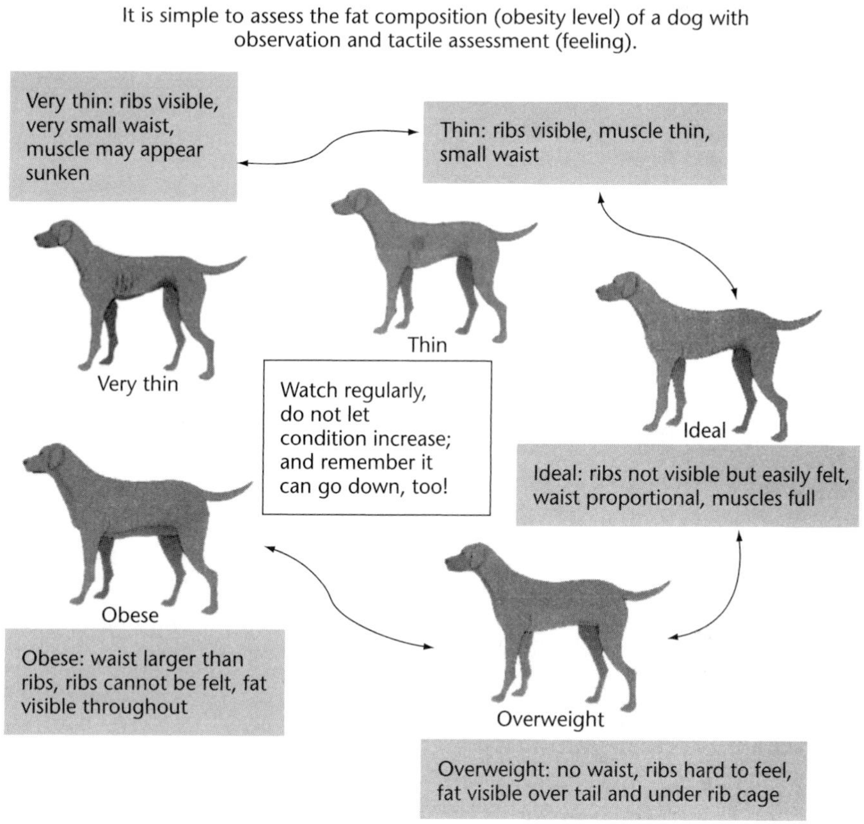

It is simple to assess the fat composition (obesity level) of a dog with observation and tactile assessment (feeling).

Very thin: ribs visible, very small waist, muscle may appear sunken

Thin: ribs visible, muscle thin, small waist

Very thin

Thin

Watch regularly, do not let condition increase; and remember it can go down, too!

Ideal

Ideal: ribs not visible but easily felt, waist proportional, muscles full

Obese

Obese: waist larger than ribs, ribs cannot be felt, fat visible throughout

Overweight

Overweight: no waist, ribs hard to feel, fat visible over tail and under rib cage

Figure 10.3   Body condition in dogs.
Source: Redrawn from Case, *The Dog: Its Behavior, Nutrition and Health,* page 332. © 1999. Blackwell Publishing Professional.

Growing too fast can put too much pressure on the growth plates in the bone, leading to minor or major joint inflammation and weakness. Too much fat (obesity), in addition to being undesirable in itself, increases the risk of almost all diseases of the skin, locomotion, respiratory, circulatory and nervous systems, and kidney, as well as increases the risk of some cancers (as does, by the way, severe underweight).

> The objective for growth is not to make it fast, but to make it right: The animal needs the *right amount and proportions* of energy and protein.

Table 10.2 lists the average requirements for the major nutrients in different stages of the life cycle. More detail is provided in Tables 10.3, 10.4 and 10.5. You may also go to other books, or to the main source for most of this information, the National Research Council's *Nutrient Requirements of Dogs* (2004), and look up such numbers. *(The student here may want to review Chapter 2 on the*

Table 10.2	Feeding Rates and Requirements		
BW of puppy, grams	150	600	3000
Milliliter milk/d	45	150	750
Energy, kcal/d	40	130	600
Protein, g/d	2.3	7.5	37
Fat, g/d	2.3	7.5	37
Lactose, g/d	2.0	6.8	33
Ca, mg/d	81	270	1330
P, mg/d	68	225	1110

Assuming the normal concentration of bitch's milk as in Table 10.1, feeding about 250 to 300 ml of milk per kg BW.

Life Cycle and Chapter 7 on Proteins and Energy.) As in Table 10.2, requirements are given in percentages of the diet. The problem here is that there never has been an animal that needed a "percentage" of anything. They need an actual amount, such as 10 grams of protein, or 1,000 kcal/d of energy. So, why do we continue to give requirements as percentages? Well, one reason is tradition—we have done it that way for a long time. Another reason is quite valid: We cannot possibly feed each animal a set amount of protein, energy, calcium, vitamin A, and so on. If we did, we would need to have all of the purified nutrients on hand and mix the diet for each individual animal. Any potential nutritional benefit from that approach is far overcome by the time and cost involved in purifying and remixing rations. But, when we formulate a proper ration and feed it in the proper amount for that animal, we actually do come very close. Table 10.2 gives an example of how two different rations, one growth and one maintenance, can be fed to dogs of varying weights to supply the proper amount of nutrients for each animal.

A label may show that a feed contains 25 percent crude protein, and that for an animal of 44 lbs (22 kg) you should feed 4 cups, or 360 grams. Doing the math, 360 grams $\times$ 0.25 = 90 grams of protein. Similarly, a chihuahua at 10 lbs (4.5 kg) only needs 82 grams of food (just under a cup); 82 $\times$ 0.25 = 20 grams of protein. In this case, the protein concentration of the diet is not as important as the amount fed. By formulating a ration for the "average" dog and feeding the amount that our specific dog needs, we can come quite close to the exact amount needed. In Table 10.3 are listed ranges of requirements for foods and in Table 10.4 are requirements for the intake of the animal. We use the information in Table 10.4 to determine that in Table 10.3. In Table 10.5 are specific guidelines for feeding amounts in grams, ounces and cups of maintenance type diets. It is critical to get this right for each animal to maintain long-term body condition properly.

Table 10.3	Nutrient Makeup of Foods Recommended During the Life Cycle					
				C18:1, %		
	Energy, Mc ME/kg	Protein, %	Fat, %	linoleic acid	Calcium, %	Phosphorous, %
Early Growth (2 to 6 months)	3.8	22	18–22.5	1.3	1–1.2	0.7 to 0.9
Late Growth (6 to 12 months)	3.7	20	10–14	1.3	same	same
Maintenance	3.5	18	8–10	1.1	0.3	0.3
Late Pregnancy	3.7	20	20–22	1.3	0.8	0.5
Lactation	3.8	22	20–22	1.3	0.8	0.5
Performance	3.8	22	15+	1.1	0.6	0.5
Aging	3.8	20+	13–18	1.1	0.5	0.4

Recommended (not minimum) requirements for nutrient content of foods, % of diet.

Source: Adapted from NRC, 2004, Nutrient Requirements of Dogs and Cats. *These are average figures to demonstrate the requirements and are not meant as absolute values. These will vary with activity, age, and weight of animal. Assumes weights of 5.5, 15, and 22 kg for puppies, adults, and bitches.*

The same principles apply to feeding energy. First, remember that we actually feed fat, carbohydrates, and protein that contain energy. Also, remember that the "ancient dogs" of evolutionary times did not eat much corn, soybeans, or wheat. They ate other animals and received all their energy from protein and fat. For the modern dog, that type of diet is far too high in protein and fat and for their energy and muscle growth needs (review Chapter 7). Prudent use of fat and carbohydrate and intake control supplies a more moderate and sensible amount of energy for our modern dogs.

Thus, as you read in Chapters 8 and 9, dog foods will contain a mix of muscle foods, animal by-products, corn, soybean, and occasionally other plants to provide that magic "3.5 kcal/g" of food with 20 to 25 percent crude protein. Remember that carbohydrate has only four-ninths of the energy of fat, so it is very easy for a dog to consume too much energy if we do not use carbohydrates from plant materials in the diets. Similarly, "pure meat" contains about 50 to 60 percent protein per dry weight, almost three times what dogs truly need.

Calcium and Phosphorous

Calcium truly is an underappreciated mineral, even though almost everybody knows it is important for bones. However, bone is not only calcium, and calcium use for bones is just one aspect of its important role in the body (review Chapter 5). It is also important to match the amount of other minerals to that of calcium. The calcium:phosphorous ratio, as well as the amount of magnesium, is important to the optimal use of calcium by the body. Bone is

Table 10.4		Average Intakes of Major Nutrients for Dogs										
BW kg	BW lbs	Life Stage	Energy kcal ME/d	Protein g/d	C18:1 g/d	DFI, G Growth Food g/d	DFI, M Maint Food g/d	DFI, G Growth Food Cups/d	DFI, M Maint Food Cups/d	Ca mg/d	P mg/d	Protein % of growth diet
5	11	Early Growth	1000	34	2.7	273	287	3.0	3.2	164	120	12
10	22	Late Growth	1125	22	3.1	306	321	3.4	3.6	184	135	7
20	44	Maintenance	1250	14	3.4	340	357	3.8	4.0	204	150	4
25	55	Pregnancy	2200	64	6.1	609	639	6.8	7.1	366	268	10
20	44	Lactation	4250	117	11.6	1160	1216	12.9	13.5	696	510	10
20	44	Performance	2350	47	6.4	644	676	7.2	7.5	387	283	7
20	44	Aging	1130	14	3.1	309	324	3.4	3.6	186	136	5

Source: Adapted from NRC, 2004, Nutrient Requirements of Dogs and Cats. Values should be considered average for the weight and will vary with activity level. They were calculated using the estimated requirement per unit metabolic body size and have been measured in many trials. However, they will vary with the energy content of the food. In practice these values can range from 30% less to 30% more. DFI = daily feed intake; Maint = maintenance food, G = Growth-type food.

Table 10.5	Feeding Rates of a Maintenance Ration to Adult Dogs of Various Sizes

Assumes ration characteristics in Table 10.4: 3.5 Mc ME/ kg food; 90 grams/cup. This is an example of a low-fat maintenance diet. For many commercial products higher in fat, less food would be fed. Note that larger dogs need slightly less food per unit BW as their metabolic rate is slower.

Weight, kg	Weight, lb	Grams Dry Food	Cups Dry Food
10	22	180	2
20	44	360	4
30	66	513	5.7
40	88	660	7.4
60	132	970	10.8

DFI, G is daily feed intake of a growth diet containing 3.67 Mc ME/kg.

DFI, M is daily feed intake of a maintenance diet containing 3.5 Mc ME/kg.

Energy Requirement assumes 1.5, 8, 4, 12, 12, 5 and 1.5 g metabolizable protein per kg MBW (BW0.67) for maintenance, early growth, late growth, pregnancy, lactation, performance, and aging.

Protein requirement assumes 300, 200, 132, 200, 450, 250, 120 Mcals ME/kg MBW for maintenance, early growth, late growth, pregnancy, lactation, performance, and aging. Late pregnancy may equal lactation in energy requirements.

Source: Compiled from NRC, 2004, Nutrient Requirements of Dogs and Cats.

Meat versus Dog's Requirements

If we feed pure meat to meet the dog's requirement for energy, it will receive twice as much protein, three times as much iron, and only 5 percent of the calcium it needs. In addition, this diet will be short in many other vitamins and minerals, and contain too much energy because of the fat. Also, if you have ever cleaned up the waste from an animal on an all-meat diet, you know it is not quite as pleasant as when we feed them a more balanced ration. We can talk all we want about "dogs evolving on meat," and "dogs need all meat," but it simply is not true. Decades of healthy, long-lived dogs consuming balanced rations from many different ingredients provide good proof.

MEAT! It's (not) what's for dinner!

	Needed for 20-lb dog	In raw beef	
	Maintenance	Amount	% Req
calories	700	700	100
protein, g	15–20	90	450–600
Ca, mg	1000	40	4
P, mg	880	700	80
Ca:P ratio	1 to 2:1	1:17	
Na, mg	530	230	40
K, mg	710	12150	175
Fe, mg	11	10	90
Cu, mg	1.5	0.3	20
Mg, mg	55	77	140
I, mg	0.3	0.03	10

made of about 1.2 parts calcium to 1 part phosphorous to 0.05 parts magnesium. Even though the amount of magnesium seems small, it is important.

If we do not have the ratio of calcium, phosphorus, and magnesium in the diet close to this ratio, bone growth is no longer optimal. It is worse if the ratio is lower (Ca:P < 1). The body simply cannot find enough calcium to make the bone, and bone cannot be made in any other proportions. It is not quite so bad if there is some extra calcium, as the body has an extensive system (see vitamin D discussion in Chapter 6) to regulate calcium absorption. At calcium:phosphorous ratios of around 2:1 or 4:1, bone will grow normally. However, this is a waste of a precious resource (calcium), and can lead to problems in old age. At ratios over about 7:1, especially if phosphorus is low in the ratio, bone formation is abnormal, and calcium excess is deposited in the tissues such as arterial walls, heart, and muscles. This leads to hypertension, loss of heart function, and diminished life span. So, again, it is important to "let mother nature take her course" and feed the animal the proper balance.

Mature Dogs: Goals and Objectives

When we go into adulthood, by definition, the composition of the body will not significantly change—we have already defined this situation as *maintenance*. We are maintaining an ideal body composition, with adjustments for normal activity and environmental situations. The basic needs are for normal organ and muscle regeneration and cell and organ functions. Growth is over, so the needs of the animal decrease dramatically. If we can keep intake to the proper level, and keep energy use (by exercise) at a reasonable level, we can prevent most of the nutritional obesity in pet populations. So why do mistakes still occur? Well, if you are wrong by 1 percent each day, and continue to make the same error in the same direction, it will not take long to be wrong by 100 percent—small errors are thus magnified over time.

Obesity

On average, animals will consume food in proportion to their energy need. If we are feeding an animal the proper diet, most animals will consume the right amount of food to maintain themselves for life. However, biology has few true constants, and there will always be exceptions. Not all animals can control their intake, and thus their weight, as well as others. Some reasons for the variation are *genetic*, as there is a wide variation in the ability of animals to control their food intake. For example, are you "always hungry," even if you eat enough; or, do you seldom feel hungry, even if you do not eat very much? This is genetic variation. It is not "willpower," although people may help to overcome their genetics through dietary discipline—it doesn't change the physiology, but is the proper *environmental* response to the genetic variation. Some animals, as well as some humans, respond differently to the normal hunger signals, and all the willpower in the world cannot make some people or dogs eat less or eat more. However, as responsible pet owners, we can help them along.

> Animals on average eat to meet their energy requirements. But, no one animal is truly statistically average! Diet, genetics, and environment can combine to overcome the normal regulatory systems leading to excess intake, decreased energy expenditure, and eventually obesity.

Figure 10.4 shows the development of obesity. As explained, many genetic and environmental factors can lead to obesity. The simplest two ways to control the situation are to limit intake and manage exercise. Note that only 100 kc/d extra is 1 kg in two months, 6 kg in one year. That is a lot of dog. One hundred kilocalories is about 1/3 to 1/2 of a cup of normal dog food, or 1/2 an ounce of fat or oil, or even less than an ounce of trimmed meat. Twenty-five grams or about 1 ounce of pure sugar is 100 kcal. Therefore, it does not take much, and again we see how small feeding mistakes can turn into major problems over time.

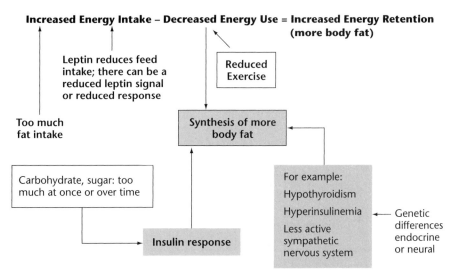

Figure 10.4 Factors of obesity.

All metabolic functions are *heritable* (they can be inherited from the parents), and with continued improvement in the detail of our knowledge of the genome, we will be able to understand this more. Some animals will be hypothyroid, and will make more fat from the same amount of food. Some animals may not make the right amount of leptin (that controls food intake), or the feeding centers in the brain will not respond to it properly, and feed intake will be greater than it really needs to be. However this is extremely difficult to identify for any one individual animal. Thus, this part of the scientific process is extremely frustrating: What we can observe for large groups of animals may be extremely difficult to measure for individuals. Additionally, change of even less than 1 percent can lead to excessive accumulations of fat over many weeks.

The other reason for variation, if not genetic, is *environmental*. Included in the environmental aspects are the type and amounts of food fed, exercise, and temperature.

Energy In − Energy Out = Energy Balance. If Energy In is greater than Energy Out, then the "balance" ends up as fat in the adipose tissue.

The environmental aspect of obesity usually consists of us feeding too much of regular food, or too much of human food high in sugar, fat, and thus energy. We have known for many decades that the normal regulatory processes can be upset and overwhelmed when animals are supplied with a diet high in fat and sugar. Additionally, feeding an animal a small amount more than it needs every day can add up over time, as noted.

It is a fact that obesity that is not caused by specific genetic imbalances can be readily prevented by feeding a properly balanced ration to requirements. Avoid excessive (> 10 percent of ration) human food or added fat and provide regular exercise. If you have an animal that eats the right amount, or less, of a proper dog food, and gets moderate exercise (15 minutes or more every day), yet still gains fat, then a veterinarian can check for some hormonal problems (hypothyroidism, diabetes) that might be at the root of the problem, but remember that these are a small percentage of the total cases.

Now, let us switch gears a minute and talk about the management of losing weight. Quite often, a person will suddenly notice (observe) that their animal is now overweight. This is not a suprise, as it is often truly difficult to notice very small changes over a short period of time. What we see gradually becomes normal, and thus our *perception* of the animal changes to adjust what we see. But now, we have finally noticed that the dog that used to be a nice trim 40 lbs is now a little rounded at 50 lbs. The first response is often "Let's start a diet and

---

Fact Box | Recommendations for Safe Rates of Weight Loss

These lists show alternative ways to calculate safe weight-loss regimens. The first, "Caloric Weight Reduction," shows that on a 400-kcal per day deficit (feeding about 60 percent of maintenance requirements), one can calculate 150 days to lose the weight, no matter what the original or target size of the dog (or cat, or horse, etc.). The second list, "Weight Reduction Management," uses an average of 1.5 percent of body weight per week weight loss, known to be able to provide weight loss without secondary deficiencies or health problems, still showing that slow, gradual weight loss is best.

Caloric Weight Reduction
➤ Moderation is the key: 2 to 3 percent of BW week 1, 1 to 2 percent per week thereafter.
➤ The weight was not gained in a month; it cannot be lost in a month.
➤ You must still feed the nervous system, organs, and muscles vitamins, protein, energy, and minerals.

Weight Reduction Management
➤ Assume 30-kg dog, 33 percent overweight = 40 kg.
➤ 1.5 percent per week is 0.6 kg/week (1.3 lbs).
➤ It will take 16.7 weeks (10 kg/0.6 kg/week) to lose the weight.

Weight Reduction, More Specific
➤ If need to lose 10 kg, at 6 Mcal/Kg of BW, that is 60 Mcal, or 60,000 kcal.
➤ Assume that 1,000 kcal/d is normal requirement.
➤ Feed for weight reduction at 60 percent of requirement.
➤ That is 400 kcal/d less.
➤ 60,000 kcal/400 kcal/d deficit = 150 days to lose the weight (21 weeks).

lose this weight fast!" Well, the sad (as well as physical, chemical, and biological) fact is it took time to add the weight and it will take time to lose it. The energy equation given earlier applies the same in either direction.

We want to avoid "crash" changes; we should decrease feed intake gradually, observe the animal, and feed to eventually reach a known target weight—not for a quick and severe weight loss. There are two major reasons to avoid extreme weight-loss diets. First, if we reduce food intake so low as to lose a lot of weight (more than 3 to 4 percent of body weight per week), we run a great risk of amino acid, vitamin, and mineral deficiencies. And, partly because of these deficiencies and partly because of hormonal survival adaptations, the body will often rebound and, if food is then supplied, the animal will eat too much again and regain much of the weight. This then goes right back to the original dangerous situation. If food is still limited, we can end up with a dog that is "always hungry," exhibiting anxiety and undesirable behaviors because we have pushed the system so far out of balance.

Performance Dogs: Goals and Objectives

Dogs, more than many species we have as companions, can be used in different situations that require some various feeding strategies. These can include ambient temperature extremes; racing dogs; hunting and field exercises; police and sentry duty; guide dogs; and the show circuit. These situations are a stress on the dog such that nutritional needs and feeding behavior are changed. Physically and psychologically stressed dogs can increase or decrease their food intake. Obviously, if activity is seriously increased, then energy needs are increased. Long-term activity may increase the need for protein, vitamins, and minerals as well.

Our feeding-management goals remain the same as in normal situations: we feed to maintain body weight and composition. We can do this with a dietary change, switching to a "high-performance" diet. We might judiciously add fat to the diet if the dog is exercising so much that normal food intake cannot keep the weight on. We will provide a highly digestible, high-energy diet with quality protein. There are "stress-type" or "high-performance" diets on the market. Read the label, looking for 3.75 to 4.0 kc/g ME; > 82 percent digestibility; > 25 percent CP; > 23 percent fat, and < 4 percent fiber (see Table 10.3). These will handle all but the most extreme situations. It is often wise to offer feed at shorter intervals if you cannot offer it constantly. Providing more meals allows the dog greater time to eat the total amount that is needed. Larger meals often slow the digestive system, limiting the total amount an animal can consume in one day.

When animals are in the cold for long periods of time, or in strenuous exercise over several hours over several days, then even intake of a high-energy ration may not be enough. Then, and only then, can extra fat be supplemented, usually simple corn oil, vegetable oil, or olive oil. Animal fat is also perfectly acceptable and is sometimes easier to handle. You usually need to ask the butcher for fat trim, or trim it yourself. Although there are differences in fatty acid content (different amounts of omega-3, omega-6, and omega-9 fatty acids), for most animals this will not make a major difference as the goal is energy, not specific fatty acids. Plan to supplement regularly, not intermittently. Intermittent consumption of large amounts of fat can cause digestive upset. Mix well with food, usually 1 tablespoon (15 ml) per cup of dry food. Be prepared for refusal in some cases; you may need to change type of fat. Feeding fat is less convenient than a high-energy diet and must be fed regularly. If we feed too much fat, we can decrease intake of the total diet and cause secondary deficiencies of protein, vitamins, or minerals (remember that the average animal eats to meet its energy requirement). We must reserve this situation for that animal that is *already* consuming all the regular ration that it can, so the fat will truly be extra, not a replacement for a portion of the ration.

For racing, endurance, and guard dogs, we still feed to maintain weight and condition. It is the total amount of exercise that is important, not just the type. Carbohydrate-loading to make a lot of glycogen (see Chapter 12) is not possible in the dog; they do not have the metabolic physiology for it. Special tricks are not necessary and can do harm. There is no benefit to feeding during exercise except if exercise is for two to three hours or longer.

Gestating and Lactating Dogs

Gestation

Continued feeding of a well-balanced adult dog food is adequate for the first several weeks of gestation. A maintenance diet could be fed in this situation. In later gestation, increased food intake should be allowed, with the average being about 10 to 15 percent more than normal, and can be 25 percent more in smaller breeds. Feed should gradually be switched to a fully adequate ration (an *all life-stages diet* that meets all NRC requirements for all stages of life) in weeks 5 to 6 of gestation. Body condition should be maintained: Ribs should be easily felt; coat and skin should be full and lustrous without rounding over bones and joints. You should allow sufficient access to food so that if the dog does need more she can consume it. Increase number of meals per day, or allow free-choice intake. As always, switch diet slowly.

> Our goal in pregnancy and lactation is to maintain condition (amount of body fat) of the bitch.

Lactation

Food intake will normally fall somewhat in the last few days before whelping (giving birth in dogs). After whelping, allow ad-libitum intake, but avoid overfeeding and increases in body condition. Increase feed intake in increments over time: first week—feed for 50 percent more than normal maintenance; second week—twice as much as maintenance; third week and later—as much as three times as much. By this time, the bitch will be consuming all she usually can. If she does not need this much, she usually will not eat it. It is difficult to

Feed intake and body weight should not increase excessively in pregnancy, only about 10 to 30 percent (more for small dogs or large litters). Feed intake will increase dramatically in lactation; the bitch should be able to eat all she wants as long as she is not gaining body condition.

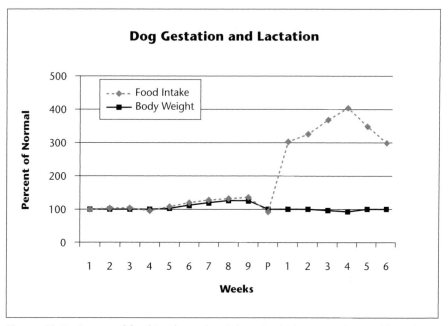

Figure 10.5   Expected feed intake and weight gain during gestation and lactation.

note changes in body condition in a week in most middle- to large-size dogs, but watch for gains in weight. They should be minimal if any.

Figure 10.5 shows intake and growth during pregnancy and lactation. Note several items: The dog does not start to gain weight, and therefore does not increase nutrient requirements, until about week 5 or 6 of gestation. Feeding management should reflect this. Note also that as soon as lactation begins (right after the pups are born) the milk production and the feed requirements increase very quickly, often doubling in two weeks. So, you should start slow and then let her eat all she wants for a few weeks. It is very unlikely that a bitch suckling more than a few pups will gain weight during lactation, and will probably lose body fat even if she is eating quite a bit. Save the weight-reduction diet, if needed, for after weaning.

Offer more meals per day or free-choice. Some dogs can handle once-a-day feeding, but twice a day is not much more work, helps prevent digestive problems, and will allow for better intake in finicky eaters. Free-choice of dry ration is usually best, but watch for overeating. Feed a high-quality ration, similar to a high-performance ration, at or above 80 percent digestibility, greater than 29 percent CP, and less than 17 percent fat. Discourage gorging by taking away food for a few minutes if needed, but over the day, allow full intake. Allow fresh water at all times.

> Lactation places the greatest demand for nutrients on animals, more than any other situation.

Feed as an individual—watch coat, condition, and health of dam. Some weight loss may occur as the uterus involutes, fluid is lost, and if milk production is quite higher (large litters or larger pups—but loss should be moderate). Do *not* try to "thin down" a fat animal during lactation—the puppies will suffer as well. Wait until after weaning. Regular exercise is always recommended throughout gestation and lactation; avoid only strenuous or high-impact exercise in last third of gestation.

Weaning from the Mama's Point of View

Earlier, we discussed the management of weaning for the litter, but what about mom? Weaning is a normal process, we must understand, that can be stressful on the bitch. The lack of suckling and milk removal will cause engorgement and discomfort. Temporarily decreasing food intake can lower milk production faster and help lessen the discomfort and shorten the period of stress. Do not feed at all on the day of weaning, but allow plenty of fresh water. Increase food intake by one-fourth of the normal daily allotment for maintenance for four days, then feed normally under maintenance conditions.

Senior Dogs: Goals and Objectives

So now time has passed and our bouncing baby puppy is now a genteel and noble elder dog. The management goals still apply: We maintain optimal body weight and condition with diet and exercise. But we also want to do all we can to prevent onset or development of clinical diseases of any kind, which are always more stressful on the older system. If problems are present, work with your veterinarian to reduce any existing problems.

> For most dogs in the adulthood and aging stage, too much is much worse than too little.

Most diseases of aging are a combination of genetics and a too-rich food supply. Deficiencies will seldom occur when feeding good-quality, nutritionally balanced rations. Deficiencies will usually occur only when too much human food is fed, which reduces the intake of vitamins and minerals from the dog food. In addition, too much human food increases the likelihood of developing obesity or renal disease.

There is confusion, if not controversy, on when an animal is considered "aged." You certainly know some older friend or relative who, though maybe an octogenarian, is as active and young looking as someone thirty years younger. You may also know some "middle-aged" person who seems older. This is, again, often, due to genetic variation. But dogs do age on a statistical average. The old "seven in dog years" is close, but not close enough. Early on, physiological maturity is faster than seven years to one. A one-year-old dog is comparable in maturity to a human at twelve to fifteen years of age (for example, they are capable of reproducing but it is not advised). After adulthood is reached, the rate of maturity slows, so that at five years, a dog may be similar to a human at thirty-four to thirty-six years; five years later, this equales to fifty-six to sixty-two human years. Giant breeds age more slowly at first, as it takes the larger body longer to mature. But as time passes, they age faster, as their physiology is such that it puts more demand on the system. This is an exception to the rule that larger animals live longer, but biology has many exceptions. Several experts consider a dog to be *senior* beginning at about seven years of age in normal breeds. However, some do not, and several experts are also in the business of selling dog food. You need to judge your dog as an individual. If you are feeding a normal maintenance ration, it may not be so critical to switch to a senior diet. However, if your dog is showing some signs of age, such as lack of enthusiasm for exercise, less activity, or perhaps a decrease in intake, that might indicate that the senior diet can be introduced.

There are several excellent products on the market, but they all follow the same basic nutritional principles described earlier: feed a high-quality protein to reduce stress on the kidneys in making urea; use a well-balanced mix of animal and plant protein, digestibility 80 percent or greater; protein concentration should be greater than 14 percent and less than 21 percent; and fat should be greater than 10 percent, less than 15 percent. Table 10.6 gives examples of several products on the market. Take a look at them and examine the differences in chemical composition and ingredient composition. Following the footnote in the table, think about the relative and relevant differences between the rations. We also feed minimal amounts of sodium and phosphorous. Too much can add to the stress on the kidneys. Sometimes you will find a diet somewhat higher in fiber (5 to 7 percent). The purpose here is to limit energy intake. These diets may be fed to an animal prone to obesity at any life stage, or to an older, less active dog.

> High-fiber diets (five to eight percent, sometimes more) can be helpful in some situations. However, some animals can develop digestive upsets and gas with high-fiber diets. Observe the individual.

As activity level decreases, you should decrease food availability. You can feed once a day for most dogs to limit intake, but feed more often if exercise level warrants it. In older dogs, more frequent meals may be necessary to stimulate intake or "even out" digestive load. Avoid periodontal disease by feeding dry diet, dog biscuits, or routine (weekly) brushing of teeth. As the animal ages, frequent dietary changes may be necessary in order to stimulate smell and taste buds. The senses deteriorate with age and an otherwise perfectly healthy dog may just not be attracted to the food. A little common sense

Table 10.6	Examples of Ingredients and Contents of Selected Commercial Dog Foods

Item		DM	Energy (ME, Kcal/g)	Protein	Fat	Crude Fiber	Ca	P
			Nutrients in percentage of DM unless noted					
Store brand all life stages	(l)	88		21	10	4		
Ground yellow corn, soybean meal, poultry by-product meal, meat and bone meal, corn gluten meal, animal fat, brewers rice								
Store brand adult	(l)	88		26	16	3		
Chicken, ground yellow corn, ground whole wheat, wheat flour, corn gluten meal, poultry by-product meal, tallow, whey								
Store brand reduced calorie	(1c)	88	3.45	16	7 to 12	6		
Ground yellow corn, corn bran, poultry by-product meal, ground brewers rice, ground whole wheat, chicken, beef tallow, corn gluten meal								
Name brand senior (weight maintenance, hi fiber)	(a)	91.5	3.33	25	6	7	0.8	0.6
Ground yellow corn, corn gluten meal, soybean meal, ground wheat, soy hulls, meat and bone meal, tallow								
Name brand puppy chow	(a)	91.5	3.89	29.5	11.5	1.8	1.31	0.94
Ground yellow corn, chicken by-product meal, brewers rice, soybean meal, tallow, pearl barley								
Name brand dog chow	(a)	88.0	3.70	21	10	4.5	1.0	0.8
Ground yellow corn, poultry by-product meal, corn gluten meal, animal fat, brewers rice, soybean meal								

Energy is metabolizable energy in kcal/g DM. It is analyzed unless noted as calculated (c), from modified Atwater values of 3.5, 3.5 and 8.5 for starch, protein and fat.

Sources include product labels (l); also analyzed data from Hand et al., pp 1074-1082, Appendix L (a). This reference contains brand information. We chose not to include them here to avoid distraction.

As a thought exercise, the student should look at ingredient composition as well as chemical composition, and determine different ways to alter chemical composition with ingredients. Why are different ingredients included? Why are many ingredient lists the same, although chemical content may vary? Which is more important (ingredients or chemical content)? Are they both important?

Calculate the energy of the first two diets from the chemical composition. Estimate the Ca and P content by comparing them to the ingredients and analysis in the others.

| Table 10.6 | Examples of Ingredients and Contents of Selected Commercial Dog Foods *continued* | | | | | | | |

Item		DM	Energy (ME, Kcal/g)	Protein	Fat	Crude Fiber	Ca	P
			Nutrients in percentage of DM unless noted					
Name brand dry snack	(l)	88.0	2.9 (c)	15	5.0	3.5	0.8-1.3	0.8
Wheat flour, soybean meal, meat and bone meal, whole milk, wheat germ, animal fat, salt, dried cheese, fish meal, dried fermented corn extractives, malted barley flour								
Name brand wet treat. % af	(l)	18.0	3.98 (c)	8.0	3.0	1.0		
(% DM)		100.0	44.4	16.7	5.6			
Water, poultry, meat by-products, wheat flour, beef, wheat gluten, pearl barley, salt								
Premium brand adult	(a)	92.0	4.22	25.0	15.4	1.7	0.72	0.65
Chicken, corn meal, ground grain sorghum, chicken by-product meal, ground whole grain barley, chicken meal, chicken fat, dried beet pulp, chicken flavor, dried egg product								
Premium brand 'lite'	(l)	91.0	3.05	20	10 to 12	5		
Corn meal, chicken, sorghum, chicken by-product meal, fish meal, ground whole grain barley, ground whole wheat, dried beet pulp, chicken flavor, dried egg product								
Premium brand puppy chow								
Large Breed	(a)	91.4	4.53	29.6	17.2	2.1	0.98	0.82
Medium Breed	(a)	93.1	4.60	32.8	20.5	2.7	1.28	1.00
Premium brand senior dry	(l)	90		20	14	2.5		
Corn, corn gluten meal, chicken by-product, poultry by-product, animal fat								
Premium brand senior wet	(l)	18	(as fed basis)	9	3 to 6	1.5	2.5	
Meat broth, meat by-products, chicken, turkey, wheat gluten, lamb, corn starch modified, soy flour, flavorings, whole rice		100	(DM basis)	50.0	16.7 to 33	8.3	13.9	

needs to be applied: "Variety is the spice of life," and if your dog has lived this long, it can probably stand a little variety!

Strategies of encouraging eating include: (1) mix canned and dry foods; (2) switch different flavors of canned or dry food; (3) warm food slightly in microwave or hot water (to no more than 40° C or 100–110° F (warm, not hot to touch), otherwise it may burn the dog; (4) feed more often; (5) substitute more "treats," but keep total diet balanced; and, of course, (6) do all of the above.

Nutritionally Related Problems of Dogs

A serious detailed description of all the nutritionally related problems of dogs is beyond this introductory text, but I think it important that you at least can tell the difference between something that is just a temporary variation, a genetic condition, or a true situation that is either caused by some nutritional situation or can be improved with nutritional management. This might seem like a minor point, but many genetic diseases can be partially alleviated by nutritional management, though they remain a *genetic* disease: The nutritional management did not cause them.

> *Hip dysplasia and bone problems:*
>
> Genetic predisposition
> Nutrition can alleviate or worsen problems
> Know your animal's genetic background
> Reduce weight gain of large breeds
> Do not supplement calcium
> Watch weight gain and condition closely

Genetic hip dysplasia is an example. Not all dogs will have bone or joint problems, but some will. The difference is genetic: Some individuals express genes differently, leading to an increase in certain problems, some of which are made better or worse through nutritional management. Larger breeds definitely have a greater incidence of bone problems. This is probably due to the rapid weight gain putting stress on the bone joints. We cover this in depth in Chapter 12 on horses, in which bone problems can be more common and worse. The basics are the same. Too-rapid growth caused by genetics and worsened by overfeeding leads to stress, inflammation, and eventually damage of the joints.

Renal Damage, Aging, and Dietary Protein

Many of us were brought up to believe that feeding excess protein is too hard on the kidneys and should be avoided. We also know that normal kidney function decreases with age. We also know that the kidney needs to filter out excess minerals that are consumed. However, more recent research has enlightened us on some of these areas. There is no doubt that consumption of more sodium, potassium, phosphorous, and magnesium causes greater blood flow through the kidneys to remove these ions from the blood and maintain the proper balance. This will, over time, increase the rate of normal loss of kidney function, and thus feeding a minimal requirement of these minerals is to be desired.

However, it may be a different story concerning the amount of protein. Excess ammonia produced from the metabolism of unneeded amino acids must be removed through urea in the urine. It was always thought, and some early research suggested, that the increased protein intake would also damage the kidney over time. However, more recent studies have not shown that reducing protein from "hi-pro" type diet levels to maintenance levels changes kidney function. So now, in the perspective of applying the scientific method, this remains an open question. Further research with more detail over longer periods of time and more levels of protein in the diet will give us a better picture. For now, in my scientific opinion, it is prudent to not feed any more protein than is truly needed, and for most older animals, we know that maintenance or senior-type diets provide a proper amount of protein. There is no good reason to feed more; it is more expensive and it wastes another precious resource.

Figure 10.6 shows the effects of nutrient use on kidney function. Most minerals and excess nitrogen must be excreted in the urine. Excess amounts of minerals cause an increased blood flow through the kidney that over time increases the normal rate of loss of kidney function. As mentioned, it is now more controversial whether excess nitrogen intake actually causes kidney damage. Yet, there is still no good reason to feed excess protein.

Gastrointestinal and Liver Diseases

Dogs can develop several gastrointestinal (GI) and liver diseases, some by genetics, others by infection, some by diet, and all of these factors interact to lead to greater incidence of diseases. Symptoms may include vomiting, regurgitation, diarrhea, constipation, bloody stools or urine, discolored (very dark or very light) stools, or persistent mucous in stool.

Vomiting is forceful emission of stomach and often proximal intestinal contents by neural control of intestinal and stomach musculature. *Regurgitation* is milder reflux of proximal stomach or esophageal contents, often due to too-rapid eating or dysfunction of the esophageal sphincter. Vomiting can be temporary or prolonged. It can be physiological, or can be caused by consumption of a poison. Sudden and persistent vomiting should mean a quick trip to the vet. Regurgitation can often simply be caused by a little irritation— a morning's taste of some sweet grass gets caught in the esophagus and causes regurgitation. Too-rapid eating can do the same. This is usually not a situation to panic over, just observe.

The kidney's ability to filter wastes from blood decreases with age; about a 75 percent loss is normal by old age (eight to fourteen years). Moderation in intake of protein and key minerals will help to keep the functional loss normal.

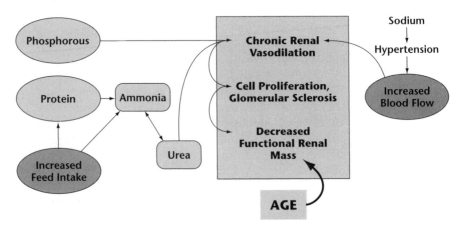

Figure 10.6 Aging and loss of renal function.

Vomiting, regurgitation, diarrhea:

1. Basic management rules apply.
2. If symptoms of persist or recur, see a veterinarian.
3. Keep things simple—rule out simple things first:
   a. Rapid change in diet?
   b. Dog got into something it should not—plant, chemical, food, garbage?
   c. History of sensitive GI tract?
   d. Temperature, physical, or social stress?

Steatorrhea is the presence of elevated levels of fat in stool. They can be light in color, usually yellow or white. This can be due to feeding too much, or indigestible fat (beef tallow). It can also be due to feeding too much milk, or garbage consumption. However, it may also be due to salmon poisoning or other infection (raw fish consumption). Sometimes, excess fat in the stool can be due to liver, pancreatic, or intestinal dysfunction—these are usually persistent and resistant to changes in diet. Again, see a vet if problems persist or are accompanied by other changes in behavior.

Food Allergies and Management

Few areas of nutrition are more misunderstood and more frustrating than food allergies, or what are more commonly referred to nowadays as *food sensitivities.* It is a fact that these sensitivities exist.

A *food sensitivity* is a definable and reversible response to some component of the diet.

It is also a fact that according to the best case studies available, these are a fraction (less than 5 percent) of problems presented at veterinary clinics. It is also a fact that when they do truly exist, they can be very painful for the animal and extremely frustrating for the owner. Signs of such sensitivities can be expressed in many ways such as diarrhea, vomiting, and pruritus (watering) of the face, eyelids, and lips (no lesions, puffiness, excess tearing). Skin signs may include pruritus, or may look like mange or eczema (dry skin) in advanced cases. Animals may scratch or bite at sensitive areas, and even in the absence of such behavior, skin under the fur may appear red or tender.

Available information would indicate that about 10 to 15 percent of all cases presented at veterinary clinics are allergies of some kind. The majority of these, 80 to 90 percent, are usually diagnosed as **seasonal allergies,** caused by contact to various materials or pests, or *inhalation allergies* such as pollens and dust. An ***allergy*** is a response by an animal to some chemical. Responsiveness varies widely among animals, most likely due to different genotypes. The foods that are reported as causing allergies in most cases are beef, corn, wheat, chicken, oats, and soybeans. Yes, basically any food component could cause an allergy in a given animal. So in one sense, we must have the combination of the wrong food for the wrong animal to have a sensitivity. So remember that these are only a small percentage, 1 to 2 percent of all cases at clinics, but when they exist, they can be a real hardship on the dog and owner.

Figure 10.7 shows the role of essential fatty acids in immune response. A truly emerging area of "immunonutrition," we now understand that the balance of certain fatty acids may help some animals in their response to the environment. In general, the omega-6 fatty acids tend to increase inflammation. Some products are now on the market with a better balance of these fatty acids. However, the student must appreciate that this is a very new area, and the fatty acids give rise to many different important regulatory molecules. We are still far from understanding it all.

Figure 10.8 shows the procedures for diagnosis and treatment of food sensitivities. Upon observation and veterinary diagnosis of a sensitivity to a food and not an environmental allergen, we can alter the ration to a product that does not contain the ingredients we were feeding (chicken to lamb, for example). Observe signs for several weeks; if they diminish, a food sensitivity may have existed. If you really want to know and, frankly, sometimes it is not worth it, you may reintroduce the old food. If symptoms recur, it was probably a food allergy. If not, it was probably not a food allergy.

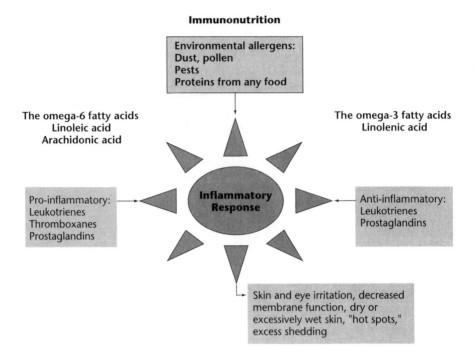

Figure 10.7 Role of essential fatty acids in immune responses.

Some animals may have mild or severe allergies to several foods of all classes. After diagnosis by a veterinarian (ruling out seasonal or parasitic allergies), the major way to treat is to remove feedstuffs from the diet and watch the signs. There are a number of dietary products on the market that tend to be lower in allergens. These are usually some mix of highly processed meat (lamb is used often as it is not usually used in dog/cat rations) and rice. Do not self-diagnose, as other nutritional problems may result.

We now know quite a bit more about the physiology of allergic (inflammatory) reactions. Essential fatty acids give rise to many different regulatory molecules we know as prostaglandins, cytokines, and thromboxanes. Some of these compounds will increase the inflammatory response to an allergen and some will decrease it. In addition, the ratio of the amount of fatty acids that have a double bond at the omega-6 position to those that have a double bond at the omega-3 position is now thought to be involved in this. Fatty acids that have a double bond in the omega-6 position tend to give rise to metabolites that cause a stronger inflammatory response *(pro-inflammatory)*, whereas those with the double bond in the omega-3 position tend to inhibit the inflammatory response *(anti-inflammatory)*.

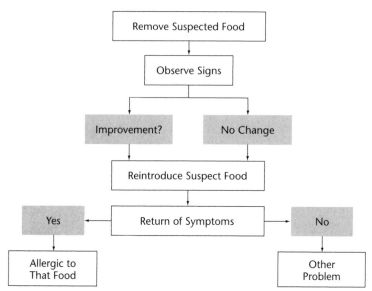

Figure 10.8   Diagnosis and treatment of food sensitivities.

Fatty acids with the first double bond from the omega carbon in the six position (arachidonic acid) tend to give rise to a greater inflammatory response; those with omega-3 double bonds tend to give rise to less inflammatory chemicals. A ratio of 5:1 to 10:1 of omega-6 to omega-3 precursors (for example: the diet would contain a ratio of linoleic acid and arachidonic acid to alpha-linolenic acid of 5:1 to 10:1) is thought to be optimal to minimize inflammation.

The omega-6 derivatives arise from arachidonic acid, and the omega-3 response is initiated with alpha-linolenic acid. Omega-9 molecules do not seem to be a major player here. There are veterinary prescription diets and a few over-the-counter diets that can be used to help reduce inflammation because they have the proper ratio of omega-6 to omega-3 fatty acids (a smaller ratio of omega-6 to omega-3, Figure 10.7). Research studies have demonstrated that a ratio of about 5:1 up to 10:1 of omega-6 to omega-3 will help to reduce inflammatory responses. In some classes more detail may be explored on this new and growing area of nutrition.

Supplementation, Special Situations, and Myths

Many "old tales" still exist in dog nutrition. A large part of this is due to the natural tendency for people to perceive the result that they desired. Also, personal life experience in one situation is often thought to apply to all other situations as well. Sometimes these "tales" are right on, but quite often they are not quite correct, apply to only a very specific situation, or in worse cases, are simply wrong and sometimes dangerous. These "tales" may be responsible for the resurgent growth in homemade diets, all-meat diets, and "natural" diets, all of which are based on the misperception that all commercial dog foods are overprocessed, contaminated, and downright despicable. Again, there is *some* truth in many ideas, but the *real* truth takes a little more work.

Danger Foods for Dogs?

Some things can be dangerous. Not all of nature is friendly. Each species of animal has some peculiarities in what affects them. For dogs, a short list includes:

Chocolate, coffee, cocoa, dry tea contain theobromine, a neurotransmitter, which leads to seizures, coma, and can be fatal. Vomiting and diarrhea may result in lower doses. Should never be fed (even if you got away with it before).

Onions: causes hemolytic anemia, liver damage.

Cat food: not toxic, but contains too much protein.

Mushrooms: contain several compounds that can cause diarrhea, vomiting, convulsions, coma, and death.

Grapes, raisins: contain polyphenolic compounds that can cause diarrhea, vomiting, and liver and kidney damage.

Plants: dozens of plants contain hundreds of compounds that in large enough doses can hurt animals. Some examples include stems and leaves of apple, almond, wild cherry, oak, philodendron, ivy, poinsettia, and aloe vera.

Supplementation with Vitamins, Minerals, or Special Supplements

If the animal is consuming a well-known store, brand, or specialty food appropriate for its life stage, no supplementation is necessary. In some cases it will do no harm except to the checking account of the owner. In some cases, especially for calcium and phosphorous, it can do harm. In the case of sodium, phosphorous, and potassium, excess amounts can increase the rate of loss of kidney function.

Supplementation with Human Food

Dogs are not humans and most human food is simply not appropriate for the digestive system of the dog or is inadequate for their metabolic rates. Meat is often given, but as noted earlier, as a major source of food it is much too high in protein and low in calcium. An occasional scrap of trim is fine and will not hurt, but one must be judicious. Large amounts (1 cup or more) at a time will more often cause digestive upset than help the dog.

Fruits and Vegetables

Even if the dogs love them, fruits and vegetable simply are mainly water, can be too high in sugar, and the fiber will be useless to the dog. Breads, pasta, and other grain foods are too concentrated with starch. Candy, donuts, and pastries are never a good idea as they are way too high in sugar and fat, and can encourage cravings and bad behavior.

Can you make a nutritionally balanced food for dogs from fresh meats, organs, and vegetables? Yes. Is it easy to miss some minerals, or vitamins, or to overfeed protein or fat? Yes. I have included a couple of good books on homemade diets for dogs in the "Further Reading" section for those truly interested in that subject. But I would certainly work with a competent veterinary nutritionist and pay close attention to detail if desiring to implement a homemade diet.

Hip Dysplasia

A common myth is that improper feeding can cause hip dysplasia. This is simply not true; hip dysplasia is a genetic disease. Some breeds, especially of larger dogs, have a genetic tendency to grow too fast, and this puts stress on the hip joints, which are genetically weak. We can help to minimize some of the problem with limited feeding of high-quality protein and limited energy to slow the growth rate. But overfeeding, even leading to increased growth rate, does not *cause* hip dysplasia. Overfeeding can cause inflammation of some joints and increase the effects of genetic problems. Feeding the proper commercial diets for the animal's life stage at the proper amount will help to avoid bone problems.

There is some evidence that increased amounts of vitamin C (ascorbic acid) and glucosamine may reduce joint inflammation and bone problems in susceptible dogs. These are two ideas based on sound science, as we know that vitamin C is required for normal collagen, connective tissue and, eventually, bone formation. Glucosamine is a component of normal connective tissue. There may certainly be animals who are genetically predisposed to make less of these compounds or use them less efficiently and may benefit from supplementation. The idea, as always, is to watch the animal, keep growth rate moderate, keep body condition average to slightly below average, have a regular, moderate exercise program, don't supplement for the suke of supplementary. There is some information that excess vitamin C may cause health problem in dogs.

Words to Know

active immunity	*food sensitivities*	*seasonal allergies*
allergy	*heritable*	*steatorrhea*
body condition	*passive immunity*	*vomiting*
colostrum	*regurgitation*	

Study Questions

1. What are the general nutrient requirements for protein, energy, calcium, and phosphorous in each stage of the life cycle?
2. Describe the practical feeding-management goal for each phase of the dog life cycle: suckling, weaning, growth, maintenance (adulthood), gestation, lactation, and aging.
3. What special situations may exist that may alter food intake and some nutrient requirements? Give some examples.
4. Describe the causes, prevention, and treatment of obesity in dogs.
5. How do different fatty acids affect the potential response to foods or other sensitivities that an animal may exhibit?
6. What is an example of a common misperception, or myth, in feeding dogs, and what should a dog owner do about it?

Further Reading

The following are primarily texts and books relating to nutrition of the dog, encompassing a wide variety of topics. Within each of them are hundreds of specific scientific references to back up the information in them, and the information in this chapter. Those interested in more specific information can find what they are looking for in this reading list. Students and teachers looking for texts for higher-level classes should certainly start looking with these texts. Also, a recent search of "dog and food" on the Internet listed about 4 million sites. Read carefully. At the "PubMed" site, a recent search for "nutrition and dogs" listed more than 40,000 articles in peer-reviewed scientific journals on a topic in that broad area. These scientific search engines usually allow you to read the abstract. Students in most universities may access the Veterinary Medicine or Science Library for most of these journals.

Allen, T. A., D. J. Polzin, L. G. Adams. 2000. Renal disease. In M. Hand, C. D. Thatcher, R. L. Remillard, and P. Roudebush, eds., *Small animal clinical nutrition*, fourth edition (chapter 19). Topeka, KS: Mark Morris Associates.

Bovee, K. C. 1991. Influence of dietary protein on renal function in dogs. *J. Nutrition*, volume 121: S128–S139.

Burkholder, W. J., and P. W. Toll. 2000. Obesity. In M. Hand, C. D. Thatcher, R. L. Remillard, and P. Roudebush, eds., *Small animal clinical nutrition*, fourth edition (chapter 13). Topeka, KS: Mark Morris Associates.

Case, L. P. 1999. *The Dog: Its behavior, nutrition and health*. Illustrated by Kerry Helms. Ames, IA: Blackwell Publishing Professional.

Case, L. P., D. P. Carey, D. A. Hirakawa, and L. Daristotle. 2000. *Canine and feline nutrition*. St. Louis, MO: Mosby, Inc.

Diez, M., P. Nguyen, I. Jeusette, C. Devois, L. Istasse, and V. Biourge. 2002. Weight loss in obese dogs: Evaluation of a high-protein, low-carbohydrate diet. *J. Nutrition*, 132: 1685S–1687S.

Hand, M., C. D. Thatcher, R. L. Remillard, and P. Roudebush, eds. 2000. *Small animal clinical nutrition*, fourth edition (appendix J). Topeka, KS: Mark Morris Associates.

Harper, E. J. 1998. Changing perspectives on aging and energy requirements: Aging, body weight and body composition in humans, dogs and cats. *J. Nutrition*, 128: 26276S–2631S.

Hincliff, K. W., J. Olson, C. Crusberg, J. Kenyon, R. Long, W. Royle, W. Weber, J. Burr. 1993. Serum biochemical changes in dogs competing in long-distance sled race. *Journal of the American Veterinary Medical Association*, 202: 401–405.

Hoenig, M. 2002. Comparative aspects of diabetes mellitus in dogs and cats. *Molecular and Cellular Endocrinology*, 197: 221–229.

Kirk, C. A., J. Debraekeleer, P. Jane Armstrong. 2000. Normal dogs. In M. Hand, C. D. Thatcher, R. L. Remillard, and P. Roudebush, eds., *Small animal clinical nutrition*, fourth edition (chapter 9). Topeka; KS: Mark Morris Associates.

Kelly, N., and J. Wills. 1996. *BSAVA manual of companion animal nutrition and feeding*. UK: British Small Animal Veterinary Association.

Lund, E. M., P. J. Armstrong, C. A. Kirk, L. M. Kolar, and J. S. Klausner. 1999. Health status and population characteristics of dogs and cats examined at private veterinary practices in the United States. *Journal of the American Veterinary Medical Association*, 214: 1336–1340.

Murray, S. M., A. R. Patil, G. C. Fahey, Jr., N. R. Merchen, and D. M. Hughes. 1998. Raw and rendered animal by-products as ingredients in dog diets. *J. Nutrition*, 128: 2112S–2815S.

National Research Council. 2004. *Nutrient requirements of dogs and cats*. Washington, D.C.: National Academy Press.

National Research Council. 1986. *Nutrient requirements of cats*, revised edition. Washington, D.C.: National Academy Press.

Richardson, D. C., J. Zentek, H. A. W. Hazewinkel, P. W. Toll, and S. C. Zicker. 2000. Development of orthopedic disease of dogs. In M. Hand, C. D. Thatcher, R. L. Remillard, and P. Roudebush, eds., *Small animal clinical nutrition*, fourth edition (chapter 17). Topeka, KS: Mark Morris Associates.

Strombeck, D. R. 1999. *Home-prepared dog and cat diets: The healthful alternative*. Ames: Iowa State University Press.

Sunvold, G. D., G. C. Fahey, Jr., N. R. Merchen, E. C. Titgemeyer, L. D. Bourquin, L. L. Bauer, and G. A. Reinhart. 1995. Dietary fiber for dogs IV: In vitro fermentation of selected fiber sources by dog fecal inoculum and in vivo digestion and metabolism of fiber-supplemented diets. *Journal of Animal Science*, 73: 1099–1109.

Warren, D. M. 2002. *Small animal care and management*, second edition. Albany, NY: Delmar Thomson Learning.

Zicker, S. C., R. B. Ford, R. W. Nelson, and C. A. Kirk. 2000. Endocrine and lipid disorders. In M. Hand, C. D. Thatcher, R. L. Remillard, and P. Roudebush, eds., *Small animal clinical nutrition*, fourth edition (chapter 24). Topeka, KS: Mark Morris Associates.

Chapter 11

Nutrition of Cats, the True Carnivores

Take Home and Summary

Cats are the only major companion species that are true nutritional carnivores that can thrive throughout the life cycle without any carbohydrate in the diet. However, they can live longer and healthier lives if the proper mix of carbohydrate, fat, and protein is supplied. Cats are not small dogs, and thus require a much different balance of nutrients. They have developed different enzyme systems to metabolize amino acids, fats, and vitamins. They require a greater amount of amino acids to supply the faster rates of amino acid metabolism. They require more taurine, a specialized amino acid, to protect against retinal degeneration and liver disease. Cats require a greater supply of linoleic acid, niacin, and vitamin A. Lower urinary tract diseases, primarily struvite crystal formation, are a function of the normally lower urinary volume, mineral content of the diet, and feeding pattern. Feeding a free-choice diet to avoid large meals and feeding a diet with the proper amounts of magnesium and other minerals can help prevent the crystallization of struvite.

Basic Characteristics of the Cat and Differences from the Dog

True Carnivore

> The cat is a true carnivore, the only mammalian companion animal that can make that claim.

A *nutritional carnivore* is an animal that can *thrive* throughout its life cycle without any carbohydrate in the diet. The animal does not just "get by"; it does well on a diet without plant material. The definition of *carnivore* in this sense is different than the phylogenetic order that they have been classified into, Carnivora (Kingdom Animalia, Phylum Chordata, Order Vertebrata,

Class Mammalia, Family Carnivora, Genus *Felidae*). This is because not all members of this family (such as dogs) are truly nutritional carnivores, even though they evolved on primarily a carnivorous diet. Although cats can *survive* on a diet very high in meat, an all-meat diet is neither *normal* nor *healthy* for cats or any other animal. The Fun Fact Box in Chapter 10 showed the problems of all-meat diets versus actual requirements for the dog, and this applies to the cat as well. Meat has too much protein and not enough calcium. The specific reasons for feeding a balanced diet of animal and plant products, with judicious use of vitamin and mineral supplements, will be covered point by point. Primarily remember that the cat is not a small dog, and will not thrive on rations meant for dogs.

Lack of Dietary Glucose

Because of the long-term development of cats as predators, they evolved consuming a diet made of other animals. This diet is primarily protein and fat, with no carbohydrate. Thus, over time, cats have evolved and adapted; cats have "given up" or "lost" several genes for the synthesis of amino acids that other animals have, because the amino acids were always present in the diet. In addition, they have evolved to a high degree the ability for *gluconeogenesis*, the synthesis (genesis) of new (neo) glucose, so much so that cats now require greater amounts of amino acids to maintain their normal rates of metabolism, even if glucose precursors such as starch are fed. The practical application of this evolutionary biochemistry is that cats require more protein in their diets than any other animal we have as companions with the sole exception of some fish and a few reptiles.

> *Gluconeogenesis* is the synthesis of new glucose. Mammals can only make glucose from amino acids, primarily glutamate, glutamine, aspartate and alanine.

Amino Acid Nutrition

Because of the differences in amino acid metabolism of the cat, there are a few other idiosyncrasies (unique characteristics) that affect their practical nutrient requirements. The cat derives substantial amounts of energy from amino acids (review Chapter 7) because it historically did not usually consume much glucose. However, because the brain of the cat still requires glucose, they must have fast rates of *transamination* and *deamination* in the liver to convert amino acids to glucose. The practical application is an increased requirement for energy and protein. The protein requirement of a cat in some parts of the life cycle can be twice that of dogs. Another practical effect of these differences in metabolism is an increase in urea synthesis and excretion, which requires more energy and increases the urea as well as ammonium ion in the urine. This rapid metabolism of amino acids also in-

FUN FACT

There is a phenomenon in scientific investigation that we call *serendipity*. Serendipity is roughly defined as "finding one thing when you are looking for something else." It was the search to determine the amino acid requirements of the cat that led to the discovery of a major difference in genotype and nutritional requirements of this animal. Dr. Quinton Rogers, Jim Morris, and colleagues at the University of California were determining the amino acid requirements of cats. Diets containing a range of amounts of a specific amino acid, starting with none, were mixed. These diets were then fed to animals for a relatively short period to observe at which minimum amount of that amino acid the animals do the best: grow properly, have the best hair coat, general health, and the like. When Dr. Rogers and his students started some cats on the arginine-free experimental diet one afternoon, and then returned in the morning, unfortunately the major observation was that many of the cats were dead or very sick. Those that were sick quickly returned to normal when fed normal protein diets. Until then no one thought that such a short period lacking one amino acid could possibly have such a major effect. However, the cat without arginine cannot make enough urea and ammonia toxicity ensues very rapidly. This is not usually a practical problem as all proteins contain arginine. It may never have been discovered had it not been for the need to define specific amino acid requirements to design proper diets. This *serendipitous* discovery led to more work and more discoveries on practical problems of amino acid nutrition that have improved the nutrition of animals and humans ever since. . . (For further information, see Morris et al., 1991; NRC, 2004.)

As mentioned, we use the term *serendipity* to define finding something (usually important) while we are really looking for something else. But where did this word come from? It has to do with world history. As European explorers were sailing around the world looking for more things to exploit, they came across an island, near India. On this island grew a strange plant, the leaves of which, when boiled in water, made a strong, stimulating, and tasty drink we now call tea. While the explorers were out looking for gold and treasure, they found a big treasure, yet it was not what they were looking for. The name of this island was *Serindip* (now it is Sri Lanka). So, from this island comes not only that drink which keeps you awake, but also the term we use for "keeping your eyes and mind open when looking for one thing, because you just might find something more important!"

creases the amount of carbon dioxide production and loss in the urine. Because CO_2 is a weak acid, it causes the pH of the urine in cats to be somewhat more acidic. This leads to potential crystallization of waste products in the urine of cats, a set of symptoms commonly referred to collectively as *feline lower urinary tract diseases (FLUTD)*.

The last interesting bit of cat nutrition lore is the strict requirement for taurine. *Taurine* is a type of amino acid called a beta-sulfonic amino acid. It is not a normally occurring amino acid in the plant world. Animals make it for

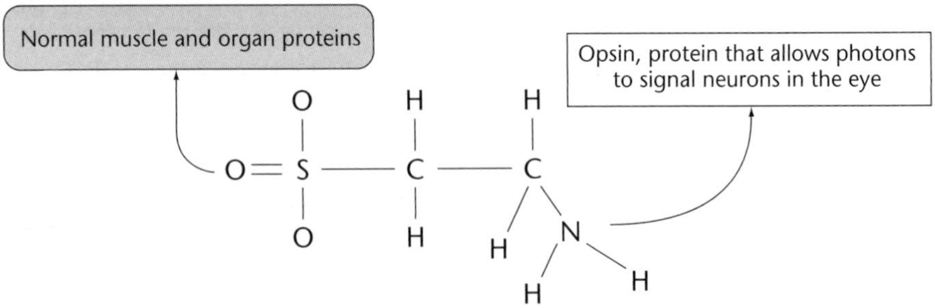

Note that taurine is not an alpha-amino acid; it is a beta-amino sulfonic acid. It is normally made in all animals. But, because cats need more amino acids overall, and they cannot make enough taurine, we have to include it in the diet, either in animal products or as a supplement to diets higher in plant protein.

Figure 11.1 Taurine, the amino acid everybody needs, but the cat needs more.

their own use. Most animals, including cats, make some taurine. Unfortunately, cats do not make enough. Because of their rapid amino acid metabolism, they must have a greater amount of taurine in the diet. Most diets for cats have most of the protein coming from animal products, and supplemental taurine is used for diets containing proportionally more plant protein.

Taurine is used in digestion of fat as taurocholic acid, a combination of part of taurine and cholic acid, from cholesterol (as discussed in Chapter 4). In addition, taurine is a critical element of opsin, a protein involved in eye function (Figure 11.1). The major symptom of taurine deficiency is reduced and then severely damaged eyesight. Signs can take many months to a couple of years to develop, and once noted are usually not completely reversible. Taurine must be included in the diet, either as contained in animal protein or by itself, at about 0.5 percent of the total diet. Taurine is often listed as an ingredient in many cat foods, especially those relying more on protein from plant sources. If it is not on the label that does not mean that the food is not adequate, but means that animal products (chicken, beef, and by-products) are supplying enough taurine.

> Taurine is a beta-sulfonic amino acid that all animals need, but cats cannot make enough, so it needs to be included in greater amounts in the diets of cats.

The information on taurine metabolism and nutrition of the cat should *not* be viewed as an argument either for or against plant proteins as a source of protein for cats. Cats do not need an all-meat diet (see discussion on this in Chapter 10), and they can derive a large part of their amino acids from plant proteins, as long as the taurine amount is correct and the overall amino acid quality (see Chapter 4) is good. Plant proteins are generally cheaper and provide a good balance to the concentrated amino acids in animal products.

Fat Nutrition

Amino acid metabolism is not the only difference between cats and other animals. Cats also evolved on a diet with significant amounts of fats. Remember that fat, especially animal fat, is also a good dietary source of fat-soluble vitamins, primarily vitamins A and D, as well as the essential fatty acids linoleic, linolenic, and arachidonic acid. So you may guess (or hypothesize) that a similar evolution has taken place in fat metabolism, and you would be correct. Cats require greater amounts of each essential fatty acid. It is not completely clear if cats can make enough arachidonic acid from linoleic acid. Thus, it is wise to increase arachidonic acid concentrations. However, some studies have shown that for maintenance, linoleic acid is sufficient.

Vitamin Nutrition

Finally, we need to discuss some adaptations in vitamin needs. Whereas many species can cleave beta-carotene to make two vitamin A molecules (Chapter 5), cats cannot do this, and must be fed pure *vitamin A* from ingredients or supplements. Supplementation of beta-carotene into the diets of cats might make for a nice yellow fat, but it will not improve their vitamin A status at all.

Niacin, better known chemically as nicotinic acid, is extremely important in all aspects of metabolism, primarily in generation of ATP for energy. Many animals can make some niacin from a related amino acid, tryptophan. However, because of the greater need for amino acids and the faster rate of amino acid metabolism the cat cannot make enough niacin from tryptophan and must have more niacin in the diet. Tryptophan has been "spared," if you will, by another genetic adaptation to help ensure that the cat has enough to supply the fast rate of protein synthesis. Evolutionarily, this was no problem as the poor little prey it caught for dinner had plenty of niacin in its organs. Nowadays, it is not a major problem to add sufficient niacin. However, this is another reason that dog food is *not* food for cats.

Finally, cats require a larger amount of the vitamin pyridoxine (B_6) because of the rapid rates of amino acid metabolism. Pyridoxine is the vitamin primarily involved in amino acid metabolism, especially the movement of the amine group from essential amino acids to nonessential ones (transamination) to make all the necessary proteins. Pyridoxine is also involved in deamination, the removal of the amine group in the conversion of amino acids to glucose or the breakdown of amino acids to supply energy.

Summary—Nutritional Uniqueness of Cats

Greater requirement for taurine

Need linoleic acid, perhaps arachidonic

Cannot make vitamin A from beta-carotene

Cannot make niacin from tryptophan

Need more pyridoxine for the greater rates of amino acid metabolism

Mineral Metabolism and Urinary Acid-Base Balance

We have already touched on the biological differences that affect mineral nutrition of cats, although you may not have realized it. But a major effect of the differences in amino acid metabolism is that the mineral content of urine is an issue for the health of cats. The first issue, discussed earlier, is the increased amount of CO_2 in the urine of cats because of the rapid breakdown and use of amino acids. This makes the urine naturally more acidic. Also, the cat has adapted to use a relatively smaller amount of water than other animals, which increases the concentration of most waste products in the urine. Therefore the cat presents a combination of problems generally referred to in the past as *feline urological syndrome*, and more recently FLUTD. These are the names used to cover any and all problems with infections, inflammations, and crystallization problems of the lower urinary tract. More specifically, we are discussing one major problem: *feline struvite urolithiasis (FUS).* This is the interaction of urinary pH, ammonia, magnesium, and phosphorous in the urine leading to the formation of a crystal called *struvite.* Struvite crystals are made up of ammonium, magnesium, and phosphorous and account for about 70 to 90 percent of cases of feline urinary tract blockages.

As an increased number of cases of problems and blockages in the urinary tract of cats were noticed by veterinarians in the last few decades, the initial focus was on the amount of magnesium in the diet. This was a valid hypothesis, as the crystals most commonly found, the struvite, contained a large amount of magnesium. It was also noted that decreasing the amount of magnesium in the diet could decrease, but did not fully prevent, the formation of these crystals. Several diets were developed with a reduced amount of magnesium, and they were somewhat effective products in reducing struvite problems. However, it was neither all of the story nor the end of the story. It was only through a good application of the full scientific process that the multifactorial nature of this problem came to be fully understood.

As more data were collected from veterinary clinics, and more directed research studies were done, it was realized that any factors that increased the pH of the urine and the concentration of magnesium, phosphorous, or ammonium ion increased the incidence of struvite crystals. What practical nutritional, environmental, and feeding-behavior problems can you think of that might increase urinary pH and concentration of chemicals in the urine?

Meanwhile, back to struvite. The crystallization of this molecule increases as the pH increases above 6.8 for any significant amount of time. Also, as noted earlier, an increased concentration of any ions prone to crystallization will increase the amount of crystallization. Now look at Figures 11.2 and 11.3 to note some of the reasons why struvite crystallization is increased in any situation leading to decreased urine volume, increased concentration of the ions, and increased pH. Thus, a well-balanced diet with no more protein than necessary, free-choice feeding to reduce large increases in pH, plenty of fresh water, and keeping a nice clean litter box will all work to keep the risk

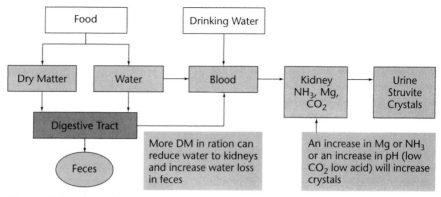

Figure 11.2   Water flux and struvite in cats.

As nutrients undergo metabolism after a large meal, a large increase in pH (alkalosis) can lead to greater crystallization in the urine. If we train or allow the animal to eat smaller meals through the day, the metabolic alkalosis is less and so is crystallization.

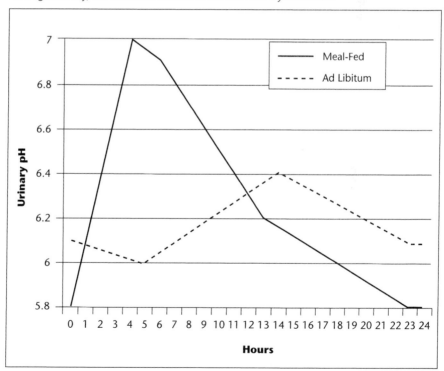

Figure 11.3   Meal pattern and urinary pH.

of feline struvite urolithiasis at a minimum. A clean litter box is important to encourage regular urination to remove these waste products and keep urine mineral concentration low and pH normal.

We also want to avoid acidic urine as well, as that can lead to excess bone mobilization and calcium crystals. Results of experiments that lowered the pH of urine showed that if the pH dropped much below 6.4 for a few hours it could increase the amount of calcium oxalate crystals because the acid draws calcium out of bone and into the urine. These occur at much lower rates than struvite (only 5 percent or so) and are usually not a major practical problem.

Finally reduction of struvite is managed by allowing free-choice feeding of diets to cats. As with any chemical reaction, if there is a greater concentration, crystallization increases. In Figure 11.3, we see a graph of the effect of meal pattern on urinary pH. As the cat eats a large meal, all the excess nutrients that are absorbed faster than the body can use them are lost in the urine. This is true for protein also: If the body absorbs enough amino acids in three hours as it needs for twelve or twenty-four hours, some will naturally be lost because the body cannot adapt to store all the extra fast enough. Therefore, in the cat, a large meal increases carbon dioxide, magnesium, and ammonia concentrations in the urine and raises urine pH above 6.8 for a few hours. So, with better-balanced rations and good nutritional feeding management (and a clean litter box) we can keep this potentially painful problem to a minimum.

> If the cat consumes free choice, it will decrease spikes in urinary pH, avoiding the big dose of waste in the urine and reducing the incidence of struvite crystal formation.

Life-Cycle Nutrition of Cats

We have spent several pages discussing the nutritional "weirdness" of cats and potential problems; now it is time to think about how to do things right.

Suckling and Weaning

As we learned in Chapter 2, the queen (mom cat) produces colostrum for the first two to three days of lactation. The colostrum is more dilute than regular milk, containing 88 percent water compared to 82 percent; however, it is enriched in the immunoglobulin proteins to provide immunity to the kittens. This is critical to their short- and long-term health, thus it is essential that the mom nurses the litter properly for the first few days. If there appears to be any problems developing (the queen will not nurse) you should contact your veterinarian as soon as possible. There are well-formulated milk replacers available. However, normally, everything is fine and the young kittens begin to get the nutrients they need and the immunity as well.

After a few days, the protein and fat content of the milk will begin to increase, from about 4 percent protein in colostrum to 6.6 percent in milk, and 3.4 percent fat to about 5.5 percent in regular milk. For the first three weeks or so, life is good and there really is not anything else you need to do other than let mom and the litter do their thing. Then, kittens should be gradually introduced to solid food around three to four weeks of age. Remember that feeding-management goals include encouraging proper eating behavior, not just providing a quality food. Follow the general directions provided for dogs in Chapter 10—feed dry rations moistened with water, decreasing the water content slowly so that by weaning they are eating fairly solid food (maybe about 20 percent water). Canned food can also be fed, but should also at first be diluted about 1:1 with clean fresh water and mixed to reduce particle size (a gentle blending or even simply mixing with a fork, depending on the consistency of the product). If you are feeding canned food, you want to reduce the particle size to avoid choking. Remember that canned products are generally quite high in protein and fat, so a smaller amount should be fed. Any diet used to introduce the kittens to food should be labeled as "complete and balanced nutritionally for all life stages," or should be kitten chow. In order to teach them good eating behavior, provide food at regular intervals, in consistent amounts. Watch what they eat, feed according to label instructions, and do not allow them to consume too much.

Growth

The biological goals of growth, as discussed in Chapters 2 and 7, are muscle accretion, good bone formation, final development of metabolic and reproductive organs, and a proper amount of body fat growth. While you are supplying the animal with the protein, energy, vitamins, and minerals, you are also managing its feeding pattern and intake to avoid problems with struvite urolithiasis and obesity. Be sure to have food available often, and feed smaller meals to reduce gorging and metabolic alkalosis (Figure 11.3). Most cats can be trained to ad-libitum dry-food feeding without allowing too much intake. Table 11.1 shows requirements and intakes for cats. Table 11.2 shows the general requirements for cats during the life cycle.

During later growth, metabolic and reproductive organs undergo final development. There is a normal amount of adipose tissue formed for fat storage. We discussed how a free-choice diet could minimize problems with struvite urolithiasis. It is also convenient for the person, and if you are paying attention to amounts, will provide the needs of the animal without risking obesity.

Feed regular amounts at regular intervals. For most animals, this means free-choice feeding.

Table 11.1	Nutrient Makeup of Foods Recommended during the Life Cycle of Cats

	Energy, Mc ME/kg	Protein, %	Fat, %	C18:1, %	Calcium, %	Phosphorous, %
Early Growth (2 to 6 months)	4.0	18–22.5	9	0.55	0.8	0.72
Maintenance	4.0	16–20	9	0.55	0.3	0.26
Late Pregnancy	4.0	17–21	15	0.55	1.1	0.76
Lactation	4.0	17–21	15	0.55	1.1	0.76
Aging	4.0	20+	13–18	0.55	0.3	0.26

Recommended (not minimum) requirements for nutrient content of foods, % of diet. These are average figures to demonstrate the requirements and are not meant as absolute values. These will vary with activity, age, and weight of animal. Assumes weights of .8 for kittens and 4 kgs for all adults. Assumes energy intake of 180, 250, and 540 kcal/d for the three stages. The column for C18.1 represents the recommended linoleic acid content of the diet.

Source: Adapted from NRC, 2004, Nutrient Requirements of Dogs and Cats.

Table 11.2	General Requirements for Cats during the Life Cycle

	Body Weight		kcals ME		Grams of food		Cups of dry food				
	kg	lbs	per day	Growth food	Adult food	Growth food	Adult food	Protein g/day	Ca mg/d	P mg/d
Kittens	0.5	1.1	113	28	32	0.31	0.40	6.7	222	167
Kittens	1	2.2	225	56	64	0.63	0.7	10.7	356	267
Adult	4	8.8	250	100	114	1.1	1.3	12.4	711	533
Pregnancy (late)	4	8.8	540	135	154	1.5	1.7	21.3	711	533
Lactation peak	4	8.8	540	135	154	1.5	1.7	32.0	1333	1067

Assumes 4.5 and 3.5 Mc ME/kg diet for growth and adult food and a minimal crude protein requirement of 24% for growth and reproduction and 14% for adulthood. Note that commercial foods will have somewhat more energy and protein to be safer for all animals. As always, individual observation of growth and body condition is important in making adjustments. Assumes 225 kcal/kg BW for kittens, and 62.5 and 135 for adults and pregnant/lactating animals.

Source: Based on data from NRC, 2004, Nutrient Requirements of Dogs and Cats.

Obesity is the most prevalent problem in the U.S. pet population. We can prevent most of this with simple observation and following the basics we have learned in previous chapters. Figure 11.4 shows body condition assessment for cats. Follow the guidelines in Table 11.1 and observe the animal over days, weeks, and months. You should watch the body condition, minimizing fat covering over the ribs, pelvic structure, and abdomen. Most tables of requirements and feeding amounts on labels are a good starting point. Adjust food amounts so that animals do not gain excessive condition or, if on the opposite side, do not have excessive hunger symptoms (vocalization, irritabil-

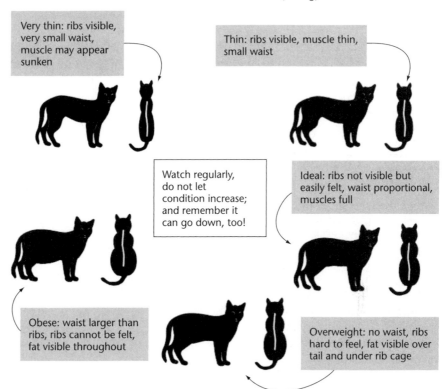

It is simple to assess the fat composition (obesity level) of a cat with observation and tactile assessment (feeling).

Very thin: ribs visible, very small waist, muscle may appear sunken

Thin: ribs visible, muscle thin, small waist

Watch regularly, do not let condition increase; and remember it can go down, too!

Ideal: ribs not visible but easily felt, waist proportional, muscles full

Obese: waist larger than ribs, ribs cannot be felt, fat visible throughout

Overweight: no waist, ribs hard to feel, fat visible over tail and under rib cage

Figure 11.4   Body condition in cats.
Source: Figures redrawn from *Canine and Feline Nutrition,* Case, Carey, Hirakawa, and Daristotle, Figure 26.3, Copyright 2000, with permission from Elsevier.

ity, searching, or begging for table scraps). A little extra fat may not hurt and may keep the animal happy, but excess fat (approximately 20 percent more than ideal body weight) is known to cause an increase in all major diseases. Even for cats, regular exercise is important. If allowing the cat to roam is possible, do so. If the cat is somewhat more confined, place the feeding area, watering area, and litter box in different areas as separated as they can be. Encourage play whenever possible. As the animal matures into adulthood, the same rules apply that we learned for dogs. The metabolic rate will slow, so watch body condition and feed accordingly.

Table 11.3 gives examples of several products on the market. Take a look at them and examine the differences in chemical composition and ingredient composition. Following the footnote in the table, think about the relative and relevant differences between the rations.

			Energy				Crude		
Table 11.3					**Examples of Ingredients and Contents of Selected Commercial Cat Foods**				

Item		DM, %	kcal ME/kg	Protein	Fat	Crude fiber	Ca	P
		Nutrient Content in Percentage of DM except as noted						
Store brand all life stages Poultry by-product meal, ground yellow corn, ground whole wheat, corn gluten meal, fish meal, dried egg product, soybean meal, brewers rice, animal fat	(l)	89	3.26	31.5	11	4.5		
Name brand maintenance Corn meal, poultry by-product meal, corn gluten meal, soy flour, tallow, cellulose, salmon meal, soy hulls, malted barley flour, brewers dried yeast	(a)	91.5	3.54	36.8	9.7	1.8	1.38	1.09
Name brand kitty Poultry by-product meal, brewers rice, corn gluten meal, soy flour, tallow, fish meal, brewers dried yeast, wheat flour	(l)	88	3.72	40.0	12.5	4.0	1.2	1.1
Name brand senior Chicken, brewers rice, poultry by-product meal, corn gluten meal, soybean meal, whole ground corn, tallow, brewers rice	(l)	91.5	3.74	36.0	13.0	5	1.0	0.9

Energy is metabolizable energy in kcal/g DM. It is analyzed unless noted as calculated (c), from modified Atwater values of 3.5, 3.5 and 8.5 for starch, protein and fat.

Sources include product labels (l); also analyzed data from Hand et al., pp 1074-1082, Appendix L (a). This reference contains brand information. We chose not to include them here to avoid distraction. This reference also includes many more products.

As a thought exercise, the student should look at ingredient composition as well as chemical composition, and determine different ways to alter chemical composition with ingredients. Why are different ingredients included? Why are many ingredient lists the same, although chemical content may vary? Which is more important (ingredients or chemical content)? Are they both important?

Female Reproduction

The principles of nutrition in the breeding queen are the same as for dogs, but with the adaptations expected because of the cat's different metabolism and smaller size. Because the average cat is smaller than the average dog, the metabolic requirement of the litter (in utero and during lactation) tends to be a greater percentage of the daily nutrient requirements than it is for the dog.

Table 11.3	Examples of Ingredients and Contents of Selected Commercial Cat Foods *continued*							

Item		DM	Energy (ME, Kcal/g)	Protein	Fat	Crude Fiber	Ca	P
			Nutrient Content in Percentage of DM except as noted					
Premium brand adult Lamb, brewers rice, chicken byproduct meal, corn meal, chicken fat, fish meal, dried egg product	(a)	92.5	4.76	33.8	23.0	1.0	0.85	0.74
Premium brand kitty Chicken, chicken by-product meal, corn gluten meal, whole ground sorghum, chicken fat, corn	(l)	90	3.88	34.0	22.0	1.0		
Premium brand senior dry Corn, corn gluten meal, chicken by-product, poultry by-product, animal fat		90	20	14	2.5	6.5		
Premium brand senior wet Meat broth, meat by-products, chicken, turkey, wheat gluten, lamb, corn starch modified, soy flour, flavorings, whole rice		18	9	3 to 6	1.5	2.5		
Premium cat treat wet	(l) as fed:	22	0.96	12	5	1		
	DM:	100	4.37	54.5	22.7	4.5		
Salmon, meat by-products, liver, water, poultry by-products, ocean fish, shrimp, guar gum, onion, salt								

The gestation period of the queen is sixty-three to sixty-five days; there may be a slight increase in this period for larger animals.

Somewhat different than for the dog, the total amount of nutrients needed by the pregnant cat starts to increase earlier in gestation (Figure 11.5). The hormones of pregnancy allow the queen to use the nutrients she consumes to meet these needs without sacrificing herself. About week 3 to 4, the size of the developing litter now starts to require more nutrients than the animal is consuming. The first adaptation in the queen is to consume more food to meet these needs. If the queen was on a maintenance ration prior to conception, she should be slowly (over one week) switched over to a ration formulated for all life stages. The extra protein, calcium, phosphorous, and other vitamins and

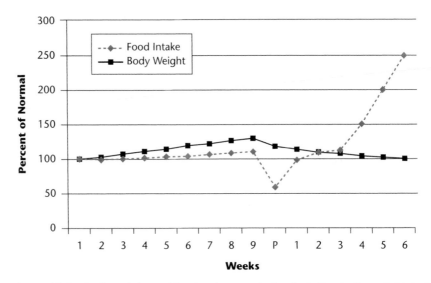

Figure 11.5 Body weight and intake changes during lactation in the cat. Note that the food intake will increase sooner in the cat than the dog, but still only increases about 10 to 15 percent.

minerals will take care of the kittens. The queen's diet is important during the third and fourth week of pregnancy because appetite loss and weight loss may occur in some animals. The amount of food needed increases gradually during gestation (about 25 percent more food is needed at parturition) as a result of a slow, steady increase in body weight of the litter. As stressed, do not supplement diets but rather feed high-quality complete and balanced rations. Table 11.3 shows the nutrient content of some common cat foods.

Lactation

Everything discussed about lactation nutritional management in the dog applies to the cat as well. When the mammary gland starts to make significant amounts of milk at day 1, food-intake controls signal hunger to the brain, and food intake increases. At this stage, a diet formulated for all life stages will supply what the queen needs. After one or two weeks, the queen can easily be consuming twice to three times as much as she normally would. For most animals, except those very small ones or those nursing especially large litters, this increase in intake can supply most or all of her needs. However, it is normal for an animal in lactation to lose some body fat—this is not a bad thing unless it goes too far. If the queen is losing too much condition, you can supplement with some higher-fat canned food, or small amounts (remember 1 tbsp of oil contains 100 kcal) of oil well mixed over the food. Maintaining proper nutritional management for the reproducing queen enhances the kittens' survival. Because of the hormones of lactation, the queen can lose substantial amounts of body fat and the milk composition and volume will not change so that the kittens will still get what they need. However, if allowed to go too far, it is potentially unhealthy for the queen. If you note that the queen's body weight and fat is dropping dramatically, and the kittens' growth rate is slowing, then it is time to check with a veterinarian about supplementing with some milk replacer, or with older litters (six weeks), beginning an earlier introduction of food.

Aging

We have covered the principle biological changes related to aging in Chapters 2 and 10. These apply to our feline friends as well. We need to think about reducing the total amount of protein, and make sure that the amino acid balance is as close to ideal as we can make it. The lowered intake of amino acids and the better balance of amino acids will reduce the urea formed and lower the metabolic load on the smaller kidney mass. Also, the content of sodium, potassium, and phosphorous should be as close to the minimal required as is safe to avoid extra stress on the kidneys.

The decision to switch to a specialty diet—a maintenance, low-calorie, or senior diet—is, again, made for each animal individually. For all but the largest (muscular, not fat) or active animals, a maintenance diet should be sufficient, and the lower amount of proteins and minerals are better for almost all animals. For animals that are reaching those ripe old ages above ten, some senior formulations will have a better protein quality, lower fiber, and a lower mineral load, and can help to reduce stress on the older organs. Other management goals such as fresh water and a clean litter box will go a long way to keeping kidney and urinary health good for as long as possible.

There are several excellent products on the market, which all follow the same basic nutritional principles: feed a high-quality protein, not high protein, to reduce stress on the kidneys in making urea; use a well-balanced mix of animal and plant protein, digestibility 80 percent or greater, protein concentration about 14 to 21 percent, and fat 10 to 15 percent. We also feed minimal amounts of sodium and phosphorous. Sometimes you will find a diet somewhat higher in fiber (5 to 7 percent). The purpose here is to limit energy intake. These diets may be fed to an animal prone to obesity at any life stage, or to an older, less active animal. There are different opinions of increasing fiber in the diet to reduce energy. Fiber can be helpful in some situations such as reducing feelings of hunger. However, some animals can develop digestive upsets and gas with this much fiber. Again, observe the individual.

As activity level decreases, you should decrease food availability. You can feed once a day to limit intake, but feed more often if activity level warrants it. In older cats, more frequent meals may be necessary to stimulate intake or "even out" digestive load. Avoid periodontal disease by feeding dry diet or by routine (weekly) brushing of teeth. As the animal ages, frequent dietary changes may be necessary in order to stimulate smell and taste buds. The senses deteriorate with age and an otherwise perfectly healthy cat may just not be attracted to the food. A little common sense needs to be applied: "Variety is the spice of life," and if your cat has lived this long, it can probably stand a little variety. You can refer to the end of the last chapter (pertaining to dogs) for tips on feeding management for the very aged cat whose appetite might be decreased. In this situation, vitamin and mineral supplements might be useful if feed intake is low.

Words to Know

deamination	*feline struvite*	*struvite*
feline lower urinary	*urolithiasis (FUS)*	*taurine*
tract diseases	*gluconeogenesis*	*transamination*
(FLUTD)	*niacin*	*vitamin A*
	nutritional carnivore	

Study Questions

1. Describe the differences in metabolism of cats versus dogs for protein and amino acids, fats, vitamins, and minerals.
2. Describe the causes, prevention, and treatment of struvite urolithiasis.

Further Reading

Allen, T. A., and J. M. Kruger. 2000. Feline lower urinary tract disease. In M. Hand, C. D. Thatcher, R. L. Remillard, and P. Roudebush, eds., *Small animal clinical nutrition*, fourth edition (chapter 11). Topeka, KS: Mark Morris Associates.

Burkholder, W. J., and P. W. Toll. 2000. Obesity. In M. Hand, C. D. Thatcher, R. L. Remillard, and P. Roudebush, eds., *Small animal clinical nutrition*, fourth edition (chapter 13). Topeka, KS: Mark Morris Associates.

Case, L. P., D. P. Carey, D. A. Hirakawa, and L. Daristotle. 2000. *Canine and feline nutrition*. St. Louis, MO: Mosby, Inc.

Hand, M., C. D. Thatcher, R. L. Remillard, and P. Roudebush, eds. 2000. *Small animal clinical nutrition*, fourth edition (appendix J). Topeka, KS: Mark Morris Associates.

Kirk, C. A., J. Debraekeleer, P. Jane Armstrong. 2000. Normal cats. In M. Hand, C. D. Thatcher, R. L. Remillard, and P. Roudebush, eds., *Small animal clinical nutrition*, fourth edition (chapter 11). Topeka, KS: Mark Morris Associates.

Morris, J. O., D. Finley, and Q. R. Rogers. 1991. Waltham International Symposium on the Nutrition of Small Companion Animals. Special supplement to *Journal of Nutrition* 121: Supplement, November 1991. (This proceeding has many research papers on all aspects of canine and feline nutrition and metabolism.)

National Research Council. 2004. *Nutrient requirements of dogs and cats*. Washington, D.C.: National Academy Press.

National Research Council. 1986. *Nutrient requirements of cats*, revised edition. Washington, D.C.: National Academy Press.

Strombeck, D. R. 1999. *Home-prepared dog and cat diets: The healthful alternative*. Ames, IA: Iowa State University Press.

Chapter 12

Nutrition of Nonruminant Herbivores: Horses

Take Home and Summary

The horse is a nonruminant herbivore, with an enlarged cecum and colon. Within this part of the digestive tract, anaerobic bacteria digest and ferment fibrous feeds (cell wall constituents), and the horse uses the energy derived from them. This process makes horses quite different than most of the other species we discuss in this book. We may feed the horse a combination of forages and grains depending on the stage of life. (Younger, more active, or reproducing, animals may be fed more grain). If we feed the horse too much energy or protein as a foal or growing animal, problems with muscle and bone development may occur. Obesity is always a problem with the less active, more "lovingly" fed horse. Feed the aging more highly digestible forages, grains only as needed, and make sure they receive regular exercise and veterinary care. The properly fed horse can be given a long and productive life indeed if the simple rules of energy, protein, and mineral nutrition we have learned are practiced.

Digestive Physiology of the Horse

The horse has a postgastric fermentation: Anaerobic bacteria in the cecum and colon break down cellulose and hemicellulose to glucose, then to volatile fatty acids absorbed for energy. As the bacteria are then lost in feces, no extra capture of nitrogen as amino acids for the horse is realized.

Horses have a digestive tract with a stomach and small intestine, as do the dog, cat, and human (Figure 12.1), with a smaller stomach per unit body weight. Behind the small intestine, there is an enlarged *cecum* and *colon*. This is where the fiber that the horse consumed, but that was not digested in the stomach and small intestine, can be acted upon by anaerobic bacteria to derive some more energy. The horse consumes forage plants such as grass, legumes, and some shrubby and bushy plants, then digests the protein and

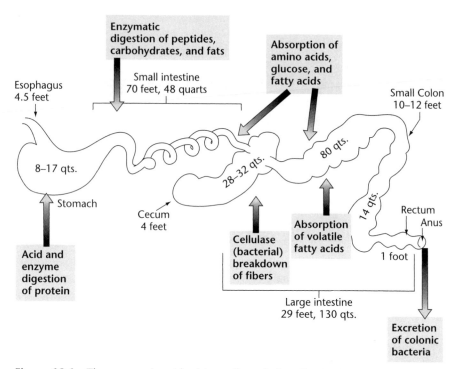

Figure 12.1 The nonruminant herbivore (horse) digestive tract.
Source: Cheeke, Peter R. *Contemporary Issues in Animal Agriculture,* 3rd edition, © 2004. Digestive track redrawn by permission of Pearson Education, Inc. Upper Saddle River, NJ.

starch (and the little bit of fat present) in the stomach and small intestine. The remaining cellulose, hemicellulose, and lignin passes to the cecum.

In the cecum, there is no oxygen, thus it is *anaerobic*. The bacteria in this environment live on plant fiber (cellulose and hemicellulose), just like in composting parts of the forest floor or your backyard (or, in a horse pen that has not lately been cleaned). The fiber, which has been only partially digested by the stomach acid and enzymes of the small intestine, is broken down to glucose by these bacteria. The major enzyme is called *cellulase*. Only anaerobic bacteria (and some fungi) make this enzyme, and these bacteria live in the rumens of ruminants or in the large intestine of some nonruminants (review discussion of fiber and carbohydrates in Chapter 3).

Once the glucose is formed by the breakdown of cell wells, it is not absorbed directly as it would be in the small intestine. First, because the large intestine does not have the transport systems to allow this. Second, because those helpful anaerobic, cellulytic (cellulose-breaking) bacteria do not work completely for free. In return for breaking down the cellulose for their own use, they ferment the glucose *(bacterial fermentation)* to three *volatile fatty acids*; acetic acid, propionic acid, and butyric acid (see Table 3.1; Figure 12.2). These fatty acids are used by bacteria to grow and reproduce. These volatile fatty acids and other carbon skeletons, when combined with ammonia,

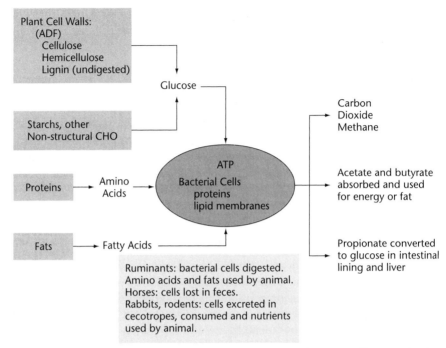

Figure 12.2 Anaerobic fermentation: using plants to make more efficient animals.

make amino acids. The amount of volatile fatty acids that the bacteria do not need are absorbed into the bloodstream of the horse through the cells of the cecum and large intestine, and the horse can use them for energy.

The first volatile fatty acid, acetic acid, can enter directly into the TCA cycle, be converted to CO_2, meanwhile generating $NADH_2$, which enters the electron transport chain to generate ATP. Butyrate, a four-carbon fatty acid, can be converted into two acetic acid molecules, which are used as described. Now the neat thing is what happens to propionate. Propionate is three carbons, and it easily can be converted back into glucose in the liver of the horse.

> From low-quality forages, the horse can derive the amino acids, obtain energy from acetic and butyric acids, and make glucose from the propionic acid—a very efficient survival strategy.

You might be thinking that this seems like a fairly convoluted and inefficient system. Well, from one point of view it is. The horse cannot be as efficient as those elegant felids and canids that consume a high-protein, high-fat diet. That is correct chemically—you just cannot deny thermodynamics and chemistry (Chapters 3–5 and 7). However, from a different perspective, the horse is efficient in that it can consume low-quality food such as grass, legumes, and other parts of flowering plants and some trees and survive quite well. They do not need to wait for some unsuspecting rodent or ruminant to happen by, nor do they need to stalk and chase their food around. They can survive year round even when forages are sparse.

Although the horse can derive the energy from the fibrous plants, it does not derive any protein from the bacteria. This is for two reasons: (1) the bacteria in the large intestine use the remaining amino acids for their own growth, and (2) the bacteria are then removed in the feces without any further digestion. The amino acids have to come from the plants through the usual methods of digestion in the stomach and small intestine. It is still an overall plus for horses. They receive the amino acids they would have anyway in digesting the plants, and then derive some extra energy from the bacterial fermentation of cellulose and semicellulose.

Fat Digestion in Horses

Because horses have evolved primarily on forages, they have not been adapted to a diet containing significant amounts of fat. That does not mean that they cannot use fat from their diet. It is similar to the case of dogs and cats: Just because their ancestors did not eat much in the way of plant material does not mean that we cannot safely and effectively feed plant material to them. Back in the dark ages when some of us went to school or grew up with horses, we learned that "horses don't have gall bladders, so they cannot eat fat." Also, it was not really necessary, and too expensive at the time, to feed extra fat to horses. There was some truth to that. The *gall bladder* is the small sac under the liver that stores *bile acids,* which help in the digestion of fat and removal of cholesterol. Although horses do not have a gall bladder horses *do* have bile acids. They just don't store them in the gall bladder; they release them directly to the digestive tract.

The bottom line is that the horse *can* use fat, even up to about 10 to 15 percent of the diet. There have now been many research trials in practical settings that have proven this fact. The real question is "Why would we ever want to feed a horse that much fat?" and the easy answer is: "Well we would not usually, but there are some situations in which this would be beneficial, so we can do it, if needed, without harm." There are a few situations, primarily for exercise or strenuous work, in which some fat can be beneficial, and we will discuss those in the section on exercise.

Life-Cycle Nutrition

The basics discussed for dogs and cats still apply, but we have the added twist of the digestive system of the horse, thus the need to feed diets with a significant amount of forages. It is a fact that many problems in horses are caused by the failure to understand that these animals evolved on a diet primarily of a wide variety of forages, and they neither *need* any grain nor *need* any one particular forage. There are certainly no magic numbers about what percentage of forage and grains we should feed, but there are some useful and practical guidelines. If we can keep the guidelines in mind, we can come up with a simple feeding-management plan to ensure proper development and a long and healthy life.

Foals

When the foal is born, its first requirement is for those immunoglobulins in colostrum. We want to make sure the foal can suckle and that the mare is providing what the foal needs. The immunoglobulins protect the foal against possible infections. The milk, of course, provides lactose, fat, and protein, just as it does for all mammals. The horse grows more slowly than the dog or cat and thus the amount of milk provided in relation to body weight is less for the horse.

> Nutritional management rules still apply: Observe and ensure that the foal is receiving all and only all it needs, that it is growing properly, and that the mare is neither losing weight too rapidly nor gaining significant amounts of body condition.

The growing foal should have access to forage in addition to the milk of mom. It is perfectly fine to allow the foal to graze good pasture, or to provide some high-quality (more protein and less fiber) hay free choice. Good pasture would include a stand of mixed grasses, or legume and grass (clover and grasses; alfalfa and orchardgrass). They will slowly nibble on it, and this helps establish the proper bacterial population in the cecum and colon. Consuming hay at a young age will not hurt the foal. Usually, the foal will regulate the intake to get a good mix of the easily digestible milk nutrients and the more-slowly digested fibrous forage. But, just as we slowly wean dogs and cats onto solid food, we need to do this for horses as well.

Now, we need to introduce another jargon term for feed meant for suckling and weaning horses. For the puppy and kitty, we just take "puppy chow" or "kitty chow" and start the animal on it. For the horse, we prepare a similar feed with a lot of energy and easily digestible components, but we call it *creep feed*. The word has nothing to do with the behavior of the foal, but with the fact that any self-respecting horse owner (historically) was not about to buy or prepare expensive, high-protein, high-energy foal feed and then let the mom eat it all. So they built a feeder with boards or fencing around it so that the foal could creep into the feeder but the mare could not reach the goodies. Thus the term *creep feed*.

Aside from deriving glucose and energy from fiber fermentation, the nutritional biology of the horse is similar to that of other mammals. We want to provide high-quality protein, so the creep feed usually contains a small amount of expensive but high-quality milk protein and legume protein such as soybean meal. This will provide high-quality protein to that developing foal. Also, some high-energy feeds such as corn, oats, and barley are included to provide starch to stimulate synthesis of the proper enzymes. Of course, a balanced vitamin and mineral mix is included to even out the variations in content in the grains. The goals are still the same here: to adapt the digestive system from wet, highly-digestible food to dry, variable food requiring a variety of enzymes to digest.

Growth

As for dogs, cats, and all other species, the principles for growth remain the same: the proper amount and balance of amino acids for muscle growth, sufficient energy to run the synthetic reactions, and sufficient amounts of vitamins and minerals for metabolic reactions and bone growth. Because the growth rate of these larger animals is slower, they tend to need less of a percentage of these nutrients in their diets during growth.

> Goals in growth include establishing good eating habits; providing optimal, not maximal, growth; and avoiding obesity and osteochondrosis.

A moderate to severe lack of protein or energy will slow growth and can lead to a permanently reduced adult size. In nutritional practice, usually only some definite mismanagement of feeding by the owner, sometimes caused by ignorance or lack of money would ever lead to such a result. Pasturing young horses on pastures with only grasses and not providing a supplement with vitamins and minerals can lead to a shortage of protein, energy, and calcium, as well as a shortage of several other vitamins and minerals. Simply providing a small amount of a grain supplement fortified with the proper mix of vitamins and minerals will provide proper nutrient needs and avoid such problems.

> Horses cannot survive well on one plant specie. Grasses are too low in calcium; several legumes are too high. A mix of plants is the best whenever possible, avoiding the use of expensive supplements.

Table 12.1 shows basic requirements and examples of feeds for young horses. Usually we would start with supplying 60 to 70 percent of the total ration as *concentrate:* the mix of grains, protein sources, and vitamins and minerals. This term has been used to describe feeds that are 'concentrated' in nutrients compared to whole forages and grains. As the animal ages, increase the forage content over time up to 60 to 100 percent for two-year-olds with minimal or no exercise training. The ultimate goal is to feed the adult horse on close to a 100 percent forage diet, with an appropriate amount of vitamin and mineral supplement in a grain carrier, for long-term weight mainte-

Table 12.1	Requirements and Feed Compositions for Young Foals								
		Alfalfa Hay		Grass Hay					Creep feed
	Requirement	early	mature	early	mature	Corn	Oats	Barley	
Energy, DE, Mc/kg	2.9	2.5	2.2	1.8–2.2	1.5–2.0	3.4	3.2	3.3	3.25–3.5
Protein, %	14–15	20–22	15–19	12	6	9	11	13	15–20
Fat, %		< 3 to 5% for all							
Calcium, %	0.8	1.7	1.3	0.45	0.30	0.05	0.1	0.05	0.8–1.0
Phosphorous, %	0.5	0.3	0.25	0.26	0.22	0.3	0.3	0.38	0.3–0.4
Zinc, ppm	40–60								
Copper, ppm	20–30								

Source: Adapted from NRC, Nutrient Requirements of Horses.

nance. Some texts may recommend to a higher ratio of grain, even up to 50 percent for two-year-old horses. This is fine *if* you remember total intake; it is just too easy to overfeed with that much grain. Most recommendations will limit grain mix to 1.5 to 2.0 kg/100 kg horse BW for a nursing foal, about 1.5 kg/100 kg for yearlings, down to 0 to 1 kg/100 kg for two-year-olds. It is in general better to have the ration consist primarily if not solely of a high-quality legume, grass, or legume/grass mixed hay and adjust the total ration with the grain mix only as needed.

The growth period, after weaning, for most horses would extend to about two years of age. Table 12.1 shows the basic requirements for weanling horses. As for dogs and cats, the percentage of nutrients in the diet will decrease during the growth phase, as will energy. The protein percentage requirement will usually decrease about 2 percent every six months, so that an adult animal only requires about 8 percent protein. A quick glance at Table 12.1 shows that very few feeds except for old grass will ever be this low. So we tend to over-feed protein to horses the same way we do to other animals. We definitely want to keep protein to a practical minimum, as the waste will just end up being a load on the environment. Similar to protein, energy density should decrease from about 2.9 Mcal DE/kg to 2 Mcal DE/kg ration dry matter from weaning to two years. This is easily done by reducing grain, and feeding more forage and, ideally, a lower-energy and protein forage such as more mature alfalfa or grasses. Calcium and phosphorous requirements will drop from 0.8 percent and 0.5 percent to 0.25 percent and 0.20 percent, respectively. Zinc and copper needs will decrease from 60 and 50 to 15 and 7, respectively, similar to what is found in normal plants. We will soon discuss why more zinc and copper appears to be helpful to reduce bone problems in growing horses. Total feed intake is always critical, and as a percentage of body weight, intake should be about 2 to 2.5 percent in weanlings, down to 1 to 1.5 percent for adults. If the grain percentage is greater, the total intake should usually be less.

Developmental Orthopedic Diseases

Our major goal here is to avoid rapid growth by the feeding of too much concentrate, and to provide the proper balance of calcium, phosphorous, zinc, and copper.

> Proper management of intake of protein, energy, and mineral balance will provide optimal growth and decrease the incidence of developmental orthopedic diseases.

Developmental orthopedic diseases rank close to obesity as potentially serious problems with growing horses. Developmental orthopedic diseases (D.O.D) is a general term used to describe various problems that can exist with bone development in young horses. Some can be due to injury but we will not cover these here. Others, primarily *epiphysitis* and *osteochondrosis*, are situations that are strongly linked to nutritional problems.

> *Epiphysitis* is an inflammation of the growing part of the bone—the *epiphyseal plate*.
> *Osteochondrosis* is improper formation of growing bone: cysts, chips, dislodged pieces.

The epiphysis is where cartilage is made and where bone crystallization starts. As the long bone grows, this constant synthesis of cartilage followed by crystallization of bone leads to the lengthening and strengthening of the bone. Any condition that can stress this area has the potential to lead to inflammation. Continued damage can lead to osteochondrosis, ranging from a mild temporary joint problem to debilitating lameness. Generally, the sooner the problem is identified and dealt with, the less permanent damage is done. The bone chip causes much pain as the weight of the animal presses on it in the joint.

Figure 12.3 shows the development of orthopedic diseases. Increased growth rate puts increased pressure on the tender developing bone near the joints, particularly of the rear legs. You should understand that any dietary factor that increases growth rate, such as too much energy, protein, or intake, would increase the potential for problems. Improper amounts of minerals (calcium, phosphorous, zinc, and copper) will also lead to malformations of the new bone. Usually there is more than one nutritional or environmental problem occurring at a time, so managing the entire situation is critical.

There are multiple causes of epiphysitis and osteochondrosis. Rapid growth can cause excess weight and pressure on the growth plate, leading to epiphysitis. If we avoid too-rapid growth, we can minimize these problems. Reducing the energy and protein content of the diet, feeding more fiber, or simply reducing intake will help. Defining *too rapid* is, as you might guess, not always an easy thing to do. Each animal has a genetically different growth potential. But some general rules would suggest that if we observe

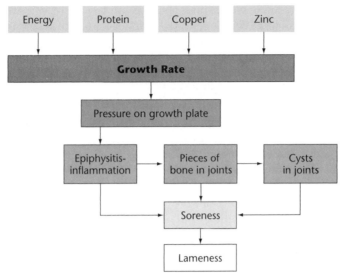

Figure 12.3   Developmental orthopedic diseases.

what a growing animal would normally eat, and we are concerned that they are growing too fast due to eating too much, we can simply reduce the amount or energy and protein content of the diet to slow the growth rate by 5 or 10 percent. An example, then, would be that a fast-growing horse may grow from 200 kg to 400 kg in 300 days; a horse fed to grow at a rate 90 percent of that would take only another month. This reduction in growth rate will not be unhealthy to the horse, certainly will help to avoid obesity, and the easing of stress on the bone joints can reduce the possibility of damage. By reducing the grain portion of the diet, feeding more mature hay, or limiting total intake, we can help prevent bone problems.

Growth rate alone is not the only cause of nutritionally related joint problems. Copper and zinc are required for normal cartilage formation, as we learned in Chapter 6. Improper cartilage formation can lead to poor structure and strength in the growth plate. Thus, most studies would recommend keeping copper and zinc at about 50 to 60 ppm of the diet, which is easily done with a commercial growing-horse supplement.

Strategies to reduce bone problems:

Observe animal daily for tenderness

Optimize growth rate at 90 to 95 percent expected breed maximum

Feed a foundation of mixed forage

Supplement with grain only as needed

Check and ensure that calcium, phosphorous, zinc, and copper are optimal

Proper calcium and phosphorous supply is critical. Either too much or too little calcium or phosphorous can lead to improperly crystallized bone. Without enough calcium or phosphorous, bone is improperly formed. With too much calcium, the same problem can occur. It is often easier to supply too much calcium than phosphorous, but there are also many situations in which it is easy to supply too little calcium. In general, cereal grains and *grass forages* (see Chapters 8 and 9) are very low in calcium (less than 0.1 percent), whereas *legume forages* such as alfalfa or clover can be high in calcium (over 1.5 percent calcium). Thus, diets that are too high in either legumes alone or grasses alone should be avoided. It is also easy to provide a nice mix of legume hay and cereal grains to balance out the lack of calcium in one type of plant with the excess in calcium in the other. By providing a ration with a good mix of plant types (legumes and cereals), a properly balanced mineral mix, and supplying just enough for proper growth rates and providing regular mild exercise, developmental orthopedic diseases can be kept to a minimum.

Obesity in horses is a problem as for dogs and cats. As for any animal, too much body fat increases stress on the circulatory, respiratory, and skeletal systems, ultimately shortening the lifespan. Excess fat during pregnancy can lead to problems foaling (see "Reproduction" section). As excess fat is accumulated, it becomes increasingly harder to lose it. Energy intake must be decreased by sufficient amounts so that the animal oxidizes fat; also, introducing an exercise regimen to a horse that carries extra fat and is not in good training condition is not always an easy task.

> It is always a good idea to prevent obesity rather than to treat it. Horses are efficient animals and we must not overwhelm them with "kindness" in feeding.

Following the same guidelines as those for reducing the incidence of orthopedic diseases will also help reduce fat gain. Feed slightly less than the animal will willingly eat. Use the feeding charts in this book and in the NRC's *Nutrient Requirements of the Horse* (1989) to provide sufficient forage and keep energy intake in line. Have a moderate and regular exercise program. Observe the animal regularly for excess build up of fat and adjust the feeding program as necessary to reduce body fat before it accumulates too severely.

Adulthood and Maintenance

Most of the lifespan of horses will be spent at *maintenance*: constant body weight and body composition. Maintenance is easily achieved by observing the animal. Maintain a good regular exercise program with fifteen to thirty minutes of regular moderate exercise (a nice ride or walk) every day or every other day. The physiological benefits of regular exercise to the circulatory and respiratory system are well documented. Provide a primarily forage diet, ideally of mixed legume and grass forages. Legumes, especially good alfalfa, are often too high

Table 12.2	Requirements for Maintenance, Growth, Reproduction, and Exercise, 500-kg Mature-Weight Horse						
Requirements **% of diet**	**Maintenance**	**Growth**	**Pregnancy**	**Lactation**	**Light work**	**Mod. work**	**Strenuous work**
Energy, Mcal DE/kg	2.0	2.9	2.25	2.4–2.6	2.45	2.65	2.85
Protein, %	8	14–15	10–11	11–13	10.0	10.5	11.5
Calcium, %	0.25	0.4–0.5	0.45	0.4–0.5	0.30	0.30	0.35
Phosphorous, %	0.20	0.25–0.35	0.35	0.35–0.5	0.25	0.25	0.25
Kg feed intake, % of BW	1.0–2	2–3	1.5–2	2–3	1.7	1.9	2.3
Kg feed intake, 500 kg	5–10	6–12	7–10	10–15	8.5	9.5	11.5
Forage: concentrate	100:0–90:10	80:20	60:40–70:30		70:30	60:40	60:40–40:60

in energy and too high in calcium by themselves for maintaining horses. Also providing some grass hay (bromegrass, orchardgrass, bermuda grass) can reduce the energy and calcium to more proper levels. You can use a little bit of a well-balanced grain supplement for adult horses primarily to relieve boredom and to provide surety of a good vitamin and mineral balance. Table 12.2 shows the feeding requirements for the life cycle of the horse.

Figure 12.4 shows body condition scoring for horses. Similar to dogs and cats, observation and palpation of fat and muscle help guide us to maintain proper health and body composition of horses. A body condition score (BCS) of 1 is extremely emaciated, bones protruding, muscles thin; a 5 is average, some fat covering, muscle conformation evident; a 9 is extremely obese, large fat deposits over all the body.

Exercise and Exercise Physiology

The days when the big old horse was the tractor for the farm are long gone, for the better not only for horses but for food production too. Nonetheless, horses often exercise extensively, be it in the form of a pleasant morning ride or a million-dollar purse prize race. There are as many or more myths about exercise and nutrition as there are in any other biological field. With some provable biological facts, you will be better able to find the few facts within the myths, identify the pure myths, and practice sound feeding management for optimum performance.

In practice, each animal is a little different, and what works for one may not work for another. There are at least as many ways of doing things right as there are ways to do it wrong.

Areas to Observe:

A = neck

B = withers

C = back crease

D = tailhead

E = ribs

**F = behind
    shoulder**

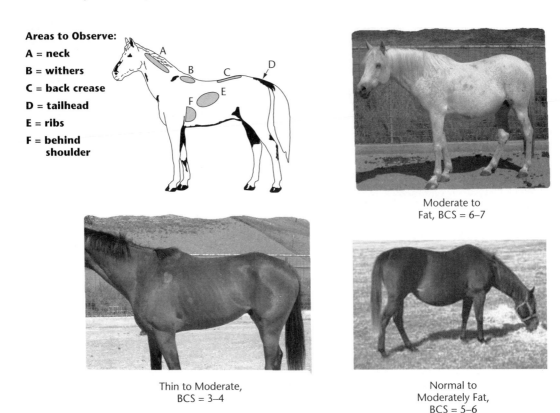

Moderate to
Fat, BCS = 6–7

Thin to Moderate,
BCS = 3–4

Normal to
Moderately Fat,
BCS = 5–6

Figure 12.4   Body condition scoring of horses. We want to shoot for BCS in the 4 to 6 range for most animals with a combination of food intake and exercise.
Source: Photos by John McNamara.

Let us take these key points of exercise nutritional physiology one at a time. First, remember that exercise is *work:* physical work, the application of force to a mass over a distance. The animal must move a mass over a given distance, whether walking to the water tank from the pasture, or racing around a 3/4-mile track, or carrying a rider or a load; in all cases, this is work. First we will cover physical work and the chemical energy that supplies it in muscles, and then we will come back to nutrition.

In the following boxed example, a 500-kg horse runs 1 km (0.61 miles) at 5 m/sec (11 mph). It takes the horse 200 seconds, and it has a force of 2,500 N, uses 2,500 KJ (600 Kcal) of energy (work) at a rate of power of 12.5 watts. That use of energy (0.6 Mcal), compared to the average daily maintenance requirement of about 10 Mcal/d of DE, shows that the energy needed to supply the work done over 1 kilometer, even at a good pace, is only about 6 percent of an increase in the total daily energy need.

Physical Work

Physical Equation, unit	*Practical Example*
Force = mass × acceleration, (kg*m/sec^2; Newton)	500 kg × 5 m/sec (11 mph) = 2,500 N
Work = Force × distance, (Newton*m; joule [or calorie])	2,500 N × 1,000 m (1 km; 0.61 miles) = 2,500 KJ (600 Kcal) this is 2.5 MJ; 0.6 Mcal, energy in 1/3 lb corn or 2.5 oz oil
Power = Work/Time; (joule/sec, watt)	2,500 KJ/200 sec = 12.5 KW

Now, you may be wondering if it is more work and energy if the horse runs faster (at the same weight and distance); the answer is *no!* There is *no* function of *time* in work. If the *same* amount of work is done *faster*, it is still the same amount of work. Only if *mass is greater or the distance is greater* is it more work. Work over time is **power**. Power is the amount of work done in a given time, such that 1 joule of work done in 1 second is one watt. One watt is 1 joule of work done in 1 second. This term is named after James Watt, who is credited with inventing the steam engine, one of the first examples of converting chemical energy (hot water) to physical energy (movement). Quite often you might hear someone say "Boy, he is really powerful!" or "That is really a powerful animal!" The speaker is correct if they are referring to the ability of the man or the horse to do the same amount of work in a shorter time, or more work in the same time—*not* just more work! That mythical horse running at 5 m/s (11.2 miles per hour) ran 1 km in 200 sec, or 2,500 J/200 seconds = 12.5 watts. Had that horse run the same distance in only 100 seconds (22.4 miles per hour, 10 m/s, or a 1:40 mile), he would have been twice as powerful, providing 25 watts of power.

Chemical Energy Use in Exercise

How do we supply energy to the muscle to do that 600 kcal worth of work, and how do we feed the horse to provide the right amount of energy to the muscle? Well, we need to review a little metabolism (see Chapter 3) and energy use. Carbohydrates, fats, and protein can all supply energy to the body. But muscle needs specific nutrients to run its own metabolism. The muscle at rest uses primarily fat, broken down by beta-oxidation (Chapter 7) and in the TCA cycle to form carbon dioxide, ATP, and $NADH_2$, which is further metabolized in the electron transport chain to form more ATP. It is the ATP that supplies chemical energy to the muscle proteins, actin and myosin, to contract (or to move against each other, shortening the muscle and thus causing movement by applying acceleration to a mass). Figure 12.5 describes some of the metabolic changes in the muscle from resting to intense exercise.

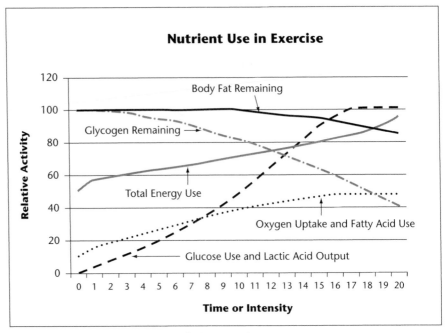

Figure 12.5 Chemical reactions during exercise.

The muscle at rest uses primarily fatty acids to supply energy; as exercise begins and increases, more glucose is used.

As exercise begins, the increased contraction rate immediately demands more ATP, and thus the rate of biochemical pathways supplying ATP increase: more glucose being oxidized, more fatty acid oxidized to the two-carbon acetate, more oxidation in the TCA cycle, more NADH$_2$ into the electron transport chain, and more ATP. If exercise continues at a moderate rate, this set of biochemical pathways can supply enough energy for hours, even days. If more fatty acids are needed, they will be released from the adipose tissue stores and moved to the muscle. If an animal is eating regularly, there really is no limit to how much total work, or exercise, can be done. There are of course many examples of horses (and people) working moderately hard every day for decades. Figure 12.6 shows the exercise metabolism process in muscle cells.

As exercise intensity (Power = Work/Time) increases (Figure 12.5), rates of the fatty-acid oxidizing pathways continue to increase, and if intensity continues to increase, eventually these reactions are proceeding at their maximal rates. Up to this point is called **aerobic exercise.** At this point, the cells are at their **aerobic threshold,** and oxygen supply is maximal. Any additional

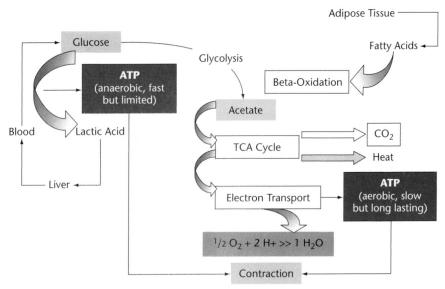

Figure 12.6 Muscle metabolism during exercise.

energy must come from anaerobic reactions, and the only major one in mammalian cells is the oxidation of glucose to two pyruvic acid in glycolysis (see Chapter 7 and Figure 12.6). These reactions provide only four ATP for every mole of glucose, or 0.2 calories per gram, only 4 percent efficiency. The advantage is that these reactions are very *fast*. So, the aerobic reactions are already proceeding as fast as possible, plus the inefficient but fast anaerobic reactions are supplying significant additional energy.

The problem with *anaerobic exercise* is that when the animal reaches its maximal ability to do work and take in oxygen, the breakdown of glucose to pyruvate forms lactic acid. Some of this lactic acid can return to the liver, be converted with almost 100 percent efficiency back to glucose, and the glucose can return to the muscle to supply four more ATP.

> The recycling of lactate from the muscle, and conversion to glucose in the liver, with the glucose returning to the muscle to supply more ATP, is known as the Cori cycle. Two married scientists, the Coris, worked diligently to elucidate this mysterious pathway. It is important to survival and, nowadays, to exercise physiology.

Sooner or later, dependent upon the genetic makeup of the animal and the amount of physical training done, either the pain stimuli from the nerves in the muscle or the buildup of lactic acid (also causing pain), or the buildup of carbon dioxide in the blood, or all three, eventually slow the animal down. If the animal is really genetically endowed and/or very well trained, this period can go on for significant intervals of time (several minutes), continuing

Table 12.3	Use of Energy by Various Activities in Horses and Energy Supplied by Feedstuffs	
Activity	DE (kc/h/kg BW)	Multiple of requirement at walking
Walking	0.5	1
Light, slow trot	5.1	10
Medium, fast trot, canter, jump	12.5	25
Heavy, canter, gallop, jump	24	48
Strenuous	39	77

Adapted from National Research Council, Nutrient Requirements of Horses, *1989.*

to use up glucose at fast rates. Additional glucose can be taken from the stores of glycogen in muscle and in the liver. However, the glycogen stores contain only a tiny percentage of the total energy available in body fat.

Within a few to several minutes, however, the usable glycogen stored in the muscle and liver is used and the rapid rate of anaerobic exercise can no longer be maintained. At this point, the muscle must slow its contraction rate, and the animal slows down. In sports training jargon, this point is called *hitting the wall*. Humans often still think they are going very fast but, in fact, they have slowed down. Even the most genetically lucky and highly trained animal must eventually succumb to the lack of oxygen, pain, and damage to muscle cells, reaching exhaustion.

Practical Feeding for Exercise

So, now that we have covered exercise physiology, let us start talking about exercise nutrition. Table 12.3 shows the amount of energy needed per unit mass for an average expenditure of energy in five different categories: walking, trotting, canter, jumping, and galloping or jumping at a more intense rate. The first part is simple multiples: As exercise (power) increases, the energy needed per unit mass increases. The second part compares this with the actual energy supplied by foods: average alfalfa, corn, oats, and fat. It doesn't take a rocket scientist, or even somebody that doesn't know a lot about horses, to tell that a little walking, or even trotting, can easily be supplied by a very small increase in intake of grain or even forage. An increase in 1 kg of forage is about a 12 percent increase in intake. If we feed fat, we only need about 300 g to supply a full hour of trotting, or only about a 4 percent increase in intake.

The Role of Fat in Exercise Feeding

For a seriously working horse, spending a total of an hour in strenuous activity, if we only provided alfalfa, it would have to almost double its feed intake (about 8 kg per day for maintenance versus an extra 7.8 for work). This is not usually practically possible. If we add in some fat, about 2.3 kg, or a 25 percent increase in intake, it would supply this energy. Now, practically,

Table 12.4	Energy Supplied by Feedstuffs for Specific Amounts of Work for a 500-kg Horse				
Activity	DE (Mcal per hour)	kg alfalfa/hr 2.5 Mc DE/kg	kg corn/hr 3.7 Mc DE/kg	kg oats/hr 2.8 Mc DE/kg	kg fat/hr 8.5 Mc DE/kg
Walking	0.25	0.1	0.07	0.089	0.029
Light (slow trot)	2.5	1.0	0.68	0.89	0.294
Medium (fast trot, canter, jump)	6.25	2.5	1.70	2.23	0.735
Heavy (canter, gallop, jump)	12	4.8	3.24	4.29	1.42
Strenuous	19.5	7.8	5.27	6.96	2.29

For strenuous exercise over time (days, weeks), you should provide approximately 25 percent more DE to maintain body weight.

> ### Nutritional Strategies for Exercising Horses, in Order
> 1. Observe condition regularly.
> 2. Supply only enough energy to maintain condition or lose at a desired rate.
> 3. Increase intake of the maintenance ration first.
> 4. Add additional maintenance ration second.
> 5. Extra grain only as needed.
> 6. Additional fat if exercise increases or intake is maximal.
> 7. Increase protein percentage in ration only if increase in muscle mass desired.

feeding 25 percent fat to a horse is too much, but with a combination of about 1 kg of fat and 3 kg of corn, we can feed what the animal needs. This would work out to about an 8 percent fat diet for a very active horse.

I would strongly recommend going with a commercial product, as mixing fat into grain is a tricky business, and it is messy and easy to go awry. If there are clumps of food with a lot of fat, the horse may not eat them, causing extra waste, or if they are eaten, that large amount of fat at once can cause some digestive problems and will not be adequately digested. If you do want to go it alone, be sure to use fresh oil, and be sure to mix thoroughly and only mix what the horse(s) will eat in one to three days. Table 12.4 shows requirements for working horses.

Early in this chapter, we touched on the difference in digestion of fat by the horse due to the fact that it does not have a gall bladder to store bile acids. Nevertheless, the liver of the horse does make bile acids and horses can digest fat, much more than we used to think. The main question is "If, when, and why" we should feed more fat than is normally present in the foods they would consume. Prolonged or strenuous exercise or work is the major situation in which horses can benefit from fat. Fat can be used judiciously to provide extra energy as needed. Using fat may also save a small amount of protein, as any grain or forage source would also contain some protein, which is not really needed by the horse and thus would be oxidized for energy and the nitrogen would have to be excreted as urea.

Another question is: Does using fat actually improve performance ability? There has been some good research conducted on this topic, with unexpected results. As glucose is used for exercise, the glycogen stores in the muscle must be used. The more glucose that is available from the glycogen, the longer the horse can maintain that intensity of exercise. It turns out that if we feed horses significant amounts of fat, usually about 10 percent of the total ration for a period of at least a few weeks, it will actually allow the muscle cells to make more glycogen. This is probably because the cells are adapted to use more fat for energy, and thus keep more glycogen in storage. Then, when the animal really needs the glycogen, in long or intense exercise, more is there to supply more glucose. This allows the animal to work harder for longer. Here is yet another example of the importance of the balance of nutrients for the use of each one.

The Role of Protein in Exercise Feeding

It is easy to understand the need for more energy (glucose and fat) during exercise, but what about protein? You may automatically think (or you may have read) that because the horse is using the muscles more, it needs more protein. Alternatively, you may think: "Well, if the horse does not need to grow, does it need more protein?" As is often the case in nutrition, all of these situations are correct for specific instances. A slight increase in exercise load does not lead to a major change in protein synthesis in the muscle; little to no additional protein is needed. As the length or intensity of exercise increases, the muscle does increase the rate of protein synthesis and the rate of protein breakdown. As we learned in Chapter 4, because of the unequal use of amino acids in the metabolic pathways in the muscle, this increases in a small but real amount, the amino acids needed to maintain muscle mass. If the animal increases food intake by 1,000 grams to meet an extra 2.5 Mcal of energy, in a normal maintenance diet, an extra 100 to 140 grams of protein will be consumed, and that is more than enough to meet the increased metabolic need. Extra protein does not need to be added to the ration.

Only if we are feeding a growing younger horse in early training, or if we wish to increase the muscle mass of the horse, will an increase in amino acids be needed to supply that growth. So usually only in these two instances (growing horses in training and adult horses that require some increase in muscle mass) should an increase in the protein concentration in the diet be considered. Again, one should be sensible; we are not talking about massive amounts, and the NRC (1989) recommends only a 1 to 2 percent increase in protein concentration. It is more important in the long run to save expensive protein, reduce the load on the environment, and reduce the nitrogen load on the kidneys than to "bulk up" a horse unnecessarily or to increase growth too quickly, risking developmental orthopedic diseases.

Mineral Balance in Exercise

Because of the rapid rise in heat production due to exercise metabolism, the horse can lose a liter of sweat for about 600 kcal of heat, or that given off by seven to eight minutes of a moderate trot. A maximal sweating rate can range from 9 to 15 liters per hour. A more thorough account of evaporative cooling, sweating, and dehydration can be found in Lewis (1995). For mineral nutrition, we need to understand that the horse can lose sodium, chloride, and potassium in significant amounts, and calcium and phosphorous in lesser amounts. A grain supplement formulated to increase the amount of these minerals should be used. Alternatively, a balanced, free-choice mineral mix is usually effective. Minerals or salt should not be added to drinking water, as they can reduce water intake. The grain mix also provides the needed carbohydrate (and fat, possibly). In more hot or strenuous situations, sodium chloride can be added at greater amount. Again, refer to more detailed texts and practitioners for more in-depth information on this subject.

Reproduction

If you need to review the basics of reproduction nutrition at this time, please review Chapters 2, 10, and 11. The same principles apply to horses. The nutrient requirements during the first trimester of pregnancy do not change significantly from the maintenance amounts, so if the horse is consuming that properly balanced ration, it should continue to be fed for the first few months. Usually, you would then note a gradual increase in voluntary intake somewhere around six to eight months. By eight to ten months, intake may increase 10 to 20 percent and should probably not increase any more. If the proportion of grain has been increased, total intake should increase less, as there will be plenty of energy in the grain. In general, a larger horse with a normal intake would not noticeably increase intake quite as much, whereas a smaller horse (thus carrying a proportionally larger foal) might show a greater percentage increase in intake. Individual observation remains the key. We want to provide the proper amount of a balanced ration to maintain the mare's body condition. If the mare is consuming a good legume forage, or early-cut grass hay, no other supplement besides trace mineral salt is needed. If lower protein (more mature cut) grass hay is fed, additional grain in the range of only 0.5 to 1 lb per 100 lb of body weight (0.5 to 1 percent of BW) should be supplied, of course with a vitamin and mineral supplement within the grain.

Pregnancy is not the time to add or subtract significant weight. If the horse seems to be thinning down, provide a slightly larger amount (5 to 10 percent more for several days). If the horse seems to be gaining fat, hold off only about 5 percent (we do not want to restrict too much to avoid vitamin and mineral shortfalls). If the dam loses too much weight, the development and growth of the foal may be limited. Too much food can lead to obesity in the mare, perhaps sometimes a foal slightly too large, and increase the incidence of *dystocia* (difficult birth). It is rare for increased intake alone to increase the size of the developing fetus, but in some instances it can happen.

More important is to avoid obesity in the mare, because the internal fat can physically restrict the movement of the foal, perhaps leading to *congenital deformities* (a situation that the animal is born with) flexure deformities and problems with labor. So, watch the body condition.

In the very last days of gestation, you may note a decrease in voluntary food intake. You should not be alarmed, as this is normal. It is a crude indication that foaling may be one or a few days away. After foaling, just as with other mammals, the mammary gland kicks into high gear and provides milk to the foal(s). Requirements of all nutrients can increase 10 to 25 percent within just a few days. The horse has a relative advantage as it usually bears a single young, who has a body mass only 10 to 15 percent of the mother's mass. Thus the increase in requirements is not as drastic as it is for the smaller and litter-bearing species such as dogs and cats. Yet the intake of a balanced ration (good-quality forage with some supplemental grain) may increase 80 to 100 percent compared to maintenance. Depending on the horse and the foal, the amount of grain may be increased. There is no hard and fast formula; continue to watch the body condition of the mare and adjust accordingly.

Aging Horses

By now, you might predict that the same general concerns that we had for dogs and cats will apply to aging horses, and you would be correct. The basal metabolic rate slows, the muscle and bones start to lose mass, and the digestive and waste-processing organs lose some of their functional ability. In most cases, activity will also be less. The same principles are applied with the same practices. We want to provide a lesser amount of a highly digestible ration. A high-quality grass or legume forage, approximately 6 to 10 percent protein for grass, or 15 to 18 percent for protein, with an ADF of 30 to 35 percent is appropriate. For some horses, continued use of a balanced vitamin and mineral mix may be all they need. For others, a small amount of grain with a balanced mineral and vitamin mix and some high-quality protein such as soybean meal might be necessary. It is usually better to provide a little grain mix to ensure proper vitamin and mineral intake; and inclusion of a little molasses as a sweetener can be used judiciously to ensure proper intake and as a little treat.

We continue to practice our individual observation, feed to maintain a good body condition score, and as much as possible, provide some regular exercise. Regular exercise in older animals helps to maintain circulatory and respiratory health, as well as helps to maintain bone and muscle mass.

Other Problems in the Nutrition of Horses

There are a few situations unique to the horse that, although not directly caused by the nutritional management of the animal, do require some special nutritional and management care. The two most important ones we will touch on here are known in horse jargon as *azoturia,* or "tying up," with its subset *polysaccharide storage myopathy (PSSM).* The other is *hyperkalemic periodic paralysis (HYPP).* These are very serious and potentially debilitating problems. Luckily, their incidence is still relatively low and can be managed with some care and foresight.

Azoturia

Azoturia is another word of Latin origin, meaning "dark urine," and is the medical term given to the condition known as "tying up." The technical term used for this condition is *exertional rhabdomyolysis.* This means damage to the muscle due to improper exercise followed by inactivity. Within this situation, we now know of a unique genetic condition: polysaccharide storage myopathy (PSSM) or equine polysaccharide storage myopathy (EPSM). It is not yet clear whether there are two genetically distinct abnormalities here, but the result is the same. Figure 12.7 shows the symptoms and effects of azoturin.

This problem was first noted by horse owners working with the horses in the field. After a hard workweek, the horses were cleaned, bedded, and left to rest for part of Saturday and Sunday. Going back to work Monday morning, it was noted that the horse would be stiff and sore (not in itself unusual, given the situation), but in many cases this was accompanied by dark colored, often sweet-smelling urine. In addition, in some cases, the stiffness was so severe that the horse was in extreme pain, often unable therefore to move, and perhaps so immobile and in pain that euthanasia was at the time the only humane course.

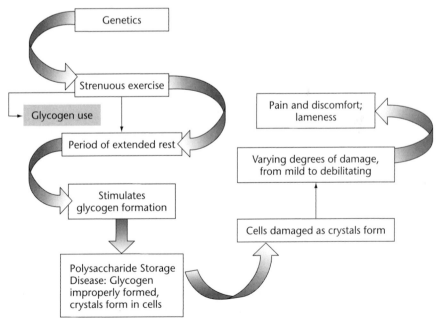

Figure 12.7   Azoturia, 'black water,' 'tying up,' exertional rhabdomyolosis, polysaccharide storage myopathy.

Although not many horses are used for farm work in the United States anymore, our exercising horses can still present with this situation.

The problem was usually diagnosed as severe muscle damage due to the hard exercise and then "forced immobility." This results in fluid retention in muscle cells. This is due to the lowered blood flow, as during even mild exercise the more rapid blood flow will help to keep water retention down. In addition, the synthesis of glycogen, after it was used up during exercise, normally draws more water into the cell. In severe cases, this can lead to cell swelling and rupture. This is accompanied, as you might guess, by pain and sometimes permanent disablement. There is also another cause, which is a temporary and acute calcium deficiency. We have already learned that calcium is needed for normal nervous transmission and muscle contraction. A calcium lack will cause improper tensing and relaxing of muscle, leading to cramping.

There is a specific mutation of glycogen formation in some lines of horses. Glycogen normally requires water interspersed with the molecule. In some horses, the glycogen is not formed properly and builds up in the cell, eventually crystallizing out like sugar crystals. This disrupts the cell membrane, causing temporary to permanent damage. The discovery of PSSM (also known as EPSM) has demonstrated that there is a genetic component present in the horse population today. It may never be known if a similar genetic problem was present earlier. PSSM has been reported to be most common in quarter horses, but is also present in thoroughbreds, especially fillies and draft breeds. However, the key ingredient was determined by modern genetic analysis. We now understand that this particular problem dates to a genetic mutation in one or two stallions alive in the 1920s and 1930s. As these horses and their offspring were used as breeding stock, the mutated gene was spread more widely around the horse population. Although not all symptoms and cases of "tying up" are directly caused by this mutation, we can avoid this serious mistreatment of animals with some simple genetic testing. If you are concerned that this might be a problem in horses with which you are familiar, you should consult with a veterinarian trained in this area about getting a genetic test done.

Treatment for tying up, once diagnosed, is absolute rest, veterinary treatment, perhaps to including sedatives, muscle relaxants, and medicines to adjust the acid balance of the blood upset by the waste products coming out of the damaged muscles. In some cases, horses can readily recover but, unfortunately, in too many cases, the severity of the muscle damage means the horse may be debilitated to a large degree for a long time, if not the rest of its life. In some severe cases, euthanasia is the only humane alternative.

This condition is easily decreased by preventative management. The prevention part comes in application of our common-sense principles: Feed a balanced ration in small meals to avoid excess glycogen storage, and exercise moderately and regularly, avoiding long periods of hard work followed by days of little or no work. Some other basic nutritional management can minimize such symptoms, whether or not the horse is a carrier of the gene. Too much dietary glucose at once can lead to too much glycogen formation, making the problem worse. More frequent, smaller meals with a minimum of grain will keep glycogen storage to a minimum. A low intake of dietary calcium can affect the muscle tone and contraction, altering glycogen use and also making the problem worse. Feed a lower-energy diet to susceptible horses, little or no grain, and added fat (25 percent of energy or 10 percent of diet). Also, feed high-quality hay such as alfalfa, which contains more calcium, but be sure to balance with phosphorous from the grain mix and supplement to avoid a calcium:phosphorous ratio that is too high.

Hyperkalemic Periodic Paralysis

Another problem known for quite awhile to be a genetically caused situation is known as hyperkalemic periodic paralysis (HYPP). Basically, that translates as paralysis caused by a temporary increase in blood potassium. It is usually reported in quarter horses about two to five years old. The symptoms include periodic episodes of muscle weakness and elevated phosphorous levels (revealed by veterinarian-performed blood tests). Specific symptoms include muscle fasciculation (quivering), ataxia, paralysis of hindquarters (dog sitting), generalized sweating, percussion myotonia (problems with the ear), prolapse of the third eyelid, and laryngeal paralysis (problems with normal breathing and swallowing). This condition is not usually fatal, but can be. It has been reported to be brought on by a meal, change in feed, or stress and is not usually exercise related. However, though it is noted that a change in feed may bring it on, it is not a nutritional problem; it is a genetic problem. Horses that do not carry this gene will not have this problem. All the nutritional management in the world cannot completely prevent the possible expression in horses carrying the gene. Figure 12.8 shows the symptoms and effects of HYPP.

A more specific explanation is provided by reviewing cell biology. Cell membrane ion transport enzymes usually keep intracellular potassium concentration high and sodium low, and blood potassium low and sodium high. But in this genetic mutation, a problem with the enzymes that transport sodium and potassium across the membranes causes an increase in blood concentrations of potassium, leading to a generalized muscle weakness—most noticeable in large hind-limb muscles. Potassium concentrations are extremely important to normal muscle function. Treatment can return plasma potassium to normal, protect heart muscle against high blood potassium, and prevent further episodes.

A practical way to decrease the dietary potassium in HYPP carriers is to feed oat hay or reduce or remove potassium from the mineral mix. Have any hay you feed tested by a competent laboratory to determine the amount of potassium. Hay grown on fields with too much potassium fertilization will have more potassium in the plant. If the horse is on pasture, altering the mineral fertilization of the pasture can reduce the potassium concentration in the plant. Consult with a nutritionist or a veterinarian well versed in this disease and in preventative nutrition. The percentage of potassium in oat hay is 0.4 percent on average compared with .3 to .5 percent in corn and barley, 3.5 percent in orchardgrass, and 1 to 2 percent in alfalfa (and beware that

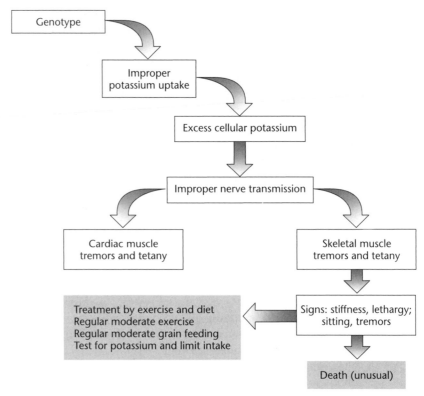

Figure 12.8 Hyperkalemic periodic paralysis.

potassium percentage of all hays can vary by 100 percent depending on season, location, and fertilization rates, so if in doubt, get it tested). The bottom line is that a simple genetic test prior to using a horse for breeding purposes can eventually lead to the removal of this serious problem from the horse population.

Myths, Secrets, and Facts

As is the case for most nutritional practices, there are a lot of so-called secrets, special tricks, and outright myths and fables out there in the nutrition of horses. Some are based in some measure of truth, others are so new that we do not really understand their true worth, and some are just outright bunk.

Sodium Bicarbonate

Sodium bicarbonate (Na_2HCO_3, known as common household baking soda) is a buffer to reduce the concentrations of acids in solutions. Sodium bicarbonate has been used to buffer the rumen of cattle to help improve the bacterial fermentation when cattle are fed grain. Somewhere along the line, someone got the idea that sodium bicarbonate might be able to buffer the metabolic acids that are made during exercise, primarily lactic acid. In this way they may decrease some of the associated pain, improve performance, and reduce recovery time. Some studies have shown that, in fact, small amounts of sodium bicarbonate given in water two to three hours before exercise can maintain a higher blood pH (less acid). However, there is no strong scientific evidence that this practice actually improves athletic performance such as speed or endurance. In addition, the increase in the carbonate ion in the blood after treating with sodium bicarbonate binds up the blood potassium, causing hypokalemia (low blood potassium). Bicarbonate can also decrease blood calcium. Both of these, even in the short term, may affect heart and skeletal muscle function. The buffering effects are highly variable with the size of the animal and only last a short time, thus managing the use (even if it were to work) is a serious challenge. Use of sodium bicarbonate is not illegal, but its use is not recommended by any serious exercise physiologist.

Myths and Facts about Nutritional Tricks for Horses

Myth	Seed of Truth	Actual Facts and Problems
Blood builders improve O_2 carrying capacity	Iron helps transport O_2	No scientific evidence Too much iron is toxic
(Di)methylsulfonylmethane beneficial effect (DMSO; MSM)	Free radical remover	No scientific proof of this
Lactobacillus beneficial effect	Establish proper intestinal bacterial population	Effectiveness has been demonstrated, especially for animals that have been ill or off feed

Probiotics

The term *probiotics* is given to a wide variety of agents thought to improve the bacterial population of the digestive tract. These may be bacterial cultures, yeast cultures, or other mixtures thought to change bacteria in the digestive tract to a more benign and helpful population. The effects certainly can exist and are based on scientific principles.

Yeast cultures are used in dairy cattle with some good effects to help prevent or reduce the severity of problems with the bacterial fermentation in the rumen. There are products on the market for horses, ranging from well-documented solutions of lactobacillus species to products not much better than snake oil. Certainly some of these products have the potential to help restore proper bacterial populations to the digestive tract after some kinds of disease or digestive upset. However, many more simply have no effect. One possible use of lactobacillus may be to reduce lactic acidosis in high-grain diets. The other side of that coin is that it is only in rare situations (primarily lactation or strenuous exercise) that diets with that much grain should be fed. Look for products containing live bacteria, not bacterial products or killed bacteria. The latter are only expensive vitamin supplements.

Potential Poisoning of Horses

Horses, as is the case for other species, can be susceptible to poisoning from different feeds and contaminants. Specifically, horses are quite susceptible to the toxins produced by several common molds. There are far too many situations to cover in this text; the more advanced reader should start with Lewis (1995, Chapter 19). The following box gives some common problems. The practical management, as ever, is to observe and inspect the feed. Watch pastures for molds, especially in wet spots and rainy seasons. Store grain dry in amounts that will not be stored for longer than a few weeks. Inspect grains that you buy (as President Reagan said: "Verite, no doverite," or "Trust but verify").

Danger Feeds for Horses

Vomitoxin, aflatoxin, mycotoxins—horses are very sensitive.

Be sure to store grain dry for short periods of time. Always buy grain "on sight."

Fescue poisoning—alkaloids from ergot molds. Always inspect pastures.

Estrogens from fusarium molds.

Slobbers—from slaframine from Rhizoctonia legumincola, in legume, especially clover, pastures with black spots on leaves.

Sweet clover disease—moldy clover from Aspergillus.

Grass staggers—excitability to tetany, from molds in bahia grass or ryegrass.

This has been a long chapter covering a wonderful, varied, and useful species. The horse has certainly been more than a companion in the history of humanity, providing economic, agricultural, and even military advantages to populations over the centuries. In too many situations, they have been reduced to a mere "backyard grass cutter," with benign neglect sometimes leading to problems in development and obesity. For those reading with serious interest in horses, I hope that the introduction provided here helps to ensure that in the future these noble animals are treated with the respect that they have earned, and are kept and fed to be healthy and long lived. There is much more to learn; readers are thus referred to the Further Reading section.

Words to Know

aerobic exercise	colon	hyperkalemic periodic
aerobic threshold	concentrate	paralysis (HYPP)
anaerobic exercise	congenital deformities	legume forage
azoturia	creep feed	osteochondrosis
bacterial fermentation	dystocia	polysaccharide storage
bile acids	epiphysitis	myopathy (PSSM)
blood builders	gall bladder	probiotics
cecum	grass forage	volatile fatty acids

Study Questions

1. Describe the differences in the digestive physiology of the horse as compared to dogs and cats. How does this provide benefits and disadvantages to the horse?
2. Describe the role of cellulose and hemicellulose (ADF, NDF) in digestion and nutrition of the horse. How does the horse derive energy from the fiber in the diet?
3. What are some fibrous feeds (pasture, hay) commonly in use by horses and what are their nutritional advantages and disadvantages?

Further Reading

Cheeke, P. R. 2004. *Contemporary Issues in Animal Agriculture*. Upper Saddle River, NJ: Prentice Hall.

Frisbie, D. D., and C. W. McIlwraith. 2000. Evaluation of gene therapy as a treatment for equine traumatic arthritis and osterarthritis. *Clinical Orthopedics* (October) 379 Suppl: S273-87.

Harris, P. 1997. Energy sources and requirements of the exercising horse. *Ann. Rev Nutr.*, 17: 185–210.

Hintz, H. F. 1994. Nutrition and equine performance. *J. Nutr.*, 124 (12 Suppl): 2723S–2729S

Hintz, H. F., and N. F. Cymbaluk. 1994. Nutrition of the horse. *Ann. Rev. Nutr.*, 14: 243–67.

Lewis, L. D. *Feeding and care of the horse*. 1995. Philadelphia, PA: Williams & Wilkins.

National Research Council. 1989. *Nutrient requirements of horses*, fifth revised edition. Washington, D.C.: National Academy Press.

Naylor, J. M. 1997. Hyperkalemic periodic paralysis. *Veterinary Clinical North American Equine Practitioner* (April) 13(1): 129-44.

Purina Mills, LLC. 2004. Horse nutrition. *(http://horse.purinamills.com)*

Spier, S. J., G. P. Carlson, D. Harrold, A. Bowling, G. Byrns, D. Bernoco. 1993. Genetic study of hyperkalemic periodic paralysis in horses. *Journal of the American Veterinary Medical Association* (March 15) 202(6): 933-7.

Valentine, B. A. 2003. Equine polysaccharide storage myopathy. *Equine Veterinary Education*, 15: 254–262.

Valentine, B. A., H. E. Hintz, K. M. Freels, A. J. Reynolds, and K. N. Thompson. 1998. Dietary control of exertional rhabdomyolysis in horses. *J. Am. Vet. Assoc.*, 212: 1588–1593.

Valentine, B. A., R. J. Van Saun, K. N. Thompson, and H. F. Hintz. 2001. Role of dietary carbohydrate and fat in horses with equine polysaccharide storage myopathy. *J. Am. Vet. Med. Assoc.*, 219 (11): 1537–44.

Chapter 13

Nutrition of the Rabbit, a Lagomorph

Take Home and Summary

Let us now turn to our friends the rabbits. The principles of nutrition and metabolism we have learned apply to these little hoppers. Yet they are unique to the point that they are their own order, the **Lagomorphs**, which includes only the pikas, hares, and rabbits, based primarily on their digestive physiology. They are nonruminants; but they are herbivores, adapted to a diet of primarily forage. Similar to the horse, they have a specialized digestive tract that requires sufficient fiber to function. The fiber content of their diet cannot vary too much without risk of problems. They have a circadian rhythm in digestion; during the night hours the bacteria in their cecum and colon make cecotropes, specialized high-protein fecal pellets, which the rabbits then consume. Thus, rabbits can capture much more nitrogen through coprophagy of the protein-enriched cecotropes, improving efficiency. They must have a high-fiber, low-energy diet to do well, and regular exercise to a rabbit is just as important as it is to horses, dogs, or cats.

The Specialized Digestive System

The lagomorph digestive system begins with sharp teeth for cutting just about any forage or woody plant, to give rabbits a wide variety of foods to choose from in the wild, even when fresh forage material is not available. The two incisors that give the animal so much power become a bit of challenge, as they need constant work to maintain ideal size. These teeth must either be provided with real forage diets to "work on," or you must pay the vet for costly rabbit dentistry work that could be avoided by feeding a proper diet. Figure 13.1 shows the unique dental and jaw structure of the lagomorph, it is similar to the rodent in structure but not in function. There are upper and lower biting incisors, a large gap to accommodate the long forages, and sharp broad grinding teeth to masticate the rough plant material.

The rest of the gastrointestinal tract of the rabbit is similar to that of the horse until we get to the large intestine (Figure 13.2). This system does not simply look different, but in fact changes the digestive physiology of the animal completely.

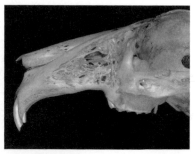

Side views of the cranial skull, note sharp incisors and large diastema, similar to rodent.

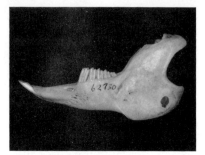

Lower jaw: note large diastema, but more forward
thrusting lower incisors and larger, sharper molars
than rabbits, adapted to grinding forage materials.

Figure 13.1   Lagomorpha: another unique adaptation.
Source: Photos courtesy of Dr. Phil Myers, Animal Diversity Web, University of Michigan, with permission *(http://animaldiversity.ummz.umich.edu).*

Similar to that of the horse, the rabbit *cecum* is replete with anaerobic, cellulytic bacteria to break down plant cell walls to cellulose, and then to glucose, which is fermented to provide energy for microbial growth. The excess volatile fatty acids are absorbed through the cecal wall to provide energy to the rabbit.

The rabbit is unique in its digestive physiology. Rabbits have a circular-structured cecum with a specialized nervous physiology.

However, the rabbit's system goes one step further. The nervous system alters the activity of the tract in a circadian rhythm, so that during the day, the microbial activity is normal and the rabbit makes normal little rabbit fecal pellets. However, during part of the circadian cycle (usually at night), the tract motility changes, and pellets called cecotropes are formed. *Cecotropes* are fecal pellets that consist of an enriched population of microbes. The microbes were busily growing off the fiber in the tract, but instead of being flushed out, they are allowed to build up to greater concentrations by the slower motility of the tract during the night hours. Then, when excreted, the cecotropes are more concentrated in crude protein and lower in fiber (because most of the fiber was digested).

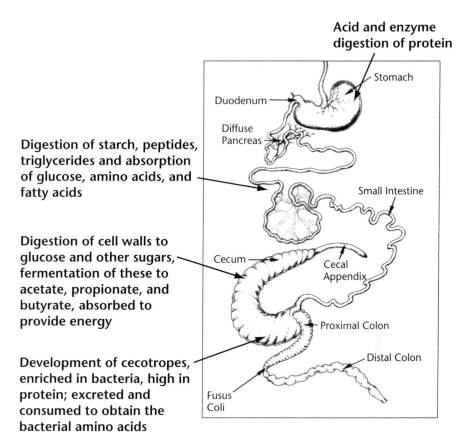

Acid and enzyme digestion of protein

Stomach

Duodenum

Diffuse Pancreas

Digestion of starch, peptides, triglycerides and absorption of glucose, amino acids, and fatty acids

Small Intestine

Digestion of cell walls to glucose and other sugars, fermentation of these to acetate, propionate, and butyrate, absorbed to provide energy

Cecum

Cecal Appendix

Proximal Colon

Distal Colon

Development of cecotropes, enriched in bacteria, high in protein; excreted and consumed to obtain the bacterial amino acids

Fusus Coli

Figure 13.2 The specialized digestive system of the lagomorph rabbit.
Source: Redrawn from *Rabbit Feeding and Nutrition,* Peter Cheeke, Figure 3.1, Copyright 1987, with permission from Elsevier.

Fact Box **Regular Feces and Rabbit Requirements**

	Cecotropes	*Normal Feces*
CP, %	38	15
Ash, %	14	15
Fat, %	1.5	1.8
Fiber, %	14.3	27.8

You may be thinking that the crude protein contained in the cecotropes is still just a waste product, and thus may be wondering what it is used for. Those of you with weak stomachs might want to be sure you are in a good mood right now, because what makes this evolutionary adaptation so efficient is that the rabbit consumes the cecotropes, and then digests them in the small intestine just like any other food. Consumption of feces is *coprophagy* and consumption of cecotropes is called *cecotrophy.* This re-cycling and capture of the bacterially produced amino acids allows the rabbit to live in environments and during seasons when only low-protein fibrous plants are available. However, this adaptation also dictates that the rabbit companion be provided with sufficient forage to allow normal digestive motility and enough fiber for the cecal microbes to use. We will discuss a serious problem in rabbits stemming from this digestive adaptation after we cover the normal life-cycle nutritional management.

> We must feed a high-fiber diet to maintain the normal circadian rhythm for cecotrope production to keep the digestive tract healthy and allow the rabbit optimal nutrition.

Life-Cycle Nutrition of the Rabbit

Suckling Phase

The litter of rabbits, after they are born, or *kindled* as the jargon goes, is supplied with food completely by the mother. However, unlike dogs and cats, which usually allow the litter to suckle several times during the day, rabbits typically allow the litter to suckle few times a day. If one stops and thinks for a minute about potential evolutionary advantages, this makes a bit of sense. Rabbits, as prey animals, had an advantage to keep the litter hidden, and thus the less they moved about, the less attention they would draw. This explanation makes as much sense as any other theory, although we really do not understand exactly how these types of things evolved. In any case, it is not unusual for the mother to ignore the litter for long periods of time. Allowing normal suckling behavior is the best management protocol for the litter.

The same principles we learned earlier also apply to preparing rabbit kits for weaning. Small amounts of high-quality forage should be provided, preferably green (not only dried hay, although dried hay is acceptable), and small amounts of a balanced grower pellet. The key is small amounts here to help the digestive tract adapt, not to provide large amounts of energy to the rabbits. Once the litter is weaned, high-quality forage (again preferably fresh, but good quality hay will work) should be supplied. Table 13.1 shows nutrient requirement for rabbit and Table 13.2 lists the nutrient content of rabbit feed. A small amount of grower pellet should be provided, primarily to provide minerals and vitamins to supplement most forages. This pellet will also supply some high-quality protein and energy, but the key (as always) is to avoid overfeeding and watch the rabbits, check for excess body fat by palpation, and provide regular exercise.

Growth

> The objectives of a growth-management plan for rabbits is the same as for any other species: Provide optimal, not maximal, amounts of protein, energy, vitamins, and minerals for organ, bone, and muscle growth and development.

Table 13.1	Nutrient Requirements for Rabbits during the Life Cycle							
Life Stage	Body Weight Kg	Energy Kc Me/ kg diet	Energy Kc/d	Feed Intake G/d	CP %	NDF/ADF %	Ca %	P %
Weanling	<1.0	2.75–3.25	200–300	130	16–18	21	0.4	0.22
Growing	1–2	2.5–3.0	250–440	170	16	21	0.4	0.22
Adult	2–4	1.8–2.5	180–250	130	12	24	0.3–0.4	0.2
Pregnancy	2–4	2.75–3.25	350–450	160	15	21	0.4–0.5	0.35–0.45
Lactation	2–4	2.75–3.25	550–750	230	17	21	0.7–0.8	0.4–0.6

The data for weanling assumes 1 kg BW; grower, 2 kg BW; adults, 4 kg.

The estimate for food intake assumes the average of the range of weights and energy needs and 2.5 kc/g DE in food for growth and pregnancy, 2.2 for maintenance, and 2.6 for lactation. Note that there is no requirement for fat other than for the essential fatty acids. Most normal rations will contain 3 to 5% fat.

Source: Adapted from P. R. Cheeke, 1987, Rabbit Feeding and Nutrition, and NRC, 1977, Nutrient Requirements of Rabbits.

As for horses, forage is the primary source of nutrients. Providing proper amounts of dietary fiber is critical, usually a minimum of 12 to 16 percent crude fiber, or 20 to 24 percent ADF. Providing most of the fiber as forage is best for teeth development and normal gastrointestinal physiology. Crude protein should be approximately 16 to 18 percent, digestible energy about 2.1 to 2.5 Mc DE/kg diet. This pattern of nutrients is similar to that for the horse, but only 60 to 70 percent of that required by cat. Cat or dog food products would provide the rabbit with far too much energy and protein and not enough fiber, a dangerous situation that should be strictly avoided.

Adult Maintenance

The rabbit will spend the majority of its lifespan at maintenance. Remember the definition of maintenance, which includes basic bodily functions, temperature, and voluntary activity. So if you have a little "cage bunny" kept indoors, his or her maintenance requirements will be quite small. If you have a nice backyard and the bunny can roam, and you have it out during the winter, the maintenance requirement can be quite a bit larger. Usually, we would feed the same types of foods to these two different animals, but alter the amount fed. In the more extreme case in which a rabbit might be quite active and out in the cold for extended periods, increased amounts of a grain supplement can be fed without the risk of obesity or digestive problems.

Reproduction

Requirements for pregnancy and lactation in the rabbit follow the same general principles we have already learned. Rabbits will generally produce four to ten kits per litter. Gestation time is about thirty-two days, and requirements do not usually increase significantly until fifteen to twenty days. Checking to

Table 13.2	Nutrient Content of Foods for Rabbits				
Food Item	**CP**	**ADF**	**Energy**	**Ca**	**P**
	% of diet	*% of diet*	*Mcal DE/kg*	*% of diet*	*% of diet*
Alfalfa					
Early bloom	21	32	2.5	1.4	0.4
Late bloom	17	39	2.2	1.2	0.24
Corn	10	4	3.8	0.05	0.3
Oats	13	16	3.2	0.09	0.4
Wheat	12–15	4.0	3.85	0.04	0.4
Barley	11–13	7.0	3.7	0.05	0.4
SBM					
Grower Ration	18	21		0.5	0.3
(40% alfalfa meal, 45% barley, 15% SBM, 0.5% trace mineralized salts)					
Maintenance Ration	14–16	35		0.4–0.8	0.3–0.5
(90% grass hay or alfalfa or alfalfa meal, 10% rabbit pellet; alfalfa-based diets will have more protein and calcium)					
Lactation Ration	16–18	21		0.7–1.0	0.5–0.8
(40% alfalfa meal, 40% barley (or corn, wheat), 14% SBM, 3% molasses, 1.5% fat, 1.25% salt and calcium supplement)					

Source: Adapted from P. R. Cheeke, 1987, Rabbit Feeding and Nutrition, *and NRC, 1977,* Nutrient Requirements of Rabbits.

make sure the ration you are already feeding is adequate for maintenance is the first step, and as the dam's requirements start to increase, she will increase her intake to meet them. Around week 3, addition of a supplement balanced for growth and lactation can be slowly introduced, primarily to ensure proper amounts of vitamins, minerals, and high-quality protein are supplied. Energy requirements will increase 10 to 25 percent, depending on the size of the litter she is carrying. A small increase in grain may be offered, but watch the weight and remember that 10 percent is not 50 percent.

Toward the last trimester of pregnancy (week 3), the animal still should be consuming primarily forage, with high-quality supplement available as needed. Nesting behavior will begin about two to four days prior to kindling and a proper nesting box or space should be provided. In the day or two just prior to kindling, feed intake will decrease. When the rabbit has the litter, lactation will commence. As for the rest of the mammals, requirements of all nutrients, including water, will increase quickly. At this time, you should still provide free-choice forage, but also allow intake of a complete supplement as needed. Let your observation of the dam be your guide—if she is always hungry and is losing weight, she should be allowed to eat a larger amount. Of course, if she is not losing but gaining weight, she need not have as much. Tables 13.1 and 13.2 give some basic guidelines. Especially for the younger owner, having a scale and measuring the weight of the dam and the litter every few days is a good learning experience, and the youngster can relate the weight of the dam and the weight gain of the litter (supplied by the milk) to the amount of food consumed. Experimental data collection will help reinforce the principles we have learned.

Old Age

The geriatric rabbit has some additional situations of which the owner should be aware. Loss of teeth can severely affect the animal's ability to forage or to masticate (chew) long forage, even if fresh or in short pieces. A high-quality balanced pellet should be offered, and it may need to be wetted slightly to help the rabbit consume it. In some animals, obesity can be an additional problem, and as the animal ages and the natural drive to exercise or forage lessens, the sedentary animal can become more obese with the natural consequences of heart weakness and kidney and skin problems that we have discussed in previous chapters.

If a high-fiber, low-energy diet is not normally supplied, the digestive physiology of the rabbit changes, and passage of nutrients through the tract is slowed. This is because the fiber consumption also increases saliva production, helping to move the material through the intestine at a proper rate, and still allowing for normal fiber amounts to reach the large intestine. If diets too high in grain (starch) are provided, the starch builds up in the small intestine, saliva production is reduced, and less fiber is delivered to the lower tract. This causes two problems. First, as digestive motility slows down, the tract can be impacted (stopped up) with large boluses of undigested feed. This is a life-threatening situation, as the tract is stopped up and the animal can be in intense pain, as well as starving. The second problem stemming from low-fiber, high-energy diets is related also to the rabbit's normal practices of grooming.

Gastric Stasis—Trichobezoars (Hairballs)

Rabbits routinely groom themselves, resulting in significant quantities of hair passing through the digestive tract. In normal situations, the saliva stimulated by the fiber in the diets simply helps to move the indigestible hair along and out. But if gastric and intestinal motility is slowed with rations containing too much grain and too little fiber, resulting in *gastric stasis,* the hair can build up into a "hairball," the official term for which being one of the best words I have ever learned in nutrition: *trichobezoar.* Fans of a young wizard will recall a meeting with him and a professor, who asked him the uses of a "bezoar." Because rabbits can't cast spells, this is a life-threatening situation, and far too many rabbits that have been given too much in the way of high-energy treats have met an early death (or at least an unnecessary trip to the vet) because of this condition.

Trichobezoars are hairballs that get stuck in the tract, usually in the stomach, because fiber and saliva flow are reduced on high-grain diets.

Rabbits cannot vomit; thus, the trichobezoar cannot be forcefully removed, and remains stuck. Feed and water intake decreases, and eventually even the least astute owner figures out something is wrong. By this time, there is often nothing that can be done. Caught early enough by responsible monitoring of the animal, a trip to the vet may still be able to save the animal. Rehydration is necessary, and mineral oil, drugs to increase gastric motility, and a change in diet to more fiber can sometimes move the bezoar through if it is not too big.

In addition to the problem of the hairball, because the cecotropes are not formed, there is instead the formation of chronic soft stools. These do not have the same composition, and are not useful to the animal, even if they are consumed. Thus the animal cannot meet its normal nutrient requirements. As with most nutritionally related problems, prevention through knowledgeable application of principles is better than trying to fix the problem after the fact. Also, as is true for most other animals, a regular exercise program, usually by construction of a larger, mobile enclosure that allows the rabbit to move around and graze, will help induce normal intestinal motility and reduce the risk of problems.

I hope this has provided a useful introduction to the basics of nutrition of rabbits. As for most animals, there is a lot of information out on the Internet, some of it good. Though you can find more detail there, be sure to be critical. The reader interested in serious data and lots of detail, especially on dietary ingredients and content, should certainly read Dr. Cheeke's (1987) seminal book. Even though, when written, it had a major audience of those growing rabbits for food, the diets of the rabbits remain the same regardless of the purpose.

Words to Know

cecotropes	*coprophagy*	*trichobezoar*
cecotrophy	*gastric stasis*	
cecum	*lagomorph*	

Study Questions

1. How are rabbits different from dogs and cats in their digestive physiology and thus their nutrient requirements?
2. How are rabbits different from horses? Ruminants?
3. Describe the basic feeding practices for rabbits during the life cycle.
4. What is gastric stasis, and how can it be prevented?
5. What are trichobezoars, and how can they be prevented? Treated for?

Further Reading

Carpenter, J. W., and C. M. Kolmstetter. 2000. Feeding small exotic mammals. In M. Hand, C. D. Thatcher, R. L. Remillard, and P. Roudebush, eds., *Small animal clinical nutrition*, fourth edition (chapter 29). Topeka, KS: Mark Morris Associates.
Cheeke, Peter R. 1987. *Rabbit feeding and nutrition*. New York: Academic Press.
National Research Council. 1977. *Nutrient requirements of rabbits*, second revised edition. Washington, D.C.: National Academy Press.

Chapter 14

Llamas and Alpacas: Ruminant Companions

Take Home and Summary

Llamas and alpacas are members of the Camelid family, which makes them ruminants. They are the only ruminants routinely kept as companions, and are used for their wool, for packing, working with children, or as with too many horses, simply for "grass-cutting." Ruminants have a large fermentation vat, made up of the rumen, reticulum, and omasum, which are prior to the gastric stomach (abomasum). In these organs, the fibrous plants are soaked and broken down by enzymes from anaerobic bacteria to glucose, amino acids, and fats. The amino acids and fats are used to make new bacterial cells, and the glucose is fermented to the volatile fatty acids acetate, propionate, and butyrate. Volatile fatty acids are absorbed and used for energy or stored as fat. This allows ruminants to live completely on forage diets, deriving energy from the fiber and amino acids from the bacteria that digest the fiber. Llamas and alpacas can thrive on low-protein diets and usually need little grain. Growing and reproducing llamas can utilize the starch but adult llamas will often become obese if even fed just a little grain. There has not been a tremendous amount of nutritional research that has been done directly with these animals; much of what we recommend is taken from another small ruminant, the sheep. Yet there is a growing, direct body of knowledge and with those data and careful extrapolation from other ruminants and nutritional principles, we can take good care of these animals.

Introduction

Llamas *(Llama glama)* and alpacas *(Llama pacos)* make up two of the four genuses in the Camelid family. The other two are vicunas *(Llama vicunga)* and guanaco *(Llama guincoe)*. They are hoofed mammals that developed solely in South America. Domesticated in what is now present-day Peru about 4,000 to 5,000 years ago, they were used for meat, wool, and work. Most references state that over 98 percent of llamas and alpacas still reside in Peru, Chile, and Bolivia. They were brought to America in the early 1980s.

Llamas are ruminants, which are members of several families of Ungulates (hoofed animals). Anaerobic bacteria and protozoa digest and ferment plant cell-wall constituents and contents to energy and bacterial protein passes to the stomach and intestines for the llamas use.

The alpaca and vicuna have no enamel on the lingual (tongue) side of their teeth, so that the teeth grow as they wear down from grazing. Llama's teeth have enamel coating the entire surface. As research is limited, and research to date shows only minor variations in nutrition of llamas and alpacas, we will discuss them together.

Ruminant Digestion and Nutrition

The ruminant animal is a pure herbivore. The tract developed to deal with very fibrous, usually low-quality feed (high fiber and low protein). *Ruminants,* by definition, *ruminate.* That is, when they eat, they generally cut and tear a mouthful of plant, and do only an initial grinding and wetting of the food with saliva prior to swallowing. The food goes into the *rumen* and *reticulum,* the first two sections of the fore stomach. There it is further wetted, and bacteria attach to food particles and secrete enzymes to break down cellulose and hemicellulose in the plant cell walls. Later, they ruminate, or regurgitate boluses of partially digested plant material and chew it again, usually ten to a hundred times. This serves to break down the tough cell-wall strands as well as to stimulate saliva production to keep the food wet and to provide buffers against the acids formed by the bacteria.

The first two compartments of the digestive tract are the rumen and reticulum. The reticulum is a small pouch that is basically part of the larger rumen. This large vat (5 to 15 L in alpacas, 10 to 20 in llamas) is where the masticated and wetted forage is kept for several hours to days. The saliva and drinking water wet the tough stems and leaves and the bacteria begin to digest the material along the breaks and tears caused by chewing and ruminating action.

Food remains in the rumen for several hours to a few days while the bacteria act upon it, food is repetitively ruminated, and eventually the size of the particles is reduced enough to pass down the tract.

The *omasum* serves as a sieve. Particles of various sizes along with liquid containing digested nutrients and bacterial cells pass into the omasum. Strong muscular folds mix this material, leaving large particles back to go back into the rumen to be worked on some more.

Llamas and alpacas do not have a fully functional omasum; there are some muscular folds in that region but it is not as developed as in other ruminants and there can be little sieving action. These animals are sometimes called pseudo-ruminants but functionally and nutritionally they are ruminants.

In the *abomasum,* the bacterial cells are killed and protein digestion begins. The remaining plant material is also acted upon by the acids.

> The *abomasum* is the true gastric stomach, in which hydrochloric acid and the enzyme pepsin begin to digest proteins and triacylglycerols.

The small intestinal tract of the ruminant is functionally similar to carnivores, though it tends to be longer per unit body size, as passage rate is slower as is digestion rate. Remaining proteins are broken down to amino acids; triacylglycerols are broken down to fatty acids for absorption (ruminant diets are usually low in fat, but it is still an important component for membrane and immune function). For most animals on normal diets, little starch reaches the intestine, as it was broken down and fermented to volatile fatty acids by the rumen bacteria. Vitamins and minerals are absorbed in the small intestine as well.

The next difference, then, is that ruminants have an enlarged cecum and colon, similar to the horse but not as developed. Here the animal has one last chance to derive as much energy from its plant diet as possible.

> Plant fibers that escaped breakdown and digestion along the tract enter into the cecum and colon, where another population of anaerobic bacteria can digest and ferment them.

The cecal bacteria hydrolyze much of the remaining cellulose and hemicellulose to make more volatile fatty acids, which are absorbed into intestinal veins, which deliver the nutrients to the liver and the rest of the animal body. Also during transit through the large intestine, most of the water left in the digesta is absorbed. Ruminants evolved to survive for long periods on low-quality plant material, and have an efficient and *symbiotic* relationship with anaerobic bacteria in order to do that.

Now we have to go back to the rumen and explain what is really going on there. The rumen bacteria are anaerobic; that is, they do not require oxygen and, in fact, to many of them oxygen is toxic. Figure 12.3 shows the basics of anaerobic fermentation. Digestive tracts of all animals have anaerobic bacteria, but in the ruminant they really are the king "bug." Rumen fluid can have a million to ten million bacteria per milliliter of fluid. There are many specialized species; some concentrate on breaking down hemicellulose and cellulose, others work mainly on starch, and others primarily on protein. All the bacteria break down the plant material, and use some of it along the way to make their own proteins and fats to form their cells. They also synthesize a large portion of the vitamins that the mammal needs, because during much of the year the plant material is low in vitamin content.

The first step, after the llama does some basic grinding, is to get the plant cell walls wet (hydrated) so that water can get in and start to swell the fibers (you have all seen this as you might soak dry plant material prior to using as food, or just in the garden). Then the bacteria can fit in the small pockets of water and release the cellulase and hemicellulase enzymes. Slowly the fibers become smaller and break apart. Repeated cycles of rumination and bacterial action make the particles smaller and smaller. Grazing ruminants swallow plant stem pieces 2 to 4 inches in length and eventually get those down to 1 inch, 1/2 inch, or smaller so that they can pass through the omasum. The average time it takes these particles to pass into the omasum is often eighteen to twenty-four hours (compare this to the total time in the digestive tract of dogs or cats of twelve to eighteen hours).

As the plant cell walls are broken down, the starch stored there swells as water enters the granules; this allows the bacteria to get into the crystals and release amylase to break off glucose. This starts in seconds and on some diets all the starch is digested in just a few hours. Next, the bacteria break the glucose, in anaerobic fermentation, to acetic acid, propionic acid, and butyric acid (a small amount of 5- and 6-carbon fatty acids—valeric, caproic, and isocaproic—is also made). Then the bacteria will use the fatty acids for energy or they will be absorbed through the cells of the rumen wall to be used by the animal. This is the first "payback" the mammal gets for carrying around billions and billions of these hungry bacteria.

Protein is broken down by bacterial proteases to amino acids, and these are used to make bacterial proteins for their membranes, cell structures, and enzymes. There is a big plus here for the animal. Bacteria can make all amino acids, even if they were not present in the plant. Most grass plants are low in lysine and legume plants are low in methionine, and in the winter, all dead plants are low in total amino acids.

> A major benefit for the ruminant animal is that bacteria and protozoa will make a higher-quality protein from the cell walls and amino acids in plant material.

The true worth of the symbiotic relationship is in the higher-quality protein that the bacteria make, and this is the second "payback" the llama receives, because the bacteria make higher-quality protein from the cellulose, starch, and hemicellulose, plant amino acids, and recycled *urea* from *saliva.* Then the bacterial cells pass to the stomach, and the amino acids are released for the animal to use.

The rumen also contains *protozoa,* which are *animals* (not bacteria) that are ten to twenty times bigger than bacteria. They are among the simplest multicellular animals. They ingest bacteria and further break them down to make proteins that have an even better amino acid balance (protein quality) than bacteria. They are much fewer in number, but yet an important part of the rumen ecosystem. Protozoa, being in the Animal kingdom, make a higher-quality protein than either plants or bacteria. In experiments that *defaunate* the rumen to remove the protozoa, the total efficiency of the digestion, especially for formation of quality protein, is diminished. This type of experiment has demonstrated the importance of protozoa in the rumen ecosystem. Without protozoa the ruminants would not be as efficient, especially in protein use.

Now we know what happens in the rumen and the rest of the digestive tract. But what else is different about ruminants? Well, remember that the ruminants have received amino acids, fatty acids, vitamins, minerals, and water just like all the other animals we have learned about. However, instead of glucose, they have received *acetic acid, propionic acid,* and *butyric acid.* Don't the ruminants' brains and other cells need glucose? Yes, they do. The ruminant has to work a little bit to finish the story off. First, acetic acid and butyric acid can be used immediately in the TCA cycle to generate energy, or if in excess, can be converted to fat for storage (see Chapters 3 and 7). Most of the propionic acid is converted back into glucose through *gluconeogenesis* (synthesis [genesis] of new [neo] glucose). See Figure 14.1. So the ruminant, through the help of its bacterial friends has derived, from low-quality plant material (or even from high-quality plant material), all the specific nutrients it needs.

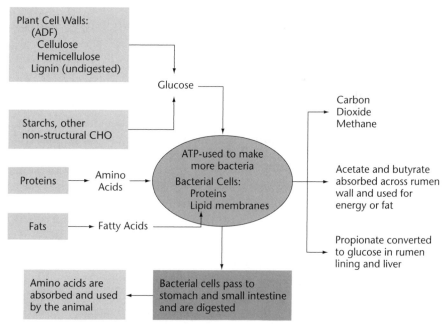

Figure 14.1 Ruminant digestion: pre-gastric anaerobic fermentation.

FUN FACT

What Is a Ruminant, Nonruminant, or Monogastric Animal?

I have alluded to the fact, from time to time, that nutritionists are just as human as any other scientists; they love to argue. You may read through literature or on Web sites and read about ruminant animals, such as cattle, elk, deer, camels, llamas, alpacas; or about nonruminants, which is pretty much anything else. But what is a "*monogastric*" animal? The simple answer is *all of them*. The term *gastric* refers to the gastric pouch, or true stomach. This organ secretes hydrochloric acid, the hormone gastrin, and the protease pepsin. In all mammals, birds, fish, and reptiles there is only one gastric pouch, or true stomach. Nonruminant animals are simply that, they are not ruminants. Ruminant animals, however, do NOT have four stomachs. Ruminants have an adaptation in their esophagus, which results in three pouches, the rumen, reticulum, and omasum, prior to the true gastric pouch, or abomasum. Birds have a similar situation with the crop, proventriculus, and ventriculus. Guinea pigs are a little like this as well. Therefore, you can be truly educated when you understand that all animals are monogastrics, but some are ruminants and some are nonruminants.

Keep in mind that these ruminants are neither like dogs and cats, nor like horses and rabbits. They are quite different. A major difference is that they specifically evolved to live on average to poor feeds, even though good quality forage will not harm them. However, inclusion of even a little grain (high in starch and or fat) in the diets of animals that are not working significantly can lead to obesity, even more easily than for other animals.

Practical Diets and Feeding

Being ruminants, llamas and alpacas have evolved to consume primarily forage materials, and we should keep this in mind when feeding. If you recall the chapter on horses, you will remember that rarely can one plant species alone provide the right balance of nutrients for an animal. Grasses (orchardgrass, brome, bermuda) tend to be low in protein, very low in calcium, and can be low in energy for rapidly growing or working animals. Legumes (alfalfa, clover) are high in protein and much too high in calcium. Although the increased calcium is not usually a problem for adult animals, it can be for growing animals. Grains of any kind are too high in energy and low in calcium. Often, as with horses, a 100 percent forage diet and a free-choice mineral mix will be sufficient for maintenance.

It is best to rely on a mix of two forages, or perhaps one forage balanced out with a properly balanced mineral and vitamin mix.

Food for Ruminants

Whole Plants

Legumes: more protein of high quality, high in calcium = alfalfa, clover

Grasses: lower in energy, less protein, less calcium = orchardgrass, fescue, timothy

Seeds

Corn: high in energy, low in calcium

Barley, oats: more fiber, more protein, low in calcium

Suckling and Growth

Right after birth, access to all of the mother's colostrum is still the first priority. Allow the *crea* full access to the dam and observe to make sure they suckle regularly. If there is any concern that the dam is not providing enough milk (will not let the crea nurse at will, the crea appears to nurse without obtaining any milk), then fresh or cleanly frozen colostrum from cows or sheep will be a good substitute. You can also test, if you are concerned, by stripping out a few times from the mammary gland after the crea has suckled (you should not obtain much material) and then after one or two hours (you should be able to get full flow). The key here is speed, you need to get that colostrum into the young within the first twenty-four to forty-eight hours, preferably the first eighteen to twenty-four hours. We aim to shoot for 8 to 10 percent of their body weight, so for average size animals of 8 to 14 kg (18 to 32 lbs) that is 2/3 to 1 1/2 L or 16 to 32 oz of colostrum in that first day. A good rule is to let them have all they can take; you will not harm them with too much (unless too much is force-fed at one time).

Growing camelids need more energy, just as do other animals. However, available knowledge from experience suggests that the camelid young grow quite a bit more slowly than the horse or even cattle.

Because of the slower metabolic rate and slow growth rate, camelids are quite prone to obesity.

Consider creep feed only for larger or more rapidly growing young or in the case when it appears the dam cannot supply enough milk. The body weight should double in six to seven weeks. If they are growing much faster than that, careful observation and palpation should determine that they are not starting to lay down too much fat. Creep feed should still consist of primarily high-quality forage with just enough grain to supply adequate vitamins and minerals. It would be prudent to not allow grain to exceed 1 percent of body weight (about 1/2 lb for a 50-lb animal). Weaning should

FACT BOX   **What Does "First Cutting" Mean? What about "Maturity"?**

We often hear about the "cutting" of the hay as a critical factor in its nutritional makeup, but what does that mean? The *cutting* refers to the number of times that stand has been cut that year. Additional cuttings, later in the growing season, are usually lower in protein and higher in fiber, but this can very tremendously with region, rainfall, and the actual number of days since the last cutting. It is the actual analysis of the hay that is critical. Generally, the *first* cutting is for high-protein, low-fiber feed for dairy cattle, so most first-cutting alfalfa is much too high in protein and energy for alpacas, llamas or horses. However, a second cutting, if the cutting actually was made on an older, more mature stand of plants, might have more fiber and less protein than a third or fourth cutting from the same field. It is much better to "buy on test," or to have a forage analysis done. Look for legumes (alfalfa, clover) in the range of 14 to 18 percent CP and 28 to 32 percent ADF (2.2 to 2.4 Mcal ME/kg) and grasses in the range of 8 to 12 percent CP and 34 to 38 percent ADF (2 to 2.2 Mcal ME/kg).

take place somewhere between four and six months of age with a weight around 100 to 150 lbs (50–80 for alpacas). Usually faster-growing larger animals can be weaned earlier, with the little ones still getting most of their food from mom for a while. Because of the use of llamas and alpacas for pleasure, work, or wool, there is no good reason to attempt to speed the growth rate.

Thus for growing llamas and alpacas, a good midbloom alfalfa with 2.8 Mc/kg DE, 18 percent CP, and 28 to 32 percent ADF would be ideal. Alternatively, grass hay at about 12 to 14 percent CP and 32 to 34 percent ADF would be fine for growth. Use a grain mix such as barley and oats, or corn and oats, with a proper balance of minerals and vitamins, only in those cases when the crea are not growing as well or when they appear too thin and bony. As for all animals, there is no substitute for individual observation and matching of the feeding rate to the ideal growth rate. Watch body condition to make sure it is growing mostly muscle and not fat.

Pastures, especially in spring or early summer, or when heavily fertilized, will usually have much more protein and energy and less fiber than cut hays, unless they have been closed off until the plant matures. In this case, your management mechanism is the time allowed to pasture: You do not want the young (or the adults) to consume so much pasture that they will grow too fast or become obese. In the summer and fall in most areas, the pasture plants will mature and have more fiber and less energy. These can still be ideal feeds for young and older animals. If pasture is available, it would only be in the late fall or winter that hay should need to be provided (again depending on all the usual caveats of rainfall, temperature, body condition score, etc.)

Maintenance and Reproduction

As for the other species, adulthood is a time to maintain body composition, muscle mass, and good aerobic circulation (Figure 14.2). For most horses and ruminants, this means a diet of almost 100 percent forage material with a mineral supplement. Depending on the quality of the forage, a vitamin supplement, carried in a minimal amount of grain, is also needed. Older forages (those that have been cut and stored for more than a few months) may lose a large amount of their vitamin potency, especially for fat-soluble vitamins. The other side of that story, however, is that for ruminants, the rumen microbes can make much of the vitamins required by the animal, especially at maintenance requirements. Earlier paragraphs, and the tables in Chapter 12, detail much of the information on feeding delivery of forage and forage composition.

Llama Body Condition Score

Frame size	Ideal BW
Small	250–275 lbs
Medium	275–300 lbs
Large	300–350 lbs
Extra Large	350–400 lbs

Condition Score 1–3	Bony, sunken musculature, bones firm and sharp to touch
Condition Score 4–6	Minimal fat cover over loin, rounded musculature Ribs may be visible, neither bony nor hidden Minimal fat in pelvic region
Condition Score 7–10	Rounded, soft, fat over loin Ribs not visible or palpable Hanging fat in pelvic region

(Adapted from Johnson, 1994.) See Figure 14.2.

Most alpacas and llamas will require 1 to 2 percent of their body weight to maintain muscle mass without excess fat tissue.

Llamas fed about 1.8 to 2 percent of their body weight from eight months to about thirty months of age may gain excess body condition. Animals that are more active will require more feed. Intact males (noncastrated) have been observed to spend more time sparring, and this exercise increases energy use and can guard against obesity. As we know for other species, castrated males will grow more slowly and accumulate less muscle and fat. However, with a simple dietary-management approach of feeding less or feeding a lower-energy diet and maintaining exercise, we can maintain lean llamas even though they

Good condition–note no excess fat between hindlegs.

Quite thin–poor condition.

Moderate condition, front view. This animal is in good condition.

Recently fleeced animal, moderate to thin condition.

Figure 14.2 Llamas of various body condition.
Individual observation is always the key; watch increase or decrease in condition and adjust feed and activity accordingly.
Source: Animals provided for photos courtesy of Jennifer Bowman, Serenity Acres, Colfax, Washington. Photos courtesy of John McNamara.

are castrated. Castration should never be done or not done based on whether or not the animal may become obese, but on safety and proper breeding concerns. The effects of castration on nutrient use can be easily managed.

Forages should be tested to determine if they are in the range given earlier. If the material available is slightly higher in protein and energy and lower in fiber, then slightly less can be fed, and vice versa. The feeder should avoid "high-quality" dairy hays, be they grasses or legumes, as they simply will be too high in protein, starch, and easily fermentable NDF for the llamas and alpacas.

> A mineral mix should be included in the diet of all llamas and alpacas, including adults.

Based on the forage fed, we can adjust the mineral mix to provide the proper calcium:phosphorous ratio and amounts, and amounts of other minerals. A commercially prepared mineral mix for sheep would usually be adequate in most situations. A custom-made mix may be made (Table 14.1). This can be adjusted based on the forage fed. If the diet is primarily good alfalfa or springtime forage, replace half of the bone meal with monosodium phosphate to reduce the calcium. If it is older forage, a vitamin A and vita-

Table 14.1	Example of a Mineral Mix for Llamas and Alpacas				

Composition:

Protein	8.5	
Fat	4	
Fiber	4.4	
Ash	63.0	

Containing:

Ca	6.30%	Mn	1025 ppm
P	3.30%	Cu	145 ppm
Mg	0.016%	Co	34 ppm
Na	14.00%	Zn	5632 ppm
S	0.70%	Fe	624 ppm
		Mo	20 ppm
		Se	4.9 ppm

This mix is formulated using equal proportions of a trace mineral salt (specifically formulated for ruminants), steamed bone meal, dry powdered molasses, and 1/5 of the amount of the first three ingredients of a zinc-methionine product (so 50, 50, 50, and 10 lbs). Nowadays it may be difficult to obtain steamed bone meal from ruminants, but bone meal from pigs is also perfectly fine.

Source: Adapted from L. W. Johnson, 1994, "Update on Llama Nutrition," Veterinary Clinics of North American Food Animal Practice, *10: 187–201.*

min E supplement at 1 to 1.5 percent of total mix can be added. Ruminants will not benefit from any water soluble vitamin supplements meant for non-ruminants, because the rumen microbes already produce most of these vitamins. If supplements are fed, the bacteria will simply break them down and the animal will get the ones the bacteria make anyway.

On pastures that are primarily grass, a problem called grass tetany can occur in spring and fall on rapidly growing pastures (either after a cut, after a rain, or after a freeze/thaw episode). This is an acute magnesium deficiency, which causes an imbalance in nervous and muscle function, with the worst case being tetany and death. This is avoided by holding animals off such pastures for several days to a week until the plant growth rate slows. Additionally, adding magnesium oxide at about 0.75 lb/100 lb of mineral mix will help. The reader looking for practical advice should definitely consult a knowledgeable local veterinarian or nutritionist.

Gestation and Lactation

The basics of nutrition for gestation and lactation are the same as for other species. If all the basic dietary-management principles are being followed, then allowing free choice (1.8 to 2 percent BW) or slightly restricted

(1 to 1.5 percent of BW) amounts of good hay and mineral mix will suffice for most of gestation. Inclusion of a vitamin mix into the mineral or included in a small amount of grain might be warranted, but there are no data to back this statement up. If llamas and alpacas are functionally similar to other ruminants, their ruminal bacteria, if provided forage and minerals, should make most vitamins except vitamins A and E.

Based solely on individual dam body condition score, a small amount of grain may be phased in during the last trimester. During lactation, again, the best approach is to allow intake of a good quality forage and mineral/ vitamin mix. Add grain only as needed to maintain weight and condition of the dam. If you note that the dam's condition is thinning, you may also note that the growth of the crea is less. Most mammals will continue to make large amounts of milk and lose their own body condition before milk production is diminished, and there is no evidence to suggest llamas and alpacas would not do the same. However, given the lack of data, good practices suggest continuing observation of the young and dam and providing only enough food to maintain condition.

Work and Exercise

We have covered the physics and nutrition of work and exercise in depth in Chapter 12 on horses. The same principles certainly apply to working ruminants. Intake or energy density of the diet is increased only as much as is needed to maintain body weight. Remember that even an hour or two of hard work only requires a few hundred grams (1/2 to 3/4 of a lb) of a normal grain mix. Pack and working animals should be offered free access to good forage, and if it is noted that body condition score is lessening, than a small amount of grain may be offered until condition stabilizes. Again, small amounts regularly fed is the key.

Certainly less is known about the exact nutritional requirements of South American camelids than for the other animals we have discussed. However, the principles and common-sense applications remain.

Words to Know

abomasum	*omasum*	*ruminate*
acetic acid	*propionic acid*	*saliva*
butyric acid	*protozoa*	*symbiotic*
crea	*reticulum*	*urea*
defaunate	*rumen*	
gluconeogenesis	*ruminant*	

Study Questions

1. How is a South American camelid different in digestion from dogs and cats?
2. How are camelids different in their digestion compared to horses?
3. Why is it not a good idea to feed only one type of plant material to camelids, or horses for that matter?
4. Is it worse to feed one type of plant to camelids, or is it worse for horses?
5. What are the problems associated with feeding grain to camelids?
6. Why do camelids not usually need vitamin supplements?
7. What basic practices should you apply to make sure that your alpacas or llamas are receiving the proper amount of energy?
8. If you were interested in feeding camelids, knowing the lack of specific data, what would you like to know the most about feeding them? Can you make a hypothesis, and construct an experiment to test it, regarding camelid nutrition?

Further Reading

Carmean, B. R., K. A. Johnson, D. E. Johnson, and L. W. Johnson. 1992. Maintenance energy requirement of llamas. *Am. J. of Vet. Research*, 53(9): 1696–1698.

Hinderer, S. W., and V. Engelhardt. 1975. Urea metabolism in the llama. *Comparative Biochemistry and Physiology*, 52(4): 619–622.

Johnson, L. W. 1994. Llama herd health. *Veterinary Clinics of North American Food Animal Practice*, 10: 248–258.

Johnson, L. W. 1994. Update. Llama nutrition. *Veterinary Clinics of North American Food Animal Practice*, 10(2): 187–201.

Johnson, L. W. 1989. Llama medicine. Nutrition. *Veterinary Clinics of North American Food Animal Practice*, 5(1): 37–54.

Lackey, M. N., E. B. Belknap, M. D. Salmon, L. Tangibly, and L. W. Johnson. 1995. Urinary indices in llamas fed different diets. *Am. J. of Vet. Research*, 56(7): 859–865.

Rubsamen, K., and W. V. Engelhardt. Water metabolism in the llama. *Comparative Biochemistry and Physiology*, 52(4): 595–598.

Van Saun, R. J., B. R. Callihan, and S. J. Tornquist. 2000. Nutritional support for treatment of hepatic lipidosis in a llama. *J. of the Am. Vet. Med. Assoc.*, 217(10): 1531–1535.

Weaver, D. M., J. W. Tyler, R. S. Marion, S. W. Casteel, and C. M. Loiacono. 1999. Subclinical copper accumulation in llamas. *Canadian Vet. J.*, 40(6): 422–424.

Chapter 15

Nutrition of Ornamental Birds

Take Home and Summary

The variation among birds kept as companions is greater than for land animals; only fish have wider variety of species. Birds come in a variety of sizes, natural habitats, and resultant nutritional requirements. Commonalities exist with mammals including most major nutrient requirements, but birds have a *much* faster metabolic rate than mammals. This faster rate of metabolism requires that the proper type and variety of food, as well as water, be present at all times. Types of foods that are normal for a given bird can range from a few seed types to dozens of seed types, leaves, stems, and for some species, animals. Birds in cages tend to have two major problems: low activity and a diet too high in fat and energy and too low in calcium. Birds in cages do not always, or even often, select the proper balance of seeds to meet their nutritional needs. These problems can easily be managed with larger cages or an aviary space and proper feeding of a commercially prepared balanced ration or a well-managed feeding plan including feed ingredients containing less energy and a balanced supply of protein, vitamins, and minerals.

Types of Birds

We will take a little different and maybe more fun approach to our discussion of birds with an introduction into the classification of the variety of orders, genera, and species in the class Aves, known as *avians.* Without listing many facts, it is germane to help us understand the special nutritional needs of various birds by understanding a little more of their family tree.

Order Psittaciformes

The Psittaciformes include the parakeets, budgerigar, lory, cockatoo, macaw, conure, amazon, parrot, and lovebirds, and are referred to as *psittacines.* Most of these birds are naturally seedeaters. However, they do not eat just one type of seed.

> Studies in the wild have shown that psittacine birds will select a diet of seeds from twenty to forty species of plants. So, providing a variety helps to make up a normal, balanced ration.

This helps them obtain a diet balanced in vitamins, minerals, starch, fat, and amino acids. Recall our discussions of the types of feedstuffs (Chapters 9 and 10) and that different seeds can vary widely in nutrient content. Grass (cereal) seeds tend to be low in protein, fat, and calcium, and are relatively low in the essential amino acid lysine. Seeds from legumes, or flowering plants and trees, tend to be higher in fat and much higher in calcium and lysine. Table 15.1 shows the diet and subfamilies of the Psittaciformes order.

Table 15.1	Psittaciformes: Parrots, Parakeets, Budgies

Diet: variety of vegetation, seeds, insects, small mammals

Subfamily Psittacinae	Several species kept as companions, require seeds and vegetative parts, sometimes insects.	
	Psittacus erithacus erithacus	African gray parrot
	Anodorynchus ararauna	Blue and gold macaw
	Many other species	
	Aratinga guarouba	Queen of Bavaria or golden conure
	Several other species	
	Psittacus erithacus	African gray parrot
	Pionus menstruus	red-vented parrot
	Pionopsitta pileata	red-capped parrot
	Poicephalus fuscicollis	brown-necked parrot
	Amazona—many species	
	Amazona ochrocephala oratrix	Mexican double yellow head
	Psittacula—larger parakeets	
	Psittacula krameri manillensis	Indian ringneck parakeet
	Psittacula cyanocephala	plum head parakeet
	Neophema elegans elegans	elegant parakeet
	Melopsittacus undulatus	budgerigar
	Agapornis species	lovebirds
Subfamily Cacutuinae	Again, require a wide variety of seeds and vegetative parts.	
	Calyptorhynchus species	black cockatoos
	Cacatua sanquinea	bare-eyed cockatoo
	Several other species	
	Nymphicus hollandicus	cockatiels

Other members of this order would include the red-capped parrots *(Purpureicephalus spurious)* and Port Lincoln parrot *(Barnardius zonarius)*. These species consume a wider variety of plants, including the fruit, nuts, and vegetative parts of plants, than do the smaller parakeets. Also included with the Psittaciformes are the subfamily Cacatuinae, including the cockatoos; the subfamily Psittacinae, including the macaws, conures, parrots, amazons, parakeets, and lovebirds; woodpeckers; also included is the family Ramphasitidae, which includes the toucans.

Order Passeriformes

Distant "cousins" to the Psittaciformes are birds from the order Passeriforme (Table 15.2). This group of birds, *Passerine* birds, takes its name from the fact that they make an annual passage every year, migrating from north to south in the autumn and back again in the spring. Included are the family Fringillidae, consisting of canaries, and the family Estrildidae, consisting of the finches and waxbills. This order also includes the waxwings, wrens, swallows, bulbuls, tits, warblers, and blackbirds. As a side note this order also includes starlings and the crow (a bird, which, unfortunately, has taken over my town and which I certainly do not enjoy as much as the songbirds).

Orders Anseriformes and Galliformes

Most companion birds are from a small number of orders and families, as we just discussed. However, there are distinct physiological and nutritional differences between birds of the orders commonly kept as companions and those of the orders making up domestic poultry, *Anseriformes and Galliformes* (Tables 15.3 and 15.4). The biggest practical distinction is that *poultry*, such as ducks, geese, chickens, turkeys, and even pigeons, have digestive systems and thus practical dietary-management practices that are quite different, primarily in the fiber-utilizing ceca and large intestine. This difference will dictate different practical rations to meet the nutrient requirements.

Order Falconiformes

This order includes birds often referred to as raptors, or birds of prey. This order includes eagles, kestrels, falcons, and hawks. These birds have gained popularity as companions. They have not been domesticated as such, although

Table 15.2	Passeriformes: Canaries, Finches		
Consume a variety of vegetation, seeds, insects, small mammals			
Family Fringillidae	*Serinus caranius domesticus*	canary	
Family Estrildidae	*Carduelis carduelis*	goldfinch	varieties of seeds, insects
	Phoephila guttata	zebra finch	invertebrates, greens
	Chloebia gouldiae	gouldian finch	
	Many other genera and species		

Table 15.3	Falconiformes: Eagles, Kestrals, Falcons

Diet: small mammals, reptiles, insects, plants (if they have to)

Falco spaverius	American kestral
Haliaetus leucocehpalus	bald eagle
Falco peregrinus	peregrin falcon
Butea Jamaicensis	red-tailed hawk

Table 15.4	Domestic Fowl

Order	Genus Species	Common Names	General Diet
Struthioniformes		ostrich	vegetable matter
Anseriformes	*Branta candensis*	Canada goose	mostly vegetation
	Anser anser	domestic goose	mostly vegetation
	Anas platyrhynchos	domestic duck	mostly vegetation
Galliformes	*Meleagris gallopavo*	turkey	omnivorous, coarse grains
	Gallus gallus	domestic chickens	mostly vegetation, grains
Columbiformes	*Columbia livia*	pigeons	mostly vegetation
	Streptopelia decaocto	collared turtle doves	mostly vegetation

they have been bred and used for hunting and sport birds. Their diet is usually carnivorous, but the prey will change with season. In addition, many specific species will prey on specific animals. This allows habitation of different falconiforme species in similar environments. Because they are primarily carnivorous, they should be fed carnivorous diets. Supplements and complete feeds containing animal products and the proper amounts and forms of vitamins may be used as needed or as treats.

> Ornamental birds are not domestic fowl, and must be fed much different amounts and types of feeds.

In addition to the obvious size difference, most birds raised for meat have a functional cecum, such that they can use significant amounts of dietary fiber. They also have slower metabolic rates related to their larger body size. Their digestive systems, because the passage rate of digesta is significantly slower and the digestive enzymes are also different, can handle more energy-dense diets with significant amounts of large grains such as corn and barley and protein supplements such as soybean meal. These ground diets would be quite detrimental to the small digestive tract of the psittacine and passerine birds. Thus, extrapolation from feeding chickens to feeding ornamental birds should never occur.

Digestive Physiology of Birds

The first obvious difference in the digestive system of birds (Figure 15.1) is the "replacement" of teeth with a beak. The other amazing thing is how widely varied those beaks will be, depending on the natural adaptations to the dietary sources for each bird. In general terms, the size and shape of the preferred seed or food source is a function of the size and shape of the *beak*, not the bird. The keratinized (similar to skin, not bone) beak is continually growing and needs to be naturally kept in proper shape, primarily by the owner providing the proper diet in the form most closely matching the actual diet of the bird. It is not appropriate to substitute grit, or even a stone, although the latter can be used safely in many situations. It is better to provide the proper type and form of diet. Regular observation and judicious trimming are also needed.

The stomach of most Aves is divided into distinct physiological sections to adapt to the lack of teeth and smaller amounts of saliva that most of our mammalian friends use. The first is the *Crop,* an enlarged part of the esophagus used for wetting and temporary storage. This is followed by the *proventriculus,* which is also referred to as the "true stomach," or acid-secreting organ. This is where the gastric acid and enzymes (pepsin) are secreted to begin chemical digestion of the food, similar to mammals. The last section of the stomach is called the *ventriculus,* better known by the term *gizzard*. This is a strong muscular organ used to grind the feed down to a smaller particle size for practical digestion, acting both as teeth and in the normal physical mixing function of the mammalian

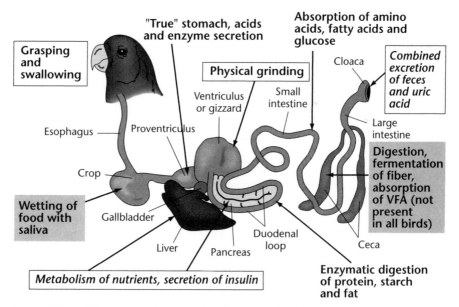

Figure 15.1   Digestive functions and metabolism of birds.
Source: Figure redrawn from *Small Animal Care and Management,* second edition by Warren. © 2002. Reprinted with permission of Delmar Learning, a division of Thomson Learning: www.thomsonrights.com. Fax 800 730-2215.

> ### Did You Know That Birds Give Milk?
>
> The order Columbiformes contains the family Columbidae, pigeons, and doves. Their crop (proventriculus) is not only used for storing and wetting foods. The cells of the crop lining can slough off and be regurgitated to the young. This *crop milk* of fluid and cells closely resembles milk in look and texture, though it contains no lactose. Secretion is also under the control of the lactational hormone prolactin. Evolution-wise, this hormone is very old, being found in reptiles, birds, fish, as well as mammals. In fish and reptiles, it controls water movement among cells. In mammals, it controls milk production and coordinates the activity of other organs (adipose, liver, muscle) to supply nutrients to the mammary gland. Here, in this specialized "milking bird," prolactin controls both water flux and crop milk. So yes, some birds can give milk.

stomach. This organ is lined with a specialized polysaccharide called *koilin* to protect the organ from the sharp food and to help aid in digestion.

In the small intestine, amino acids, vitamins, minerals, glucose, and fatty acids are absorbed much the same as in other animals. However, after this things do get different. The cecum and colon of most companion birds is small or even absent, as in the budgerigar. This severely limits the use of dietary fiber by most of these birds. Because of their small size, constant food intake rate, and lack of a cecum, the transit time is quite fast. ***Transit time*** is the time from the ingestion of food to the time when the undigested food from that meal is excreted. In large ruminants such as dairy cattle this can be 24 to 72 hours for different foods. In humans it is usually 18 to 24 hours, dogs and cats 12 to 18 hours. The transit time will vary with the type of food and rate of intake. In most caged birds, transit times are never more than 12 hours, in budgies and finches it is three to six. The practical message is that the digestive tract of small birds can be empty six hours after a meal; the bird is then fasting. Some small birds can be in a state of starvation within three days or shorter. Thus, food must be available at all times, unless you are in the process of weaning the bird from one diet to another, and even then, 12 hours without food is long enough.

The final structures of the avian digestive tract are radically different than that of any other mammals. The urinary tract and digestive tract 'reunite' in the structure called the ***cloaca***. The cloaca is the final structure of the digestive tract that includes both urinary and fecal material, and all undigested material is excreted together.

Highly concentrated urine and feces are mingled together and excreted together (as just about everyone has observed directly, even if they don't own a bird). The reduction in water volume is a tremendous efficiency advantage to the bird. Birds excrete excess nitrogen as uric acid, and this compound does not need to be as dilute as urea for reducing toxicity. The evolutionary advantage of reduced water volume for flight should be obvious. It has, however, reduced our ability to define strict chemical nutrient requirements for birds, as it is impossible to collect urine and feces separately. But, we can still measure total waste and uric acid excretion, so with proper measures we can determine energy and amino acid requirements of birds.

Life-Cycle Feeding Management

General

A general practice has already been noted: You have to feed the bird the type and form of diet that it will naturally get in the wild. However, we do know that many birds will not actually consume all of the variety we offer them. They may pick and choose, and although you think that you are providing a balanced mix, the bird is not really consuming the overall balance. A simple management rule is to leave the entire amount of food there until the bird has actually eaten all of it—even the parts it "doesn't like." As our knowledge and experience grows, most professionals recommend feeding a well-balanced commercial ration that provides the proper balance of nutrients and physical form that the birds need.

There are exceptions and newer information, however, that provide management alternatives. Recent research has shown that normal foraging behavior is important to the caged bird. In the lack of foraging opportunities, there will often be increased feather picking and other destructive behaviors. Studies demonstrated that if different feeds were placed in various parts of the cage, the birds would move around and forage and the incidence of destructive behaviors decreased. This is something that intuitively makes sense, provides a good opportunity for the owner and bird to interact more, and allows the bird to obtain much-needed exercise. I would recommend this as long as the owner carefully observes that the bird is in fact eating all the diet, and would still recommend including some commercial mix to help ensure the proper mineral balance.

> Most passerine and psitterine birds may feed primarily on seeds for a large part of the year, but on a wide variety of seeds, some high in oil and some high in protein or carbohydrate.

Several seeds in seed mixes, especially those from grasses (corn, oats, barley, wheat, and any grass seed) do not contain sufficient calcium for the bird. Many seeds (sunflower, safflower) are often high in fat energy in relation to vitamins and minerals and can cause secondary deficiencies of calcium and magnesium. Birds in the wild will also adapt during various seasons to consume vegetative parts of plants and even insects. During breeding and moult, their nutritional needs, primarily protein, also increase and they will adapt their diet. It is difficult to match this in a captive situation, which is why the major consensus among nutritionists and veterinarians is to recommend a complete feeding mix in many situations. Alternatively, feed a variety of seeds and vegetative parts of plants.

Growth and Reproduction

Growth is very rapid; young birds can increase body size ten-fold in ten days. This rapid rate of growth requires a quite high amount of dietary protein, about 25 percent protein early in growth, decreasing to about 20 percent for most of the growth curve (one week to five to eight weeks).

> Feather growth also requires great amounts of amino acids, especially proline, glycine, methionine, and cystine.

Thus, in terms of the amount of protein required per body weight, the bird requires more than any other animal, including cats. When birds are laying eggs, the protein requirement is about 20 percent.

There are some clear metabolic differences between birds and mammals in their adaptation to environments. In mammals, excess nitrogen is secreted as urea (Chapters 4 and 7). However, birds secrete excess ammonia in a more complex compound called *uric acid*, which is derived from the amino acid glycine and that simple fatty acid called acetic acid. Remember that in mammals, the urea cycle transfers nitrogen from different amino acids through the amino acid arginine to make urea. The trick with mammals that makes this work is that a large volume of urine dilutes out the very toxic urea. Birds do not have the luxury of carrying around large volumes of water (which would make it difficult to fly). Thus, birds have developed a concentrated urine using uric acid. This compound does not need to be diluted out as much as urea, so the bird can conserve water. Uric acid, however, takes more energy to make than urea, which increases the energy requirement of the animal.

There are a few other small but important differences. You may remember from Chapter 5 that glycine is not normally an essential amino acid. But, because of the fast rates of synthesis of uric acid, which requires glycine, this amino acid becomes an essential amino acid for growing and reproducing birds.

> *The nutritional differences of the bird:*
>
> Faster metabolic rate, so must have food present most or all of the time
>
> Rapid digestive passage rate
>
> Must have "animal" vitamin D_3, cholecalciferol
>
> Require proline and glycine
>
> Require increased amounts of sulfur amino acids
>
> Excrete nitrogen as uric acid; extremely low urine volume
>
> Require more essential fatty acids
>
> Most, especially smaller, psittacines and passerines do not utilize fiber

Energy

Small birds require about 80 to 140 kcal per $Kg^{.75}$, or about 10 to 25 kcal per bird per day. Most seed mixes are about 4 kcal per gram, so 3 to 6 grams of seed is average. This can be a few hundred to 1,000 individual seeds, depending on the type of plant. Obviously we are not going to count the seeds, but we are going to observe what we feed and what the bird eats. We want to minimize sorting as much as we can and provide every opportunity for the bird to only have those seeds that are proper in amount of fat and calcium.

Essential Fatty Acids

Birds need a relatively greater amount per body weight of the essential fatty acids: linoleic, linolenic, and arachidonic acid for a healthy immune system and membrane and skin function. Because of the more rapid metabolic rate of all organs, including skin, this is not surprising. Many plant seeds have sufficient amounts of the essential fatty acids, so that a deficiency is not a problem unless the bird is only being fed seeds from grasses (canarygrass, for example). These grass seeds tend to be very low in fat. Signs of deficiencies included loss of feathering and scaling of skin. This leads to excess water loss from the bird, followed by increased water intake. The immune system may also be suppressed, and with the skin weakened, the likelihood of a dangerous infection and inflammatory response can increase.

Vitamins and Minerals

Vitamin and mineral nutrition of birds can be complex given the different bone structure, need for fast rates of protein synthesis, and the tremendous amount of variety in seeds and plant structures fed. The simplest way to supply the proper amount and balance of vitamins and minerals is to provide a mix of different types of seeds and vegetative parts. A commercial ration usually contains supplemental vitamins and minerals to ensure adequacy, but be sure to read the label.

Calcium and Nutritional Secondary Hyperparathyroidism

Calcium is such an important mineral for nervous transmission and muscle contraction that there exist hormonal systems to regulate calcium metabolism. We have learned about some of this regulation by vitamin D in Chapter 6. If birds are consuming a large amount of cereal seeds, or eating a large amount of seeds high in fat, then a calcium deficiency can occur. The cereal grass seeds are low in calcium. The fat from the high-fat (legume) seeds also binds up calcium so it is not digestible. When there is not enough calcium in the blood to supply the nerve and muscle cells the normal amounts they need, the body draws on the major source of calcium in the body: the bone. Many hormonal systems come into play to take calcium from bone for the organs to use, so eventually the bone will lose enough calcium and phosphorous to become brittle and weak.

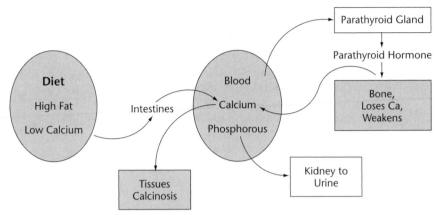

Figure 15.2 Nutritional secondary hyperparathyroidism (metabolic bone disease).

Because of this hormonal control, the system can overreact and in fact take too much calcium from bone. The excess calcium mobilization from bones is common in parrots and other psittaciformes. This condition is *nutritional secondary hyperparathyroidism* (Figure 15.2). The parathyroid gland, which is in charge of helping to get calcium out of bones when it is needed, overreacts and causes too much calcium to be released. Signs include loss of appetite, feather chewing, weakness, bone fractures, and tetany (muscles contract to the point that they "lock up" and the animal cannot move). Quite often you may see a broken leg and think of an accident, when in fact it is due to brittle, calcium-poor bones. Or the animal may simply die of a heart attack due to large amounts of calcium leaving the bone and being deposited in the arteries and heart. This makes the heart work harder (this is similar but not the same as atherosclerosis in humans) leading to early heart attack.

If the animal is also consuming a large amount of fat from high-fat seeds, then a more serious atherosclerosis occurs. Both fat and calcium are deposited in the arterial wall and heart, leading to early heart attack. The hard part is that signs can often take several years to develop. Birds should be supplied with calcium in the range of 0.5 to 1 percent total calcium in the diet, more during egg laying. This amount cannot be supplied by a diet primarily of seeds, especially cereal grains such as corn, oats, canarygrass, and Kentucky bluegrass (as examples). Oilseeds and seeds higher in protein and calcium such as legumes and vegetative parts of plants must be provided.

Vitamins D and A

Aves require vitamin D, but the story is a little more complicated (you may want to review the section on vitamin D in Chapter 6). Ergosterol is the provitamin (precursor to the vitamin) contained in plant tissue. This can be converted to ergocalciferol (vitamin D_2) upon exposure to sunlight (in dead plants). Cholecalciferol (vitamin D_3) is the provitamin found in animal tissues made as sunlight converts 7-dehydrocholesterol to cholecalciferol. For some reason, most animals can use either form of the provitamin, but birds require vitamin D_3 from animal products. What this means is that birds do not obtain a lot of vitamin D from plants, but will make the active vitamin from ergosterol in their own tissues. If vitamin D is to be supplemented, it should be the animal form, vitamin D_3, 7-dehydrocholesterol. Animals should be exposed to sunlight regularly and fed a good balance of seeds, some oilseeds and some not. Oversupplementation with high-fat seeds actually leads to a vitamin D problem, as fatty liver and buildup of D_2 can occur, leading to lethargy and liver disease (jaundice, yellow color of skin).

Vitamin A, as we learned in Chapter 5, is derived from the cleavage of beta-carotene into two vitamin A molecules. Vitamin A itself only occurs in animal tissues, not in plants. Plants contain beta-carotene and many other carotenoid components that may have some vitamin A activity. Beta-carotene can be lost with drying, heating, and storage of plants, so most commercial mixes are supplemented with vitamin A. Use of fresh green vegetative parts of plants and fresh seeds can supply active beta-carotene to supply vitamin A. Use of dried and prepared foods would include supplementation with beta-carotene, or active vitamin A, the chemical name of which is retinol. If you see retinyl acetate or retinyl palmitate on a label, that is vitamin A.

Moult

As part of their normal life cycle, birds occasionally *moult*, or lose their feathers and replace them with new ones. This cycle may begin about four months of age (budgerigar), and then usually on an annual basis in the fall. During this time, the potential for problems can increase. Fast rates of feather regeneration requires a large amount of well-balanced amino acids, especially methionine, cystine, proline, and glycine. At least half of the sulfur amino acids should come from cystine. Seeds such as rapeseed and white millet are good sources. If a bird is defeathered, both protein and energy requirements will increase (energy may increase by as much as 80 to 90 percent). Seeds high in energy such as rapeseed (canola) or groats (germ) should be fed.

Carbohydrates

We know that birds can use starches from a wide variety of seeds; these are their main sources of glucose. Yet too much starch as is present in a diet consisting solely of grass seeds provides more energy than the average bird needs. Also, many small ornamental birds, different from poultry, basically cannot derive much benefit from fibrous carbohydrate. They do not have the developed cecum and colon, and associated bacteria, to break down the fiber. In addition, their normal diets do not contain much of the vegetative part of the plant. This is the major physiological reason why it does not make a lot of sense to feed too much fresh fruit and vegetables to most birds. In addition to these sources being primarily water, the large amounts of fiber they contain will not be available to the bird for energy. Fruits and vegetables are poor sources of protein. Fruits and vegetables such as berries, dark green or yellow vegetables, peas, and kidney and pinto beans will supply a mix of carbohydrate, fats, vitamins, and minerals. Most modern recommendations are no fruits such as apples and bananas as they have little in the way of protein or useful carbohydrate for birds.

Some of the larger psitterines can use fiber as part of their diet; however, it needs to be balanced, and it is important that they receive sufficient protein and the proper amounts of vitamins and minerals. The point is that a pile of carrots, apples, and lettuce won't do the trick. Nutrient contents of various seeds and other ingredients show the wide variety of protein, fiber, starch, and minerals (Table 15.5). Table 18.3 also has the chemical content of some greens fed to reptiles and birds. It is difficult but not impossible to put together some homemade rations to meet the needs of different birds (Table 15.5), but it is certainly easier to let the "pros" do it for you (Table 15.6). Remember that not all birds will select the proper balance of nutrients from a smorgasbord-type arrangement. It is usually better to mix the ingredients into a ration so that you are sure they are always getting the proper amount and balance.

Grit

The gizzard, or more technically correct, the *ventriculus* (Figure 15.1) is the large muscular grinding organ. As humans came to domesticate poultry, somehow, somewhere, somebody decided that inclusion of some fine *grit*—hard particles of quartz, sand, or calcium carbonate (oyster shell)—increased

Table 15.5	A Variety of Nutrient Needs and Feeds for a Variety of Birds				
	Protein, %	Linoleic Acid, %	Fat %	Ca, %	P, %
Average Psittacine Requirement	12	1.0	2–5%	0.3–1.2	0.3
Average Passerine Requirement	14	1.0	2–5%	0.5–1.2	0.5
Seed Type					
Legumes—*too high in fat by themselves, need to be part of the total ration*					
Canola	35–40	6–8	15–20	0.2–0.3	0.5–0.6
Soybean	35–40	7–9	18	0.25	0.6
Peas	24	trace	1–2	0.13	0.1
Grasses—*too low in calcium by themselves, need to be part of the total ration*					
Canarygrass	9–14	0.5–1.0	2 to 5%	low	0.3
Millet	9	0.5–1.0	2 to 5%	low	0.3
Corn	8.5–9.5	0.5–1.0	2 to 5%	0.03–0.05 (low)	0.3
Oats	11–13	0.5–1.0	2 to 5%	low	0.3
Sunflower	15–20	6–8	15 to 25%	0.45	1.0
Safflower	15–20	6–8	15 to 25%	0.45	1.0

Table 15.6	Diets for Various Birds

Psittacine Birds (budgies, parakeets, parrots)

One example of a home-prepared mix:

20 to 30% seeds and nuts (safflower, canarygrass, milo, peanuts)

20 to 30% dark green, yellow, and orange vegetables

10–15% fruit: berries, melons, citrus (not apples, bananas)

20 to 30% pelleted or extruded psittacine foods, mixed in well to the mixture

Examples of commercial mixes for:

Large and medium psittacines

    Kaytee Products Exact Original All Parrot and Large Conure Daily Diet

        Extruded; 10% moisture, 16% protein, 6% fat, 5% fiber

    Lefeber's Premium Daily Diet

        Pelleted; 10.5% moisture, 14% protein, 4% fat, 2% fiber

Small psittacines

    Kaytee Products Exact Original Parakeet Daily Diet

        Extruded; 12% moisture, 14% protein, 5% fat, 5% fiber

    Premium Nutritional Products ZuPreem Avian Maintenance

        Extruded: 10% moisture, 14% protein, 4% fat, 2.5% fiber

Passerine Birds (finches, canaries)

    Kaytee Products Exact Canary/Finch Rainbow

        Extruded; 12% moisture, 15% protein, 6% fat, 5% fiber

    Premium Nutritional Products ZuPreem Avian Maintenance

        Extruded; 10% moisture, 14% protein, 4% fat, 2.5% fiber

digestion of dry foods (grains, primarily) of these food-producing species. This is because as we fed them coarser, harder grains such as corn and oats, they needed more physical help to grind the particles. As we came to have more birds as companions, it was easy for people to simply extrapolate and say "birds need grit." One can still purchase grit, and there is disagreement among nutritionists on whether or not it is truly needed in the psittacine and passerine birds. These birds, such as canaries, parrots, budgies, and parakeets are not anseriformes and galliformes (chicken, ducks, and turkeys). The former have little capability for eating the harder-grain seeds such as corn; they have little to no ability to ferment fiber in their hindgut (before fiber can be fermented it needs to be ground up a little bit to increase the surface area).

    Grit can be insoluble (quartz) or soluble (oyster shell). Insoluble grit stays in the gizzard for a long period of time. It is now known through many case studies (reports at veterinary clinics) that in many ornamental birds, the grit actually builds up in the ventriculus, and can lead to impaction. Over the last twenty years or so, recommendations have changed from: "well grandpa always fed chickens that way" to "grit must be supplied" to "grit may be beneficial in some situations" to the more reasonable, scientifically sound recommendation that it is simply not necessary for most of the birds we keep as companions, and can lead to problems more than benefits. If you are supplying a commercial ration, or a mix of appropriate seeds and vegetative parts, grit probably will not be needed.

Practical Diets and Feeding Management

I have been chastising you about all the things we might do wrong. But what is more important is how to do it right. The best course of action for many of the birds we keep is to research, find, and feed a good commercial mix. The second best is probably to make, from products available in the stores, a balanced ration for individual species of birds. The extreme variety of diets appropriate to each species is far too broad to include here. One example of a homemade mix is shown in Table 15.6. Be sure that you start with the information here and use the end-of-chapter references to formulate and mix a diet with the right balance of carbohydrate, protein, fat, fiber, calcium, phosphorous, and other minerals and vitamins. Consult with a knowledgeable pet supply house or avian veterinarian. Diets can be made ahead and frozen, then thawed completely right before use. Fresh fruits and vegetables can be offered but should be changed daily. Apples are primarily water and their fiber is not usable by most birds. Fruits may include berries, melon, and citrus; vegetables may include carrots and squash. Care must be taken to ensure that the bird consumes all of the mix before more is fed, so that they are truly obtaining the balance of nutrients. Allowing them to select parts of the diet will almost certainly lead to some of the problems discussed earlier.

Birds usually feed as a learned behavior from the parent. A bird not trained to a commercial mix will require some coaxing, even to the point of you showing the bird that it can be eaten. As you try to wean the bird from a seed or other mix to a commercial ration, start by offering the commercial ration in the morning. Do not offer any other food. For small birds, later in the day, if the bird has not eaten, supply some of the seed mix. Larger birds can be fed every other day to induce intake. Continue this until the bird is comfortable eating the commercial product. Some may take quite some time or may never adapt. In these cases, it is imperative that you are providing a balanced mix of seeds, fruits and vegetables, and vegetative parts (dark greens such as romaine lettuce). As always, observe feed intake, do not overfeed, and allow exercise as much as possible.

Along with the information on nutrition in the first several chapters, it is hoped that this chapter has provided some helpful practical guidelines for feeding birds for a long and healthy life. Remember the basics, be sure to read the label, and be sure to provide a complete mix of foods and plenty of exercise.

Words to Know

anseriformes and	*crop*	*psittacine*
galliformes	*grit*	*transit time*
avian	*moult*	*uric acid*
beak	*passerine*	*ventriculus*
cloaca	*proventriculus*	

Study Questions

1. Describe the differences in the digestive tract of avians compared with that of nonruminant mammals.
2. Describe the differences in amino acid metabolism and requirements of birds.
3. Describe the calcium and water requirements and metabolism of birds.
4. What is nutritional secondary hyperparathyroidism, and how can it be avoided?
5. How does the amount of calcium and fat in the diet of birds combine to cause some long-term health problems?
6. Describe different feeds for birds, including their nutritional strengths and problems.
7. Why is it so important to provide birds with a wide variety of food materials, and what are some practical feeding-management practices to ensure that the bird actually consumes what is fed?
8. Is exercise important for birds? Why or why not?
9. How does metabolism of vitamin D differ in the bird compared to most animals, and what must we do to avoid a deficiency?

Further Reading

Burger, I. 1993. *The Waltham book of companion animal nutrition.* Oxford: Pergamon Press.

Butcher, G. D., and R. D. Miles. 1993, revised 1996. *Understanding pet bird nutrition.* Institute of Food and Agricultural Sciences, University of Florida, Cooperative Extension Service. Circular 1082.

Harper, E. J., L. Lambert, and N. Moodle. 1998. The comparative nutrition of two passerine species: the canary (Serinus canaries) and the zebra finch (Poephila guttata). *Journal of Nutrition*, 128: 2684S–2685S.

Kollias, G. V., and H. W. Kollias. 2000. Feeding passerine and psittacine birds. In M. Hand, C. D. Thatcher, R. L. Remillard, and P. Roudebush, eds., *Small animal clinical nutrition*, fourth edition (chapter 30). Topeka, KS: Mark Morris Associates.

Warren, Dean M. 2002. *Small animal care and management*, second edition. Albany, NY: Delmar, Thomson Learning.

Wissman, M. A. 2000. *The importance of avian nutrition.* Retrieved Nov 14, 2004 from *www.exoticpetvet.net*

Chapter 16

Nutrition of Aquarium Fish

Take Home and Summary

The tremendous variety of fish species now kept as companions exceeds even that of the avian class. We do not have detailed information for exact nutrient requirements for the members of the class Osteichthyes. We must rely on a balanced application of nutritional principles, analysis of the data that are available, and a reasoned extrapolation from similar species used in aquaculture. Fish are ectotherms. They do not attempt to maintain a body temperature different from their environment. This has a major impact on their energy requirements, as the energy requirement will vary exactly in relation to water temperature. Fish excrete excess nitrogen from amino acids as ammonia; they do not need to make uric acid or urea. This saves energy, but puts a strong chemical back out into the environment. In the oceans, lakes, and rivers, the ammonia is used efficiently by bacteria, algae, and plants. This also happens in an aquarium, so we want to make sure not to overfeed, which leads to too much growth of bacteria and algae in the tank. Fish have a strict requirement for vitamin C, and as this vitamin is soluble in water (and then diluted out in the water), it must be heavily supplemented. Fish have a stricter requirement for the essential fatty acids (this is one reason why fish oil is in demand as human food, because of the ratio of essential fatty acids and their metabolic derivatives, which have a role in membrane function, coronary health, and the immune system; review Chapters 3 and 10). Carnivorous, omnivorous, and herbivorous fish can be fed with a few simple diets balanced to meet their basic nutrient needs.

What Kind of Fish?

There are over 5,000 species of the class *Osteichthyes,* or fish with a bony skeleton, kept in aquariums. There are about 300 popular species. It is neither reasonable nor possible to think that we need to have 300 different sets of requirements, just as we do not need a specific set of requirements for each breed of cat or dog. Within this class, there are just a few orders that make up most of the aquarium fish.

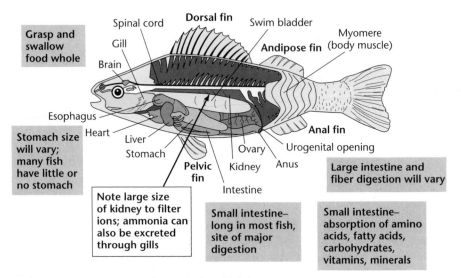

Figure 16.1 Digestion and metabolism in fish.
Source: Redrawn from *Small Animal Care and Management,* second edition by Warren. © 2002. Reprinted with permission of Delmar Learning, a division of Thomson Learning: www.thomsonrights.com. Fax 800 730-2215.

> Given the genetic diversity of aquarium fish, and the metabolic similarities among them, we need to stick to some of our basic principles and make practical nutritional management decisions.

Fish require amino acids, essential fatty acids, vitamins, and minerals. They can and do use carbohydrate in limited amounts; many fish are complete herbivores, many are omnivores, and a few are carnivores; thus their digestive tracts vary as such (Figure 16.1). It is likely we will not truly have a quantitative knowledge of the requirements of each of these thousands of species, so we need to think of requirements of groups of fish. Major classifications include warm (tropical) or cold (temperate) water; fresh or marine water; carnivore, omnivore, or herbivore; top-, bottom-, or middle-feeding; and night- or day-feeding. It is also an issue if we have a mixed population in an aquarium. We can do this if we pay attention to the feeding needs of the total population.

The koi carp, including goldfish *(Carassius auratus),* are temperate omnivores; they have no defined stomach. The coral butterfly *(Chaetodontidae spp.)* is a tropical omnivore that eats live coral; it has fine protruding teeth. The marine coral parrot fish of the family Scaridae have a beak with fused teeth to scrape polyps. The marine coral triggerfish *(Balistidae)* has a strong protruding beak. The Cichlidae are tropical freshwater carnivores and omnivores. The moonlight gourami *(Trichogaster microlepis)* is a carnivore with a low ability to digest carbohydrate.

Digestion of Fishes

First, we need to think about the digestive systems of fish (Figure 16.1). Few species of fish have teeth, but possess instead a bony mouth that can grasp. The food is usually eaten whole or in large particles. In most species the stomach is generally small, but can be quite large in omnivores (goldfish, catfish) or herbivores. In many small fish, there is no stomach; the food passes directly into the intestinal digestive system. The neon tetras (*Paracheiron innesi*) have no stomach and a fast metabolic rate; they can lose 3 percent of their body weight per day if not fed. Fish have the same basic organs as all chordata: liver, kidney, spleen, and pancreas. They make bile salts and have a gall bladder. Their pancreas makes digestive enzymes to digest food. The pancreatic duct enters at the proximal end of the ileum where the stomach would be, and the enzymes then can digest the newly consumed food. The pancreas also makes insulin to control glucose use, just as in mammals. If there is one major difference in digestive physiology for most of the fish kept in aquariums, it is that they must eat small meals more continuously. This will vary, as the more carnivorous fish can handle larger meals such as other small fish.

> It is better for the fish to feed frequent small meals. This also helps in tank health.

Metabolism of Fishes

Energy Metabolism and Temperature

The big difference between Osteichthyes, Aves, and Mammals is that fish (osteichthyes) are *ectotherms;* they do not maintain a constant body temperature (we sometimes call this "cold blooded"). They live and carry out their chemical reactions at the whim of the environmental temperature that they live in. This is a major survival benefit, as they require less energy to maintain normal life. They do not need to expend a large amount of energy to keep the body systems warm, thus they do not need as much food. Their energy requirement is a direct correlate of water temperature and will rise and fall with the seasons, or in our case, with the temperature of the house or aquarium. It is also a problem in management, as we should not feed fish at 60°F the same as at 75°F. Thus nutritional requirements will vary in the same animal depending on temperature conditions.

> The *Arrhenius equation*: as the temperature increases 10°C, the rates of all chemical reactions double; if the temperature decreases 10°C, reaction rates will be cut in half.

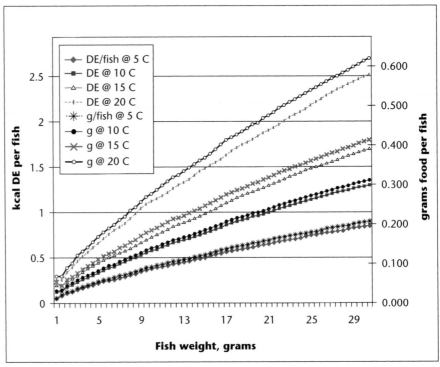

Figure 16.2 Effect of temperature on energy and food requirements of fish. Compare this graphic with the tank health chart shown in Figure 16.3. Note that both increasing fish size and tank temperature will increase feeding rate and require more frequent water turnover in tanks.
Source: John McNamara.

At a temperature of 20°C (68°F) the metabolic requirement will be twice what it was at 10°C (50°F); at 30°C (86°F) it will be four times what it was at 10°C. Does the temperature of your dwelling vary between 55° and 85°F during the year? If so, then the requirements of the fish in that small, uncontrolled tank can easily vary 200 to 400 percent during the year. This definitely affects how much we should be feeding. At a temperature of about 5°C (41°F), metabolism basically stops. Figure 16.2 shows the relationship between temperature and rates of metabolic reactions. Note that fish in an aquarium at 72°F will have a metabolic rate, and thus a nutritional requirement, only half of that at 89°F.

Nitrogen and Amino Acid Metabolism

The other major difference between the fishes and birds and mammals is that fish excrete the excess nitrogen from amino acid metabolism directly in the form of ammonia. It makes perfect sense from a chemical efficiency point of view. Animals and birds make urea and uric acid to turn the toxic ammonia into a safer compound while they are waiting to void it in the urine. Fish, however, are swimming around right in a big dilution system. So why pay the cost in energy of making that urea or uric acid (remember, that energy cost is about 10 percent of the energy value of the protein to begin with) when you can just "wash" the ammonia right out into the water? This saves them about another 10 percent on their energy requirement compared to mammals and birds. Because they do not have the full urea cycle or uric acid pathways, they have a reduced energy requirement. Also, note from Figure 16.2 that they do have kidneys rather large in relation to the body. This is because they must continually balance the ion and mineral content of their bodies against that of the water (freshwater fish would generally concentrate most minerals, whereas marine fish may need to filter out some minerals such as potassium, sodium, and chloride).

In the aquarium, ammonia can build up just like any other toxic material. The ammonia, along with any uneaten food or undigested protein that is lost in the feces, is now there in the tank. These compounds will be used by any plants you may have in the tank. But any excess will also be used by bacteria and algae. So the more extra protein you give the fish (over what they truly need), the faster the buildup of wastes and bacterial contamination in the tank. So, just as it makes no sense to overfeed protein to mammals and birds, it makes no sense for fish either. The less we feed, the less often we will need to be concerned with water replacement and tank cleaning. Figure 16.3 shows how the feeding rate affects water change and tank-cleaning practices.

Because of their general reliance more on protein than on carbohydrate, the fish do have a faster rate of protein turnover in the muscles, liver, and other organs than mammals. This is similar to but not as extreme as in the cat. Because of this fast rate of amino acid metabolism, fish do require more protein than most mammals, about 30 to 40 percent in most stages.

Nutritional Differences of Fish

Can be carnivorous, omnivorous, or herbivorous

Ectotherms: metabolic rate and nutrient requirement will vary with ambient temperature

Excrete excess nitrogen as ammonia

Require more specific essential fatty acid ratio

Another biological difference between fish and land or air animals is that they are not always growth-limited. Several fish can grow to large sizes if environment and nutrition allow. Your average goldfish dumped out in the city park pond will grow to a greater size than the one correctly left in the nice little bowl or aquarium. Conversely, the final adult size of the fish can and will be limited by environment and nutrition. This can be used to fit some fish to the situation at hand if space is limited. This is not a strict rule, nor can it be applied in the extreme—a great white shark will not stay a little tetra-sized fish in a small bowl; it will continue to grow according to its genetic code until there is no further space or nutrients provided, and then it will slowly starve to death. So this is not a major issue, but another difference that does apply in some situations for matching fish to a specific environment. Table 16.1 lists ingredients common to fish diets and explains their nutrient content.

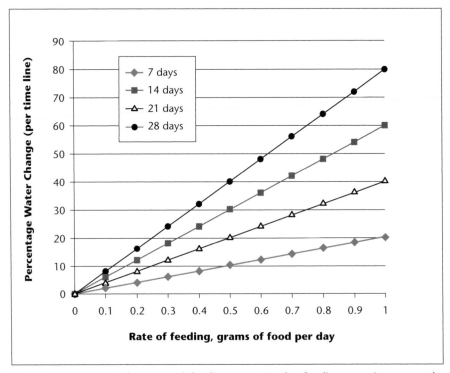

Figure 16.3 Water change and feeding rate. As the feeding rate increases, the amount of water that should be changed also increases (as a simple function of chemistry). Thus, for a 100-L tank a feeding rate of 0.5 grams per day (that is a lot of food), about 10 percent of the water should be changed every week, or 20 percent every two weeks, 60 percent every three weeks, and 40 percent every month.
Source: Adapted from I. Burger, 1993, *The Waltham Book of Companion Animal Nutrition,* figure 7.1.

Omnivorous and carnivorous fish, similar to cats, can utilize starch in commercial foods derived from plant seeds. The utilization is much better if the starch is cooked first to start the gelatinization process—allowing water into the starch granules to separate the spaces between the glucose molecules. This allows the enzymes to act more quickly and the fish will digest more of the starch. Fish food manufacturers know this and the basic fish flakes and pellets come from cooked food for both the safety aspect and for increased utilization of the plant starch.

The fish digestive tract, especially for herbivores, is not adapted to large meals. Fish in the wild often will feed for most of the day, but also they may—at different times of the year or during different stages of their life cycle—go without eating for a long period of time. So for most of the fish we deal with, three or four small meals per day are usually much better than one. This also makes sense from a basic tank-management viewpoint. The less often we feed the more likely we are to provide more food at one time than the fish can readily eat. That unused food will dissolve in the water and eventually build up as toxic by-products or be used by bacteria or plants to grow. This growth uses up more oxygen in the water, leads to more waste products, and in-

Table 16.1	Composition of Ingredients Common to Fish Diets (Also Used for Other Animals)				
Feed Ingredient	**DE, kcal/kg**	**CP, %**	**Fat, %**	**C. Fiber, %**	**Ash, %**
alfalfa meal	500–800	17–18	3	24	9–10
blood meal	1800–4200	85–90	<1	<1	2–4
corn gluten meal	3500–4200	60	<2	<2	1–3
crab meal, process residue	3000–3500	30–34	<3	11	41
fish solubles, dehydrated	3300–3400	62–65	7–9	<2	<5
fishmeal, anchovy	4200–4800	62–66	7–9	1	13–14
fishmeal, catfish	no value	48–52	9–10	<1	18
fishmeal, menhaden	~4000	62–66	9–11	<1	18–20
meat meal	>4000	54–58	8–10	<2	25–30
meat and bone meal	2900–3200	49–51	9–11	2–3	28–32
poultry by-product meal	2900–3600	58–60	13–15	< 2.5	14–16
poultry feather meal	3200–3400	80–85	<6	<1.5	<3
rice bran	2000–2500	12–14	13–14	10–12	10–12
shrimp meal, process residue	3500–4500	38–40	<4	12–13	26–28
sunflower meal	2500–3500	45–47	<3	11–12	7–8
wheat	~2400	12–13	<2	<3	<2
casein (milk protein)	~4400	83–85	<1	none	2–3

Source: Adapted from NRC, 1993, Nutrient Requirements of Fish.

creases the frequency at which we need to clean the tank. So again, observe what the fish eat and do not feed more than they want. Usually most fish will eat all they are going to in three to five minutes, so stop and look at and enjoy your fish while you feed them and you won't be tempted to feed as much. Fish have "taste sensors" over their skin, so they respond to amino acids and betaine (an amino acid breakdown product) to help them find the food.

An interesting study asked the question if goldfish would control their intake and select a proper diet if given the opportunity. When provided three dietary choices from self-feeders with different protein amounts (and the same vitamin and minerals): "Goldfish made their dietary selections on the basis of energy content, so that food demand increased to compensate for changes in the digestible energy density of the diets" (Sanchez-Vasquez, Yamamoto, Akiyama, Madrid, & Tabata, 1998). This fits with our basic concept of food intake control and suggests that perhaps the requirement for protein for this fish is not as great as is currently thought (the scientific method in action).

"Feed the fish, not the tank!" Observe the fish and offer only what they will eat in five minutes at a time. Increase number or size of meals only as needed.

Special Problems/Management of Fish

Fish have some unique metabolic requirements due to their water environment that we should note.

Minerals and Vitamins in Fish Nutrition

Fish live in an environment that contains many minerals dissolved in solution. Fish can absorb these from the intestine and gills and use them. However, the concentration in normal environments can vary significantly from place to place and from time to time (weather patterns, rainfall). In aquariums, even though the concentrations may be steadier, we cannot assume that there is a sufficient concentration of any mineral to meet the requirements of the fish, and we should plan to supply the minerals in the food.

Calcium and Phosphorous

Fish need calcium and phosphorous for the normal cellular metabolism, nervous transmission, and muscular contraction, just as any other animal. They do not need as much for the bones, as the bones are not as dense in the apetite crystallization (hard bone) as are those of birds or mammals. Bone is not an important source of calcium during periods of deficit, so calcium should be supplied regularly. Calcium and phosphorous excretion and uptake are also partially regulated through the gills—they can use the free minerals in the water. However, the concentrations are not usually sufficient to supply the needs of the fish. In marine water, there may be some problems with calcium uptake being reduced by magnesium, zinc, or copper. This is simply another reason to make sure it is supplied in the diet.

Iron

Iron is required by fish just as it is for birds or mammals. In some situations, extremely high concentrations of iron (> 200 ppm) can interfere with oxygen transfer. Most areas of the world do not have naturally occurring amounts of iron that concentrated, but if you live in a high-iron water area, you may want to check with a local aquarium veterinarian to see if it is a problem. Iron can be partially removed from water through different types of filters, or can simply be diluted with distilled water, to an extent.

Minerals for Ionic Control

Fish do require sodium, potassium, and chloride and must maintain a general balance of pH (acids or bases), just the same as other animals. Fish must continually adjust the concentration of minerals in their bodies, as the water concentration of these minerals can vary from next-to-none to very strong in marine water. Fish can thrive in a wide range of mineral concentrations, but just as for other animals, it is not a good idea to change water very abruptly to avoid stress to the homeostatic systems that keep minerals in balance. A regular change of a percentage of the water (Figure 16.3) is the best idea.

Fluorine

There is often confusion or misperception about the mineral *fluorine*. Remember that (Chapter 6) fluorine is required for proper teeth enamel and to some extent for bone growth. There is no evidence that fluoridated water is not appropriate for fish. The amounts used are in the range of 0.5 to 2 ppm and are designed to supply only what is needed and no more. There is no evidence that fluoridated water is any problem for fish.

Vitamins

There are few differences between fish and other animals in basic vitamin biology and nutrition. The optimal delivery of vitamins in a water environment provides some easily solved feeding challenges. Water-soluble vitamins are quickly solubilized from the diet to the water. Vitamin C can leach out of food at a rate of loss of up to two-thirds the amount in the food in thirty seconds. Some water-soluble vitamins can leach out at rates of loss up to 75 to 90 percent. These will be diluted and will quickly break down or be used by bacteria. Vitamins floating free in the water are in too low of a concentration to be of any benefit to the fish. Thus, most foods are oversupplemented with amounts of vitamins more than the fish actually need if they were to eat all the food. But by the time they eat what they are going to eat, there are still sufficient vitamins. As there is no vitamin toxicity for any water-soluble vitamins, if they do consume too much, the excess will be lost in the feces and urine.

Vitamin D metabolism, as in birds and mammals, is important in cartilage and bone formation. A deficiency similar to rickets is called scoliosis. *Scoliosis* is a sideways curvature of the spine such that the fish will appear bent in a V-shape if viewed from the dorsal surface.

Vitamin C is a required vitamin for fish, as they cannot synthesize it. As for mammals and birds; it is important for antioxidation and in cartilage formation. A deficiency of vitamin C is seen as *lordosis,* an upward and downward curvature of spine, so that the head and tail are down and the midsection is up if viewed from the side; scoliosis is also possible. Generalized poor health, lethargy, and aphagia (lack of eating) are also signs of Vitamin C deficiency. In order to reduce the amount of vitamin C loss in the water, it is often fed as phosphate or sulfate-stabilized salt.

Fish Pigmentation

Skin color in some specialized fish is a function of the intake of certain dietary components. As for coloring in birds, it can also be a function of the acid and base balance of the foodstuffs consumed. Most normal colors are derived from the class of chemical compounds known as **carotenoids,** and over 600 different molecules are known. These will cover the spectrum from blues and greens to yellows, oranges, and reds. The two major compounds fed are astaxanthin and canthaxanthin.

> Manipulation of the amount and type of carotenoids can bring out or subdue different colors in many fish.

Carotenoids can be supplied naturally from flower seeds and petals or, as is more likely in recent history, by supplying the compounds directly. As the interest in carotenoids in coloring and for antioxidant use grows, they are now often raised by algae with genes for specific carotenoids inserted into their genome.

The vitamin A precursor, beta-carotene, can be used in mammals and birds and many fish, but some fish such as goldfish cannot convert beta-carotene into canthaxanthin for coloration.

Fatty Acids

There is a fair amount known about essential fatty acid requirements and content of fish because of the great interest in the potential health benefits of certain fish oils in human food. Remember from Chapters 3 and 10 that the different essential fatty acids (linoleic, linolenic, and arachidonic) can give rise to different chemicals that regulate immune function and inflammation. The omega-3 series (linolenic) can help reduce inflammation, while the omega-6 fatty acids (linoleic and arachidonic) generally give rise to compounds that increase inflammation, but may suppress immune function. This is an oversimplification of an extremely complex system. Yet practically speaking, over-the-counter dog and cat diets as well as veterinary prescription diets use the amounts of omega-3 and omega-6 fatty acids to help reduce inflammation in sensitive animals.

Both categories of polyunsaturated fatty acids also affect membrane function, providing a more fluid lipid bilayer membrane that is thought to improve arterial and coronary (heart) health. Fish naturally have a greater amount of the omega-3 fatty acids such as linolenic acid than do other animals. Because of the tendency to reduce inflammation and their different properties in membranes, the omega-3 fatty acids are generally associated with a lower incidence of heart and artery problems in humans.

> Fish have strict dietary requirements for the omega-3 fatty acid, linolenic acid.

Fish floating (swimming) around in water that can vary tremendously in temperature would have a greater requirement for linolenic acid to keep the membranes fluid even when temperatures are cool and other fats become more solid. At temperatures of about 5 to 10°C, the ratio of omega-3 to omega-6 should be about 2, while at 15 to 20°C, it only need be about 0.5. The ratio should be changed by adding omega-3 fats, not by increasing the omega-6. Food manufacturers will do this by altering the percentages of ingredients coming from fish (higher in omega-3 fatty acids), other animals, and plants. Unsaturated fats oxidize rapidly and lose some of their chemical effectiveness, so this is another reason to keep feed fresh. Saturated fats such as beef fat and hydrogenated vegetable will solidify and may clog the digestive system; these are not appropriate fats for fish.

Feeding Management for Water Quality

Because fish live and excrete in water, and because of all of the other organisms in water, maintenance of water quality is a top priority. In both aquaculture and in ornamental fish, often overnitrification from too much ammonia leads to rapid bacterial and plant growth, which uses up too much of the oxygen, and then everything suffers. In aquariums, the same problem exists only on a smaller scale. Water management should be simple: Provide good, fresh water and a system of oxygenation constantly, either with a "bubbler," or by the batch by replacing a percentage of water on regular basis. In general, the higher the feeding rate, the more often the water should be changed. Remember our general rule: Feed only what fish will eat in a few minutes; if the tank has bottom-feeders, let some sink to the bottom.

Feed Quality and Selection

The similar requirements among several species of fish suggest that only a few general rations are likely to be available. Further definition by species is simply too costly and probably not necessary. Good, documented previous experience will be the best guide. Be aware (learn) whether or not the fish are top- or bottom-feeding and feed accordingly. Because of rapid spoilage, dry feeds are best for most aquarium fish. Most veterinarians and nutritionists today are recommending that if you are keeping meat eaters (piranhas, oscars), it is best to provide a complete ration instead of actual fish. This does provide more quality control as well as reduce the number of fish grown simply as food for other fish. This, at present, is a personal and not a legal choice. If you are feeding fish, be sure you are providing healthy, well-fed fish that themselves do not have any deficiencies. Be sure they are from a disease-free source to prevent the spread of disease.

When you are looking at products in the store, try to find one that refers to the type of fish you have. Look for a balance of fish and other animal and plant sources, not too heavily weighted toward any one. Also, be sure that vitamins and minerals are added. If a pelleted product, be sure that the pellet size is appropriate for the size of fish you have. A tiny neon at 0.5 cm long will not look at a big pellet as food, but as an enemy! Have fun, observe, and enjoy a long and healthy relationship with your fish.

Words to Know

arrhenius equation	*ectotherms*	*osteichthyes*
carotenoids	*lordosis*	*scoliosis*

Study Questions

1. What are the major metabolic differences in fish compared to mammals for energy metabolism? Nitrogen metabolism?
2. What is the Arrhenius equation and why does it apply to practical nutrition of fish?
3. How does vitamin C metabolism and requirement differ for fish? How does essential fatty acid requirement and metabolism differ?
4. How can coloration of some fish be altered nutritionally?
5. Describe an effective management strategy for water cleanliness in aquariums.

Further Reading

It may seem like I am giving short consideration to further reading for fish. However, the fact is that very little published, refereed, scientific research has been done on aquarium fish. Most products and recommendations have been made from careful extrapolation from aquacultural research.

Buller, N. B. 2004. *Bacteria from fish and other aquatic animals: A practical identification manual*. London, UK: CABI Publishing.

Burger, I. 1993. *The Waltham book of companion animal nutrition*. Oxford, UK: Pergamon Press.

Dunham, R. A. 2004. *Aquaculture and fisheries biotechnology: Genetic approaches*. London, UK: CABI Publishing.

Lakdawalla, P. 2003. "Sea-ing" results. An aquarium can help boost Alzheimer's patients' appetites. *Contemporary Longterm Care*, July, 26(7): 28.

National Research Council. 1993. *Nutrient requirements of fish*. Washington, D.C.: National Academy Press.

Sanchez-Vazquez F. J., T. Yamamoto, T. Akiyama, J. A. Madrid, and M. Tabata. 1998. Selection of macronutrients by goldfish operating self-feeders. *Physiology and Behavior*, 65(2): 211–218.

Stickney, R. R., and J. P. McVey, eds. 2002. *Responsible marine aquaculture*. London, UK: CABI Publishing.

Warren, D. M. 2002. *Small animal care and management*, second edition. Albany, NY: Delmar Learning.

Webster, C. D., and C. E. Lim. 2002. *Nutrient requirements and feeding of finfish for aquaculture*. London, UK: CABI Publishing.

Wedemeyer, G. A. 2002. *Fish hatchery management*, second edition. London, UK: CABI Publishing.

Yanong, R. P. 1999. Nutrition of ornamental fish. Veterinary Clinics of North America Exotic Animal Practice, 2(1): 19–42.

Chapter 17

Nutrition of Rodents

Take Home and Summary

The order Rodentia contains about 1,700 species that are similar in nutritional needs. However, many companion species have different needs. The term *rodent* derives from the Latin *rodere*, "to gnaw." Their general jaw and teeth structure is such that they can and do chew in both a lateral, side-to-side (gnawing) and back-and-forth (chewing) motion depending on what they are chewing and when. In general, rodents are fairly hardy and adaptable. They are usually omnivorous, switching from plant to animal sources of food as the seasons and conditions change. Similar to the rabbit, they are *coprophagic,* selectively consuming fecal pellets that are enriched in protein and B vitamins from the intestinal bacteria. Many species have been used as "lab animals" for research in a spectrum of biological sciences: cancer studies, aging, nutrition, reproduction, immunology, organ function, and basic metabolism. We have a good understanding of most of their nutritional needs, especially for rats. In fact, much of what we understand on vitamin and mineral use in humans has come from studies on rodents. Rodents as companions can be a tremendous amount of fun if they are given the proper attention and care. As with many of our companions, feeding recommendations are to feed a mix of ingredients to supply protein, energy, vitamins, and minerals, or "feed a good commercial diet." However, a good, balanced ration can easily be prepared at home, perhaps for less money. The rodents supply a number of interesting unique situations of biology and care.

> The term *rodent* comes from the Latin *rodere*, to gnaw.

The Order Rodentia

The teeth and jaw structure relate directly to how we should feed rodents. For rats, mice, guinea pigs, and gerbils, requirements are fairly well known and several complete diets are available. There are rat chows and mice chows and many different commercial diets. However, rodents have varied facial and teeth structures and must be fed in different ways to fully optimize their diets. Rodents have a fast metabolic rate and a relatively short lifespan, in keeping with the biological observation that the smaller the animal, the faster the metabolic rate and the shorter the life span. The fact that they are named for their specific eating mechanism should not be lost on the reader. The way we present the food and house the animals is important to their overall health and well-being.

> The rodent holds the food with its forepaws, while the lower incisor teeth move back and forth against the immobile upper incisor teeth. During this process, the lower jaw moves forward so that the top and bottom teeth line up.

Rodents have a large *diastema,* or gap between the front teeth (incisors) and cheek teeth. They can draw the cheek into this gap, and keep the food in the front part of the mouth; thus they can chew at length without having to swallow. When they continue chewing, the lower jaw moves *caudally* (backward) so that the side (cheek) teeth now can chew. The variations in jaw and facial structure among the species are taken into account in the foods fed and the presentation of the food (the bottom line is that a gerbil is not a guinea pig and a guinea pig is not a big rat).

Figure 17.1 shows the basic structure of the jaw and skull of the rodentia. Note the large gap, or diastema, between the incisors and the molars, also referred to as "side teeth" or "cheek teeth." Large muscles insert on these jaws

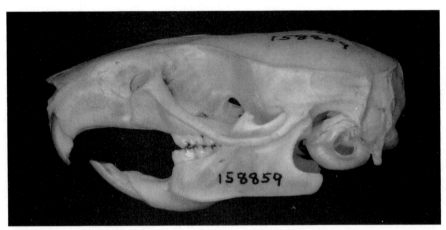

Figure 17.1 Example of rodent skull structure. Note sharp incisors and large diastema (gap) between incisors and molars. Note that the incisors are presently not lined up. When the animal eats, the lower jaw is moved forward so that the incisors align and they can gnaw off food. They store it in the large check pouches, which are where the large gaps are. Then they move the lower jaw caudally (rearward) so that the molars (cheek teeth) align and they can grind the food prior to swallowing. This adaptation allows a large amount of food to be gathered in a short time, a great advantage for a prey animal. *Source:* Photos courtesy of Dr. Phil Myers, Animal Diversity Web, University of Michigan, with permission *(http://animaldiversity.ummz.umich.edu).*

to move the bottom one forward to gnaw at food, and then backward to line up the molars (as depicted in Figure 17.1) to grind.

Rodents are often coprophagic (see Chapter 13), ingesting fecal pellets directly from the anus. These pellets are high in protein and B vitamins made by the bacteria in the cecum. Thus the rodents can be efficient on poor diets. Table 17.1 shows the classification of different types of rodents and Table 17.2 shows some life-cycle characteristics of rodents.

Table 17.1	The Wonderful World of Rodents	
	Order Rodentia	
Family	**Genus Species**	**Common Names**
Sciuridae	Several	Squirrels, chipmunks, marmots
Muridae	*Rattus Norwegus*	Old world rats and mice
Cricetidae	*Mesocricetus auratus*	Golden or Syrian hamster
	Cricetulus griseus	Chinese hamster
	Phodopus sungorus	Dwarf hamster
	Meriones unguiculatus	Mongolian gerbils, also voles, muskrats
Heteromyidae	*Mus musculus*	Mice, pocket mice
Chinchillidae	*Chinchilla laniger*	Chinchillas, viscachas
Dasyproctidae		Agoutis
Caviidae	*Cavia porcellus*	Cavies, guinea pigs, maras
Hystricidae		Old world porcupines

Table 17.2	Some Characteristics of the Rodent Life Cycle				
	Rat	**Mouse**	**Guinea Pig**	**Hamster**	**Gerbil**
Age at maturity, d (reproductive)	60 to 90	25 to 30	60 to 90	60 to 70	70 to 90
Gestation length, d	20 to 22	20 to 22	59 to 72	15 to 18	24 to 26
Litter size, range; mean	4 to 20; 10	4 to 20; 10	1 to 8; 4	2 to 16; 11	1 to 12; 5
Lactation length, d (weaning age)	21	14 to 18	21	21	21 to 24
Adult body size, g	200 to 500	20 to 50	f. 700 to 850 m. 900 to 1,200	Golden 80 to 150 Chinese 15 to 18 Syrian 30 to 50	70 to 140
Lifespan, d	400 to 1,500	350 to 1,000	700 to 1,400	350 to 1,000	1,400+

Housing

The housing requirements vary and are important to the general health of the animal. First, of course, sufficient space must be provided so that the animal has room to exercise and play. The cages must be escape proof and predator proof (those of us who have simultaneously kept cats and mice can attest to this). Constant attention is critical. Cages can be metal, plastic, or glass. It is not a good idea to use wood, as the rodents will gnaw it and it is difficult to keep clean. Wheels, tunnels, boxes, and the like should be provided in keeping with the specific animal to provide opportunities for exercise and to relieve boredom.

Most professionals will advise against shavings from trees such as cedar or pine. They contain resins that affect the health of the animals when they are exposed to the shavings. Hardwood shavings (oak, maple, beech) are much better. In the "old days" it was a concern using newspaper because the inks were made with a large concentration of metals (such as lead, zinc, and copper) that could affect the metabolism of the animals. Because almost all inks today are soy based, this is not an issue. However, newspaper and other paper may be eaten by the animals, leading to possible impaction in the digestive tract. Other types of cellulose fiber bedding are available and are neater and more efficient than newspaper.

Specific Families of Rodents

Rats and Mice *(Rattus Norwegus, Mus Musculus)*

These species can thrive on "rat chow" and "mouse chow" that has been developed from thousands of studies on nutrient needs. A small amount of treats such as small pieces of vegetable or cooked meat may be provided, but only as a special treat. Animals should not be allowed to come to expect these at the risk of not eating the properly balanced diets provided. These animals and other rodents can use beta-carotene for vitamin A and can use either the plant or animal form of vitamin A (Chapter 6).

Protein requirements are fairly low:
- 12 to 13 percent protein for growth
- 18 percent for reproduction
- 4 to 6 percent for maintenance

Problems relating to overfeeding protein to rats and mice in terms of kidney function are not a concern. Rats and mice will increase feed intake dramatically in lactation, up to four or five times maintenance rates. They will usually produce litters of six to ten offspring, but can have more. Care must be taken in group housing as the pups can be attacked and eaten by other rats, and occasionally by the dam herself.

Table 17.3	Nutritional Needs and Foods for Rodents			
Nutrient	Rats	Mice	Gerbils	Hamsters
Protein, %	5 to 15	12 to 15	12 to 18	15 to 17
Fat, %	5	5	5 to 20	5
Energy, DE kcal/g	3.8	3.8	3.8 to 4.5	4.2
Calcium, %	0.5	0.4	0.6 to 0.8	0.6
Phosphorous, %	0.4	0.4	0.3 to 0.4	0.3

Source: Adapted from NRC, 1995, Requirements of Laboratory Animals, *fourth revised edition, and I. Berger, 1993,* The Waltham Book of Companion Animal Nutrition.

Gerbils

There are just a few species of gerbils; all are members of the subfamily Gerbillinae: *Meriones unguiculatus* is the Mongolian gerbil; and the European or Tunisian gerbil is *Meriones libicus* or *Meriones shawi*. They make great small companions and are relatively easy to care for. Their protein requirement is about 16 percent on average, lower for adult animals and higher during lactation. Several good commercial chows provide proper fiber, protein, vitamins, and minerals. General recommendations are given in Table 17.3 for nutrient content and Table 17.4 for dietary ingredients that will provide the proper nutrients.

The gerbil is susceptible to magnesium deficiency on diets made up only of fruits and vegetables. The low amount of magnesium does not allow normal neural function and is associated with an increase in seizures. A seizure is a temporary loss of muscle control, in which the muscles can twitch uncontrollably for a period of several seconds. They are not usually fatal but can be. The magnesium deficiency also is associated with *alopecia*, or loss of hair. An interesting bit of nutritional trivia for gerbils is that on diets with a normal amount of fat (about 2 to 4 percent), they do not require the compound *inositol.* Inositol is a carbohydrate somewhat similar to glucose, which is used to make a lipid called phosphatidyl inositol, which in turn is used in membranes and control of metabolism in cells. Mammals usually make enough of this that it is not a required part of the diet. But if a higher-fat diet is fed, about 5 to 7 percent, the requirement increases from 20 ppm to 7 ppm to help metabolize the extra fat. As with many of the idiosyncrasies of species nutritional requirements, this is not really a practical problem on normal diets.

Guinea Pigs and Cavies *(Cavia Porcellus)*

The guinea pig is in the *Caviidae* family. They are thought to have originated in South America, where they have been bred for at least 400 years and still are raised for food as well as for research. As for other rodents, they have no canine teeth and have a large diastema (space) between the incisors and molars. They are a rodent, but in some ways are similar to the rabbit from the Order

Table 17.4	Examples of Foods and Diets for Rodents			
	Data are in percent of the total ration			
Ingredient	Rat	Mouse	Guinea Pig	Hamster and Gerbil
Alfalfa meal (17% CP)	5	8	35	20
Soybean meal (49% CP)	24	26	15	22
Ground yellow corn	30	28		31
Ground whole oats			22	20
Ground whole wheat	30	27	21	
Soybean (or other) oil	3	3	2	1
Dicalcium phosphate	1.5	1.5	0.5	2
Calcium carbonate	0.5	0.5	1.0	0.5
Salt (NaCl)	1	1	1.0	1.5
Brewers dried yeast	1	1		
Molasses (dry weight)	2	2		
Mineral and vitamin mix	2	2	2.5	2
Crude protein, %	18.6	19.6	18.3	19.4
Calcium, %	0.68	0.73	1.0	1.0

Note: The protein and calcium compositions of these rations is somewhat greater (about 10%) than what has been determined in experimental conditions to be adequate. This is in keeping with the "safety" philosophy that a little more is prudent to account for less-than-ideal conditions. More calcium could safely be added but may not be necessary in the rat and mouse rations. Also note that the ration for the guinea pig is markedly higher in alfalfa. Why do you think this is? You might want to review Table 13.2 on the nutrient content of these foods.

Source: Adapted from NRC, 1995, Nutrient Requirements of Laboratory Animals, *fourth revised edition.*

Lagomorpha. They are herbivores and have an enlarged cecum and colon similar to that of the rabbit and the horse. The cecum is large and semicircular with numerous lateral pouches similar to the rabbit. All rodents are coprophagic, but this practice is much more important for the guinea pigs and cavies. They have a somewhat greater requirement for protein, about 18 to 20 percent. Their gestation length is fifty-nine to seventy-two days and they usually give birth to one to three pups.

Their natural diet is green vegetation and fruits, and some of the older chows tended to be higher in energy and lower in fiber. This created a problem with poor formation of cecotropes and impaction similar to that which can occur in rabbits. It is best to allow many small meals a day. Because they have a relatively sensitive digestive system, care should be taken to make dietary and environmental changes gradually. If food is available for long periods, be sure to keep the total amount no more than 1 to 2 percent of body weight a day, depending on activity level. Observe and gently palpate the abdomen, ribs, and rump area to monitor body fat regularly.

Cavies, Guinea Pigs
  More like a lagomorph
  Enlarged cecum and colon
  Need more fiber
  Higher protein requirement
  Cecotrophy more important
  Cannot make vitamin C

The unique nutritional attribute of this family is that they have an absolute requirement for dietary vitamin C. They cannot make any ascorbic acid (vitamin C), and this is unique among animals. Commercial rations provide sufficient vitamin C, as well as do fresh dark greens. However, this metabolic difference is one reason why the name "guinea pig" has come to be used interchangeably with "experimental animal." Because they do not make ascorbic acid, it is easy to create a deficiency. Remember from Chapter 6 that vitamin C is used in normal collagen and connective tissue synthesis. Thus, guinea pigs have been used as experimental animals in the study of connective tissue synthesis and wound healing. Some of the major advances in post-surgical care and wound care have come from studying connective tissue and skin synthesis in guinea pigs.

Hamsters

Hamsters are close cousins to the gerbils, also members of the family Cricetidae, subfamily Cicetinae. There are three major species: the Golden or Syrian *(Mesocricetus auratus)*; the European *(Cricetus cricetus)* and the Chinese or Grey *(Cricetulus griseus)*. Their protein requirement is about 15 percent up to 20 percent for reproduction. This is greater than many grains, so it is important that they receive some type of ingredients with more protein, such as legumes (soybean) or oilseeds such as sunflower and safflower. As for all the other rodents, there are several well-balanced chows available.

Hamsters are somewhat different in their digestive tract than the other rodents. Their stomachs are separated into the first (cardiac) region, which does not secrete any digestive acids or enzymes, and a caudal portion (pyloric) that secretes stomach acid and enzymes. Although they are not ruminants, this fore stomach does act in a similar way. Ingested food is wetted, bacteria grow and apparently perform some digestion and fermentation, and then the food and bacteria pass to the primary stomach for acid and enzymatic digestion. As a supplement to the normal chow, dark greens and yellow vegetables can be fed, preferably in pieces small enough to be bite size to encourage eating and diminish spoiling. Remember to avoid things like iceberg lettuce, which is mainly water with little fiber, and apples, which are too high in water as well as in unusable fiber.

> Hamsters have a pouch before the stomach, similar to a small rumen, so they:
>
> can use more fiber and generate volatile fatty acids
> need more fiber in diet

There are a couple of nutritional idiosyncrasies in this family as well. They do not require vitamin D in the diet *if* the calcium to phosphorous ratio is 2:1 and dietary calcium is above 0.6 percent. At that level, they can make active vitamin D from cholesterol and sunlight. All mammals can do this (Chapter 6) but most others require some extra in the diet.

Their other uniqueness is in cholesterol metabolism. They are sensitive to dietary cholesterol and develop atherosclerosis quite easily—more easily than humans. *Atherosclerosis* is defined as plaques made up of fats, foam cells, and calcium that can build up in arteries. If they become large enough they can restrict or even block the flow of blood. In humans this can lead to heart attacks or strokes, situations in which the heart muscle or brain do not receive enough oxygen and cells start to die. We now know that atherosclerosis is a complex disease that has a genetic component (it is hereditary); an anatomy component (arteries with many bends generally get more plaques, because the turbulence in the blood flow around the bend damages the arterial wall and starts plaque formation); and a dietary component (too much fat, saturated fat, and sugar). Other factors include inactivity (less blood flow allows more plaque buildup) and smoking (nicotine, tar, and other chemicals can enhance plaque formation). Although hamsters do not usually smoke, the point of all this "human nutrition" information is that atherosclerosis is a complex problem requiring some serious scientific research to study.

So, hamsters and rabbits (Chapter 14) are often used as models of cholesterol metabolism and atherosclerosis. We have learned a lot about this disease from this research. However, as mentioned in Chapter 1, it is too simplistic to extrapolate all the findings from hamsters to humans. Now, because of some of the findings from research in hamsters, we, with better scientific understanding and better technology, can now study many such diseases directly in humans and do not need to rely solely on animal models. The other note is to be careful about the information you read in the paper or some magazines, understanding how much information from research done in animals can be extrapolated to humans—some can and some cannot, and usually the more complex the situation the worse idea it is to extrapolate from animal models to humans.

Rodents can and do make excellent companions, often being the animal of choice in urban settings or for smaller children who may not yet be able to properly care for a dog, cat, or horse. As is the case for most of our companions, many excellent commercial products are available to make life easy. The information in this chapter, I hope, introduces the basic nutritional biology of rodents. It may be a starting point for more in-depth research for a class project or for more advanced classes. The same nutritional principles we learned in Chapters 1 through 10 apply, and remember that not all rodents are the same.

Words to Know

atherosclerosis	diastema	rodent
Caviidae	inositol	rodere
coprophagic		

Study Questions

1. Describe the basic feeding process for rodents, detailing their unique jaw function.
2. For each of the specific genuses of rodents, describe one unique metabolic characteristic and how we alter ration formulation and feeding management to deal with it. Do this for rats, mice, hamsters, gerbils, and guinea pigs as well.
3. What are your thoughts on using rodents as laboratory animals—for what reasons should we use them, or for what reasons do you think we never should?

Further Reading

There are hundreds of thousands of research papers on rodents, especially rats and mice. It is probably a safe statement that we understand more about the nutrition of rats and mice than any other animals. I have stuck with my philosophy and listed a few key, recent references that will take you into the realm of rodent nutrition. On June 9, 2004, a query of "rodents AND nutrition" at pubmed's site (*http://www.ncbi.nlm.nih.gov*; but type in "pubmed.com", it is faster) listed 30,671 articles.

Battles, A. H. 1991. The biology, care, and diseases of the Syrian hamster. In D. E. Johnson, ed., *Exotic animal medicine in practice. The compendium collection*, volume 1 (pp. 15–27). Trenton, NJ: Veterinary Learning Systems.

Carpenter, J. W., and C. M. Kolmstetter. 2000. Feeding small exotic mammals. In M. Hand, C. D. Thatcher, R. L. Remillard and P. Roudebush, eds. *Small animal clinical nutrition*, fourth edition. Topeka, KS: Mark Morris Associates.

Harkness, J. E. 1993. Biology and husbandry—Gerbils. In *A practitioner's guide to domestic rodents*. Lakewood, CO: AAHA Professional Library Series.

Harkness, J. E. 1993. Biology and husbandry—Rats. In *A practitioner's guide to domestic rodents*. Lakewood, CO: AAHA Professional Library Series.

National Research Council. 1995. *Nutrient requirements of laboratory animals*, fourth revised edition. Washington, DC: National Academy Press.

Peters, L. J. 1991. The guinea pig: An overview. In D. E. Johnson, ed., *Exotic animal medicine in practice. The compendium collection*, volume 1 (pp. 15–27). Trenton, NJ: Veterinary Learning Systems.

Chapter 18

Nutrition of Reptiles

Take Home and Summary

The eating habits of the class Reptilia are as diverse as any animals we have discussed so far. The metabolic pathways of reptiles are similar to those of birds and their sources of nutrition are equally diverse. The various families include carnivores, omnivores, and herbivores. Their normal environments run from desert to aquatic, temperate to tropical. When kept as companions, misperceptions about what they normally eat have lead to various nutritional deficiencies and imbalances, primarily that of too much energy and not enough of vitamins and minerals. For most situations, the safest way to feed most of these animals is with well-designed and prepared commercial foods, or with a variety of natural sources that provide the full balance of nutrients that the animal would have in the wild, not just a part of them. Providing sufficient water in ways similar to the animal's normal environment is important to their overall health. Dusts and powders to supply vitamins and minerals cannot take the place of a well-balanced, normally formulated ration and can lead to spikes of deficiencies and toxicities. Most veterinarians and nutritionists today recommend that it is best for many reasons to feed complete rations, or in the case of some carnivores to supply killed prey and not live prey. Live prey in a captive situation may harm the reptile, and concerns about raising animals to be killed as live prey are slowly moving most specialists toward humane killing prior to being fed as food.

General Saurian Digestion, Metabolism, and Nutrition

I can do no better here than to quote from the specialists, Scott Stahl and Susan Donoghue (2000): "with the exception of field studies using free-living reptiles, nutritional research is limited. Thus, dietary recommendations for reptiles are based on knowledge of natural diets and feeding histories, clinical experience, and principles of comparative nutrition" (p. 961).

Table 18.1 shows the classification of the order Squamata.

Table 18.1	Our Reptilian Companions: Squamata, Sauria		
Order, Suborder, Family	Genus	Common Name	Dietary Information
Squamata Suborder Sauria			
	Iguana	iguana	herbivore
	Uromastyx	spiny tailed lizard	herbivore
	Sauromalus	chuckwalla	herbivore
	Cyclura	rock iguana	herbivore
	Corucia	skinks	herbivore
	Pogona	bearded dragons	omnivore
	Tiliqua	blue-tongued skinks	omnivore
	Physignathus	water dragons	omnivore
	Gerrhosauras	plated lizard	omnivore
	Varanus	monitors	carnivore
	Heloderma	gila monster, bearded lizard	carnivore
	Tupinambis	tegus	carnivore

Carnivorous reptiles have strictly adapted to prey containing:

protein

fat

calcium

vitamin D_3

This includes primarily vertebrate animals with some invertebrates.

I hope that the previous chapters, and recounting the basic history of nutritional principles, will help you to investigate and decide the best practices for your particular companion from the huge number (over 6,500) of potential reptilian species. Some idiosyncrasies we can deal with, but the take-home message is still the same advice of making sure to feed a fully balanced diet with only sufficient, not excessive, energy and protein, and the proper amount of vitamins and minerals. This is done either through feeding a prepared formula, or the normal and complete variety of plants and animals that the reptile would encounter in the wild. There are many complete feeds, treats, and supplements on the market. Many of these are very good and designed as best as possible for reptiles. Others are too low in calcium if the animals are only fed things like light salad greens or invertebrate prey. If you are feeding omnivorous or carnivorous reptiles a mix of healthy vertebrate prey and they are eating all you feed, supplementation with vitamins and

Table 18.2	Our Reptilian Companions: Squamata, Serpentes		
Order, Suborder, Family	Genus	Common Name	Dietary Information
Suborder Serpentes			
	Elaphe	rat snake, corn snake	mammalian prey
	Boid	pythons, boas	mammalian prey
	Pituophis	bull and pine snakes	mammalian prey
	Lampropeltis	king snake	amphibians
	Thamnophis	garter snakes	amphibians, fish, reptiles
	Diadophis	ring-necked	amphibians, fish, reptiles
	Opheodrys	green	insectivore
	Carphophis	worm snake	insectivore
	Storeria	brown snake	insectivore

> For herbivorous reptiles, common problems include the feeding of only a minimal variety of plant matter.

minerals is probably not necessary. As in all else we have discussed, research and plan first, feed the total needed ration, and supplement only as necessary.

Because most readily available fruits, vegetables, and greens (apples, carrots, head lettuce) that are fed are too high in water and fiber, and too low in minerals such as calcium and protein, than the normal diet of these animals, plants, whole plants, and seeds must be provided in a full variety to supply the right amount of protein, vitamins, and minerals. Normal foods would include dark green lettuce, kale, chard, and alfalfa. It is important to have a variety. Metabolic bone disease is similar to nutritional secondary hypoparathyroidism in birds: Calcium deficiency, vitamin D deficiency, and excess of energy and phosphorous lead to malformed and weakened bones. Table 18.2 shows the classification of the suborder Serpentes.

The digestive system of the herbivores is somewhat different than that of other mammals. There are numerous folds and partitions that probably slow passage through the tract, increasing the amount of time for the bacteria and protozoa to digest and ferment the food. Recall the digestion system of the horse; this is similar in that the herbivorous reptile will obtain volatile fatty acids (acetate, propionate, and butyrate) from the fermentation. Most herbivorous species are larger than carnivorous ones, generally because a larger body size and digestive tract size will allow more of the bulky vegetative matter to be consumed. This size probably also limits their habitat and optimal temperature range. They take longer to heat up and cool down. There are few desert-type herbivores (desert iguana is an example).

The direct relationship of temperature and chemical reaction rate applies to reptiles as well. As temperature increases, metabolic rate also increases and the animal will require more digestible foods (less fiber). As temperatures decrease, their energy requirement decreases, and they can use higher-fiber feeds with more efficiency. This is a reason to try to keep temperatures relatively constant.

> Misperception leads to feeding carnivorous reptiles a plant-based diet or an animal diet with a too-narrow range of prey, especially invertebrates.

Invertebrate prey (worms, larvae) do not have enough calcium. Worse yet, similar to some misperceptions for dogs and cats, the feeding of only "meat" (muscle tissue) leads to excess of protein and energy and deficiencies of calcium and other minerals. A wide variety of prey, or at least the right prey class (mammals, insects), should be given, and the prey should themselves be well fed to avoid deficiencies.

For mammals, we know that in late pregnancy and lactation they require more protein, energy, vitamins, and minerals. However, many animals, including aquatic mammals such as whales and sea lions, some Aves, Osteicthyes, and Reptilia actually decrease or cease feeding during egg laying (or lactation in sea mammals). There is simply little known on the nutrient requirements of egg-laying reptiles. Simple management and observation, as always, is the key. It is likely that requirements increase, allow an increase in intake, but an additional 10 to 20 percent will probably be sufficient.

Water

Water delivery to reptiles is a little more challenging than to dogs and cats. Reptiles obtain water in various ways. Anoles and geckos, for example, lick up drops of water from leaves. Iguanas and monitors usually can learn to drink from *shallow* bowls or lids (like a butter tub lid—only 1/4 inch or so tall). Smell and taste are critically important, and many specialists recommend using glass instead of plastic, which can absorb odors. Obviously, keeping water clean is important. Dehydration can be a problem if water is not managed correctly. A lack of water in some animals can lead to gout, a crystallization of purines in joints (purines are breakdown products of genetic material, DNA, and RNA). Gout can be caused by a combination of feeding certain high-purine foods and low water intake. This causes crystallization of purines in joints, causing pain and in serious cases, immobility. Foods that contain a lot of purines would include anchovies, sardines, organ meats, and veggies such as asparagus and mushrooms. It would probably be prudent to avoid these foods.

Water Delivery

Must fit the environment of the animal

Proper depth

Proper interval

Proper humidity

It is also important to be aware of the humidity level in the reptilian environment. If the humidity is too low, it may lead to dehydration, and a condition called *dyscedysis,* which is an excess drying of skin and the inability to shed the skin correctly. Small pieces of old skin may adhere and will need to be gently removed (after soaking to help soften). This condition may lead to open sores and infections. If the humidity is too high, it might lead to an increase in skin problems such as *hyperkeratinization* (an excess of growth and scaling of skin) or infections.

Nitrogen and Energy

Reptiles resemble Aves in many aspects of metabolism, and resemble Osteichthyes in other respects. They excrete nitrogen as uric acid, similar to birds. This saves both energy and water as the urea made by mammals requires a lot of water to dilute it to safe concentrations. Similar to fish, they are poikilotherms, more commonly referred to as *ectotherms;* they do not maintain a constant body temperature. Instead, their body temperature can fluctuate with that of the environment. A temperature that is too low will cause a decrease in food intake, nutrient requirements, and growth. Temperatures too warm will lead to excessive metabolism of nutrients, also leading to decreased feed intake (due to metabolic toxins made in the body) and reduced growth.

Key Metabolic Differences

Slower metabolic rate, lower energy requirement

Energy requirement varies with ambient temperature

Nitrogen excreted as uric acid

Require vitamin D_3 and sunlight or proper UV light

On average, reptiles only require about one-quarter of the energy of similar-sized mammals. Also, there is a strong inverse relation between metabolic rate and body size: The larger the reptile, the slower the metabolic rate. This means, in practice, that the smaller reptiles need more food per unit body weight and need to be fed more often. Similar to nutrition of mammals and birds, excess energy supply will lead to obesity. Obesity is harder to notice in

Table 18.3	Foods for Reptiles, Be They Herbivores, Omnivores, or Carnivores	
Carnivorous Reptiles	**Foods**	**Nutrient Content**
Snakes, aquatic turtles, monitors, most lizards, geckos, chameleons	mealworms, flies, crickets, mice, fish, rats	25 to 60% protein 30 to 60% fat
Omnivorous Reptiles		
Box turtles, day geckos, anoles, bearded dragons, arboreal tortoises	slugs, snails, crickets, fruits, greens, vegetables	15 to 40% protein 5 to 40% fat 20 to 75% carbohydrate
Herbivorous Reptiles		
Most tortoises, iguanas, spiny-tailed lizards	greens, fruits, vegetables, clover, dandelions, grasses	15 to 35% protein less than 10% fat 55 to 75% carbohydrate

birds and reptiles compared to dogs, so you must take care to supply just the right amount of energy and provide an environment that allows good activity. Using a well-balanced ration, providing exercise, and watching intake all apply to reptiles as much as they do for birds and mammals.

Carnivorous reptiles may require as much as 30 to 60 percent crude protein, with herbivores somewhat less. Green iguanas have been tested to require about 28 percent for growth. These high requirements for protein require use of animal proteins or high-protein seeds and nuts. Table 18.3 shows the nutrient content of certain reptite foods.

> Some carnivorous reptiles can be fed commercial dog and cat food along with insects or small mammals. For herbivores, dog and cat foods are too high in protein, fat, and vitamins.

Vitamins and Minerals

Because so little work has been done on vitamin and mineral nutrition of reptiles, and because the vitamin and mineral content of natural foods can vary so widely, this is a serious problem area for reptiles. Metabolic bone disease is known to be caused by a calcium deficiency. On the opposite end of the spectrum, green iguanas have been reported to have vitamin A and vitamin D toxicities when fed commercial dog and cat foods. Vitamin A can be limiting when reptiles are fed only produce and insects without supplementation. The technical term is **hypovitaminosis A,** or vitamin A deficiency. *Chelonia* (turtles) are especially susceptible as they are often fed only iceberg lettuce, hamburger, or other vitamin A-deficient foods.

You may remember from Chapter 6 that there are two types of vitamin D: *plant vitamin D*, ergocalciferol, or D_2, and *animal vitamin D*, cholecalciferol, or D_3. Reptiles, like birds, must have vitamin D_3; the D_2 form is not usable by them. Under proper sunlight they can make a significant amount of active vi-

| Table 18.4 | Nutrient Content of Some Vegetable Foods for Reptiles |

Weights are 100 grams of food as fed (1/2 to 2/3 cups loose) except as noted. Data are in percent of dry matter or in kcal/gram for energy (multiply by 4.184 for kJoule). NFE is *nitrogen free extract*, which is 60 to 80% starch. AF = as-fed basis (with water); DMB = dry-matter basis.

	Water			Protein	Fat	NFE	Fiber	Calcium	Phosphorous
	Energy	AF	DMB			(% of Dry Matter)			
Romaine lettuce	94	0.18	3.0	36	7	50	11	1.1	0.4
Spinach	91	0.26	2.9	36	3	48	7	1.0	0.6
Dandelion	86	0.44	3.1	18	5	61	11	1.2	0.4
Mixed frozen vegetables	83	0.47	2.8	16	2	68	7	0.1	0.3
Apple, 1 medium, no skin, 130 g	84	0.51	3.2	1	2	86	4	basically none	
Strawberries	92	0.28	3.5	6	4	77	6	0.2	0.2

Note that all of these foods are low in fat, most of them get the primary energy from carbohydrate; only greens are adequate in protein and calcium and have the proper Ca:P ratio for reptiles, while the vegetable and fruits are inadequate in protein, calcium, and phosphorous, have an inverted Ca:P ratio (1 or less) and get way too much energy from carbohydrate.

Source: Adapted from S. Stahl and S. Donoghue, 2000, "Feeding Reptiles," in M. Hand, C. D. Thatcher, R. L. Remillard, and P. Roudebush, eds., Small Animal Clinical Nutrition, *fourth edition.*

tamin D_3. A judicious and not excessive supplementation with vitamin D_3 should be considered. If you are feeding well-fed, whole vertebrate prey, vitamin D_3 amounts should be adequate. Table 18.4 gives the nutrient content of some vegetable foods for reptiles.

Metabolic Bone Disease

Calcium deficiency is the other serious problem, often occurring simultaneously with deficiencies of vitamin D_3. Diets consisting of light-green produce, fruits, meats, and grains do not supply enough calcium. Coupled also with use of high-fat seeds or fat meat, the fatty acids bind calcium in the diet and prevent absorption. As noted for birds, it is probably safe to say that a high-fat, high-energy, low-calcium diet has been the death of most reptiles kept as companions. We refer to this syndrome of calcium deficiency in reptiles as *metabolic bone disease.* Metabolic bone disease is also potentially caused by an excess of phosphorous, a deficiency of vitamin D_3 or, most likely, a combination of all three factors. This problem is metabolically similar to *secondary hyperparathyroidism* in birds (Chapter 15) and has been reported in lizards, tortoises, and turtles. Signs would include soft or malformed bones, shells, and fractured limbs. Once diagnosed, supplementation and use of sunlight or full spectrum ultraviolet light to encourage vitamin D_3 synthesis is the best course. But even better is to prevent this by feeding a mix of plant and whole-animal products and using calcium and vitamin D_3 supplements judiciously.

Table 18.5 shows the nutrient content of certain animal foods for reptiles.

Table 18.5	Nutrient Content of Some Animal Foods for Reptiles

Animal Matter				Data are in percentages on a dry-matter basis					
		Energy		Protein	Fat	NFE	Fiber	Calcium	Phosphorous
	Water	AF	DMB			(% of Dry Matter)			
Vertebrates									
Mouse, adult	65	1.7	4.8	65	28	7	5.0		3.6
Mouse, pup	81	0.8	4.2	73	23	4	3.8		3.7
Rat, adult	66	1.6	4.7	72	25	3	4.4		3.2
Chick, day old	73	1.3	4.8	69	26	5	2.7		2.0
Herring	69	1.8	5.7	58	38	4	1.4		1.4
Smelt	77	1.0	4.3	76	17	7		3.2	4.4
Invertebrates									
House cricket (*Gryllus domesticus*)	68	1.0	3.1	57	34	9		0.3	2.7
Commercial cricket (*Achet domestica*)	62	1.9	4.8	66	26	8		0.2	2.6
Mealworm larvae (*Tenebrio molitor*)	58	2.1	5.0	56	40	5		0.1	1.2
Fly larvae (*Musca domestica*)	70	1.5	4.9	64	26	11		0.1	1.0
Earthworm (*Lumbricus terrestris*)	84	0.5	3.1	79	6	15		<1.5	<1

Note: Be sure that you note the numbers and patterns here. Older animals will tend to be higher in fat and calcium. Invertebrates are almost always too low in calcium and have an inverse (<1) Ca:P ratio. This will inevitably lead to poor growth and calcium deficiency. Note that these two examples of fish that are often used as food will be higher in fat, but what the table does not show is that the fat will vary with season—tending to be higher in fish caught in summer and fall and lower in fish caught in winter and spring. Use of these foods should be monitored closely and if feeding invertebrates at more than 50% of the diet, a calcium dust should be applied, without phosphorous, such as oyster shell or calcium carbonate.

> Metabolic bone disease is similar to nutritional secondary hyperparathyroidism:
>
> Calcium deficiency causes excess parathyroid hormone
>
> Too much bone mobilization
>
> Excess loss of bone calcium *and* phosphorus
>
> Calcium goes into soft tissue, phosphorus is lost in urine

Feeding Herbivores

Diets for herbivorous lizards should be a mix of dark greens such as romaine, collard, mustard, spinach, dandelion, and clover. Some vegetables such as carrots, zucchini, yellow squash, and mixed vegetable preparations from the store can be used at about 10 percent up to 20 percent of the diet. More than that and you risk too much energy and not enough calcium. Greens such as spinach, cabbage, beet greens, peas, and potatoes contain increased amounts of oxalates that can inhibit mineral absorption, so use these sparingly. Cruciferous plants include broccoli, brussels sprouts, cabbage, kale, and mustard. These can contain goitrogens (glucosinolates) that might negatively affect the thyroid gland if fed as more than 10 or 20 percent of the diet. Use fruits only sparingly as treats.

Commercially prepared mixes can (and some would say *should*) be used in total or about 50 percent of the diet. This helps ensure that the vitamins and minerals are adequate and balanced. If feeding only vegetable matter, even significant amounts of dark greens, a balanced vitamin and mineral mix with sufficient calcium should be used.

Feeding Omnivores and Carnivores

When feeding the omnivorous and carnivorous reptiles, we still must think of variety and balance. No free-living animal, including reptiles, eats only one thing all of the time (a few might, but not many). In these cases, it is quite easy to get into the vitamin and mineral problems discussed earlier, especially for calcium. It is simply wise to feed a trusted and analyzed commercial mix that contains at least 30 to 40 percent protein, 10 to 20 percent fat, and has at least 1 percent calcium and a Ca:P ratio of 1 at the very least to about 2. The remaining part of the diet should include a mix of invertebrates such as crickets, meal worms, earthworms, snails, and slugs; and vertebrates such as mice or rat pups, adult mice or rats or chicks, and even dry, cooked eggs. If you choose to feed homemade preparations, use the tables as a guide, consider a vitamin and mineral supplement, and practice close observation on intake, behavior, and health.

There is always the philosophical question of whether to raise animals as food for other animals. One viewpoint (whether you are a "strict creationist" or a "natural selectionist") is that animals eat other animals in nature. The other side is that we can feed other materials carefully to meet the nutrient requirements of the animal. However, feeding a strictly vegetarian diet to carnivorous and omnivorous reptiles is wrought with peril, almost always leading to protein deficiency, and deficiencies of calcium and fat-soluble vitamins. I personally do not agree with such a vegetarian approach, but should you choose to do such, be sure to include dark green leafy vegetables, high-protein seeds, and use a good calcium and vitamin supplement.

To Feed or Not to Feed

Carnivores eat other animals alive in the wild. Should we?

    Pros: Natural

    Cons: Dangerous in confinement (prey can fight back);

    not "nice" to grow animals and feed live to others

It is your choice, but you should think first—philosophy is action, so is a lack thereof.

In the case of carnivorous and omnivorous reptiles, the use of commercial cat food, balanced for all life stages, can be appropriate. It has enough protein and the right amount of minerals and vitamins (including vitamin D₃). For omnivores, only about 50 percent of the diet should contain this; the rest should be vegetables to avoid too much fat. For carnivores, more may be used. If about 50 percent or more of the diet is cat food, further vitamin and mineral supplementation is not warranted. If none is used, then supplements should be used and invertebrate prey should be lightly dusted with a mineral mix (high in calcium) prior to being fed. Remember that "dusting" with a supplement means to dust the food that we are feeding (vegetables, invertebrate prey), *not* the animal we are feeding the food to.

Feeding carnivorous snakes presents another physical problem: whether or not to feed live prey, and how to "wean" an animal already used to live prey onto killed prey. Most veterinarians and nutritionists now agree for practical as well as ethical reasons to encourage the consumption of killed prey. Live prey can traumatize a snake, even seriously injure it when put into an enclosed space (cage life is *not* "real life," in which a snake usually and very effectively "surprises" a prey animal and can capture and consume it without harm to itself).

Rat and mice pups can be humanely killed and frozen for later use. I would recommend that you consult with a veterinarian trained in herpetology and in killing and storing prey animals before you do this. The frozen prey should be warmed quickly in hot water (or a microwave if you wish—but be careful to avoid overheating, you will not like the result). We want to warm quickly so that the bacteria in the intestinal tract do not have time to grow and ferment. It is important to feed the whole animal, as muscle itself is too high in protein and too low in calcium and many vitamins and minerals. From the intestinal tract the reptile will obtain the full complement of B vitamins and vitamin K. From the bones, it will obtain calcium, phosphorous, and magnesium. It is also important that the prey themselves were fed properly (see Chapter 17) so that they do have a balanced content in their digestive tract and do not suffer from any deficiencies. The warm, dead animal can be wiggled and moved with a rod or long forceps to trick the snake into thinking it is alive. Snakes should be housed or at least fed separately to avoid inadvertent injury or even consumption of one snake by the other.

Several snakes such as king *(Lampropeltis)*, indigo *(Drymarchon)*, and water snakes *(Nerodia)* often normally consume amphibians, fish, and other reptiles. If feeding these prey, they should be frozen at −20°C (a home freezer set cold) in order to minimize potential infections from protozoa, nematodes, and bacteria in the prey. When raising these prey, regular cultures of feces should be done and deworming as necessary. It is probably best, overall, to wean these animals onto rodent prey, for safety and economy reasons. One trick for a finicky snake may be to rub the rodent with a toad or frog to impart the scent. If feeding whole vertebrate prey, supplementation with vitamins and minerals is probably not necessary.

Insectivorous snakes include green *(Opheodrys)*, worm *(Carphophis)*, ring-necked *(Diadophus)*, and brown *(Storeria)*. Again, variety is the key; you should offer mealworms, crickets, earthworms, nightcrawlers, and larvae. The insects and larvae should be fed on a complete diet and probably dusted with a calcium and vitamin supplement weekly, as they will be low in calcium.

Chelonia: Turtles and Tortoises

The last order of reptiles we will discuss is the Chelonia, turtles. They can also be carnivorous, herbivorous, and omnivorous. Snapping turtles *(Chelydra)*, mata mata *(Chelys)*, and alligator snapping turtles *(Macrochelys)* usually only eat when they are in water. They can be fed in their own tank, or a separate feeding tank. The latter is a good option if you can manage to keep the primary environment clean. You might predict from earlier chapters that the best and simplest feeding practice is to feed a well-balanced commercial mix. Look for 30 to 50 percent protein and 1 percent calcium with a Ca:P ratio of at least 1. You may want to get trout food or other aquacultural fish food, if you can get it in small enough amounts. I do not mean "fish food" for aquarium fish, as it is not balanced the same as for trout and other aquacultural species. Small fish such as goldfish, minnows, and guppies can be fed. Fish such as mackerel or smelt may be too high in fat to feed routinely—this can lead to liver damage (steatitis) and Vitamin E deficiency. Table 18.6 shows the classification of the order Chelonia.

Omnivorous aquatic turtles include red-eared sliders *(Trachemys)*, reeves *(Chinemys)*, diamondback terrapins *(Malaclemys)*, map turtles *(Graptemys)*, and

Table 18.6	Order Chelonia		
	Carnivorous Turtles		
	Genus	**Names**	**Diets**
Chelydridae	*Chelydra*	snapping turtle	fish
Chelyidae	*Chelys*	mata mata turtle	
	Macrochelys	alligator snapping turtle	
	Omnivorous Turtles		
	Trachemys	red-eared slider	mixed
	Chrysemys	painted turtle	
	Chinemys	reeves turtle	
	Malaclemys	diamondback terrapins	
	Graptemys	map turtle	
	Pseudemys	river cooter	
	Terrapene carolina	eastern box turtle	

painted turtles *(Chrysemys)*. These fine folks can be fed the same type of mix of greens as discussed for omnivorous and herbivorous Squamata: romaine, collard, endive, chard, kale, and small pieces of floating veggies such as carrots and yellow squash. These species tend to be more carnivorous as juveniles so should be fed as noted for the carnivorous turtles, with about 25 percent vegetable matter in the diet, increasing the vegetable portion to 50 percent in adulthood.

> **Nutrition and Environment**
>
> Chelonia are extremely adapted to specific environments.
>
> We must feed and house them in situations that they have become adapted to.

Box turtles such as the eastern box *(Terrapene carolina carolina)*, ornate box *(Terrapene ornata ornata)*, three-toed box *(Terrapene carolina triunguis)*, and the Asian box *(Cuora)* fall into the aquatic turtle category. Often these are fed commercial box turtle food, low-fat dog food (maintenance chow), trout food, or small amounts of cat food. Fruits and vegetables should be offered and kept fresh. Some experiences reported in the literature suggest that they prefer red, orange, and yellow foods. Provide a large and shallow container for water so that the animal can enter it without spilling it.

Herbivorous tortoises include California desert *(Gopherus agassizi)*, South American red-footed *(Geochelone carbonaria)*, leopard *(Geochelne pardalis)*, yellow-footed *(Geochelone denticulata)*, Greek *(Testudo graeza)*, gopher *(Gopherus polyphemus)*, and hingeback *(Kinexys)*. Thanks for reading the book, I hope you enjoy your reptiles and all your animal friends with a long and healthy life.

Words to Know

dyscedysis	*hyperkeratinization*	*metabolic bone disease*
ectotherms	*hypovitaminosis A*	

Study Questions

1. Describe a simple life-cycle feeding-management strategy for carnivorous serpentes; omnivorous serpentes; carnivorous sauria; omnivorous sauria; and carnivorous and omnivorous chelonia.
2. What are the major practical problems in feeding reptiles as companions? Be sure to consider water management.
3. What is metabolic bone disease, and to which similar disease of birds does it relate? How can it be prevented?
4. How do vitamin D metabolism and requirements differ among Reptilia, Aves, and Mammalia? A simple practical way to deal with this difference is:_____.

Further Reading

Donoghue, S., and S. McKeown. 1999. Nutrition of captive reptiles. *Veterinary Clinics of North America Exotic Animal Practice*, 2(1): 69–91, vi.

Karasov W. H., E. Petrossian, E. L. Rosenberg, and J. M. Diamond. 1986. How do food passage rate and assimilation differ between herbivorous lizards and nonruminant mammals? *Journal of Comparative Physiology* [B], 156(4): 599.

Kik, M. J., and A. C. Beynen, 2003. Evaluation of a number of commercial diets for iguana *(Iguana iguana)*, bearded dragons *(Pogona vitticeps)*, and land and marsh tortoises. *Tijdschr Diergeneeskd*, Sep. 15, 128(18): 550–554.

Stahl, S., and S. Donoghue. 2000. Feeding reptiles. In M. Hand, C. D. Thatcher, R. L. Remillard, and P. Roudebush, eds., *Small animal clinical nutrition*, fourth edition. Topeka, KS: Mark Morris Associates.

Sulcata Station. 2003. *How to take care of your new sulcata tortoise.* Retrieved from *http://www.sulcata-station.org.*

Tosney, K. W. Professor of Biology, The University of Michigan. *http://biology.lsa.umich.edu/research/labs/ktosney/index.html.*

Zwart, P. 2001. Assessment of the husbandry problems of reptiles on the basis of pathophysiological findings: A review. *Veterinary Quarterly*, Nov. 23 (4): 140–147.

INDEX

('b' indicates boxed material;
'i' indicates an illustration;
't' indicates a table)